AF575452

Account Determination in SAP S/4HANA®

SAP PRESS is a joint initiative of SAP and Rheinwerk Publishing. The know-how offered by SAP specialists combined with the expertise of Rheinwerk Publishing offers the reader expert books in the field. SAP PRESS features first-hand information and expert advice, and provides useful skills for professional decision-making.

SAP PRESS offers a variety of books on technical and business-related topics for the SAP user. For further information, please visit our website: *www.sap-press.com*.

Stoil Jotev
Configuring SAP S/4HANA Finance (2nd Edition)
2021, 738 pages, hardcover and e-book
www.sap-press.com/5361

Anand Seetharaju, Mayank Sharma
General Ledger Accounting with SAP S/4HANA
2023, 886 pages, hardcover and e-book
www.sap-press.com/5630

Tritschler, Walz, Rupp, Mucka
Financial Accounting with SAP S/4HANA: Business User Guide (2nd Edition)
2023, 571 pages, paperback and e-book
www.sap-press.com/5698

Stefan Pougkas
SAP S/4HANA Financial Accounting Certification Guide (3rd Edition)
2021, 449 pages, paperback and e-book
www.sap-press.com/5310

Mehta, Aijaz, Parikh, Chattopadhyay
SAP S/4HANA Finance: An Introduction (2nd Edition)
2023, 394 pages, hardcover and e-book
www.sap-press.com/5606

Abdullah Galal, Jonas Tritschler

Account Determination in SAP S/4HANA®

Business Processes and Configuration

Editor Rachel Gibson
Acquisitions Editor Megan Fuerst
Copyeditor Melinda Rankin, Yvette Chin
Cover Design Graham Geary
Photo Credit iStockphoto.com: 493334040/© Sezeryadigar
Layout Design Vera Brauner
Production Kyrsten Coleman
Typesetting III-satz, Germany
Printed and bound in the United States of America, on paper from sustainable sources

ISBN 978-1-4932-2416-6

1st edition 2024

Library of Congress Cataloging-in-Publication Data:
Names: Ali, Abdullah Ali Ahmed Galal, author. | Tritschler, Jonas, author.
Title: Account determination in SAP S/4HANA : business processes and configuration / by Abdullah Ali Ahmed Galal Ali and Jonas Tritschler.
Description: 1st edition. | Bonn ; Boston : Rheinwerk Publishing, [2024] | Includes index.
Identifiers: LCCN 2024001736 | ISBN 9781493224166 (hardcover) | ISBN 9781493224173 (ebook)
Subjects: LCSH: SAP HANA (Electronic resource) | Accounts--Data processing. | Accounts payable--Data processing. | Managerial accounting--Data processing.
Classification: LCC HF5681.A2 A45 2024 | DDC 657.0285/53--dc23/eng/20240125
LC record available at https://lccn.loc.gov/2024001736

Contents at a Glance

Contents

Preface

This book, *Account Determination in SAP S/4HANA: Business Process and Configuration*, offers a detailed journey through the multifaceted capabilities of SAP S/4HANA, a leading-edge enterprise resource planning (ERP) system. SAP S/4HANA is designed as a vital resource for understanding and implementing key business processes through configuration within the system, which is pivotal for the seamless functioning of modern businesses.

SAP S/4HANA stands out for its ability to integrate various business processes, from materials management and sales to production planning and financial transactions. This integration is crucial in today's fast-paced business environment, where decisions must be informed by data from multiple departments. This book meticulously explores how each business area within SAP S/4HANA not only functions in isolation but also interacts with others, providing a holistic view of business operations.

Objective of This Book

The primary objective of this book is to impart a thorough understanding of the SAP S/4HANA core business transactions, focusing on their accounting impact and configuration. Readers will learn how to configure account determination in purchasing, sales, manufacturing, accounts receivables (AR) and accounts payable (AP) as well as for fixed assets. Each chapter is crafted to enhance your proficiency in navigating and optimizing core business configuration to achieve efficient business management.

Target Audience

This comprehensive book is an invaluable asset for a wide range of professionals, including SAP consultants, logistics experts, sales executives, finance specialists, and accounting practitioners, especially those who are currently using or planning to transition to SAP S/4HANA. Its detailed content makes it a critical tool for understanding and applying the complex functionalities of SAP S/4HANA in various business contexts. Additionally, this book serves as a rich educational resource, beneficial not only for students pursuing studies in business and information systems but also for educators and trainers seeking to impart practical knowledge and insights into these fields. Its relevance extends to academic curricula, professional training programs, and self-learning endeavors, making it a versatile guide for anyone interested in the intricacies of SAP S/4HANA and its application in contemporary business environments.

Structure of This Book

This book is structured into six chapters, each dedicated to a key area of configuration in SAP S/4HANA. Following an introductory chapter to set the stage, the subsequent chapters delve deeply into specific areas, providing detailed insights and guidance. Here's a preview of what each chapter encompasses:

- **Chapter 1**
  This chapter focuses on integrating accounting impacts within purchasing and inventory management processes. We begin by outlining common purchasing procedures, emphasizing accounting entries and general ledger accounts involved at each step. This chapter then delves into the detailed configuration steps in SAP S/4HANA for determining these general ledger accounts, providing insights into the nature and critical fields of every general ledger account used.

  We also examine the normal purchasing process, including the stages of ordering, receiving items, posting supplier invoices, and making payments, highlighting how each step results in distinct accounting entries and impacts inventory management. This chapter further explores the accounting viewpoint of goods receipt (GR), where a company accepts ownership of items, and the importance of understanding the legal and contractual nuances related to these transactions.

  A discussion of invoice receipt (IR) and payment processes will underline the variations in accounting entries that may arise due to discrepancies between the purchase order and the supplier invoice. This chapter concludes with a focus on configuring account determination for these processes in SAP S/4HANA, thereby offering a comprehensive guide to the materials management module and its integral role in SAP S/4HANA's accounting framework.

- **Chapter 2**
  This chapter focuses on the sales and distribution module in SAP S/4HANA, particularly emphasizing the accounting impacts of various sales processes. The chapter's primary goal is to educate readers on the main business processes in sales and distribution and the configuration of account determination for these processes within SAP S/4HANA. We detail the standard sales process from stock, also known as the order-to-cash process, which is commonly used in organizations for selling manufactured or procured items. This process includes several steps starting from a customer inquiry to the final payment receipt, each with its own accounting implications.

  The chapter comprehensively covers each step of this process, explaining the accounting entries involved, and delves into the configuration of general ledger accounts for these entries. We also address the relationships between materials management and sales, illustrating shared areas like inventory management. Through detailed examination, you'll understand the nature of general ledger accounts and learn the important aspects of maintaining general ledger account master data. This chapter thus serves as an insightful guide for professionals seeking to understand and configure the sales and distribution module in SAP S/4HANA.

- **Chapter 3**
  In this chapter, we'll dive into the complex processes of manufacturing management within SAP S/4HANA. Focusing on the make-to-stock discrete manufacturing process, we'll highlight how different accounting entries are configured and determined. This chapter covers three main components: product cost planning, cost object controlling, and actual costing/material ledger. We'll emphasize estimating product costs, tracking actual production costs, and analyzing variances. You'll learn the steps involved in the production process, from estimating costs using various components like raw materials and manufacturing activities to the final steps of posting accounting entries for stock revaluation and managing cost variances and work-in-process. This content is essential for understanding the financial integration and cost management in SAP's production planning module.
- **Chapter 4**
  This chapter provides an in-depth exploration of customer and vendor transaction management, emphasizing the transition from SAP ERP to SAP S/4HANA, focusing on the new framework for AR and AP under the unified business partner model. We'll cover various facets including the integration of business partner transactions into the general ledger, configurations for value-added tax (VAT), and both manual and automated payment processing methods. We'll elaborate on the goods receipt/invoice receipt (GR/IR) account, a critical component in managing the timing differences between GR and IR in the system. Additionally, you'll learn all about complex topics like withholding taxes, interest calculations, and dunning, which are essential for comprehensive financial management. This detailed exploration is crucial for understanding the financial nuances and operational efficiencies achievable through SAP S/4HANA's AR and AP functionalities.
- **Chapter 5**
  In this chapter, we'll focus on the integral aspects of cash management and banking operations and their impact on accounting practices. We begin by exploring core business procedures within these domains, emphasizing the importance of general ledger account master data and setting up automated determination for all accounting entries. This chapter provides a comprehensive guide to manage and align financial and accounting processes effectively in the context of cash management and banking operations. Key topics include setting up house banks and bank subaccounts; the intricate process of account determination; and handling various banking transactions such as bank statements, lockbox transactions, and bills of exchange (BoEs). This chapter concludes by discussing the setup of cash journals and the management of transactions, highlighting the relevance of cash transactions in daily business operations and the need for accurate accounting and financial tracking in SAP S/4HANA.

- **Chapter 6**
  This chapter is an in-depth exploration of the asset accounting module, crucial for managing both tangible and intangible assets. We'll focus on the comprehensive process for determining general ledger accounts within asset accounting, highlighting functionalities like managing machinery, computers, and office equipment, along with tracking capital costs and calculating depreciation. We'll also delve into handling intangible assets, reevaluating assets in inflation-prone countries, and managing assets in multiple ledgers for various accounting purposes like International Financial Reporting Standards (IFRS) and US Generally Accepted Accounting Principles (GAAP). We provide an extensive breakdown of various aspects of fixed asset accounting, including account determination logic, parallel valuation, and detailed transaction types for asset accounting, making this chapter a crucial guide for understanding and implementing asset accounting in SAP S/4HANA.

Acknowledgments

We'd like to express our heartfelt gratitude to the dedicated team at SAP PRESS, with special thanks to Rachel Gibson, our esteemed editor. Her steadfast dedication and invaluable support have been instrumental for the successful completion of this project.

Additionally, we extend our appreciation to our friends, colleagues, and family members for their unwavering support and encouragement, which have been essential in bringing this book to completion.

Chapter 1
Materials Management

Materials management is a crucial part of SAP S/4HANA, encompassing a broad spectrum of functionalities essential for effective enterprise resource planning. In this chapter, we will concentrate on the purchasing and inventory management processes that have a significant accounting impact and thus are integral to the efficient operation of any organization.

Our goal in this chapter is to provide a comprehensive understanding of the key business processes within the realm of materials management, particularly focusing on their accounting implications. We will embark on a detailed exploration of each step within these processes, examining the associated accounting entries and the usage of general ledger accounts. A critical part of this journey involves diving into the detailed configuration steps in SAP S/4HANA to ensure accurate account determination.

As SAP S/4HANA is a vast enterprise resource planning (ERP) system tailored to accommodate diverse organizational needs, it is essential to recognize that while we may not cover every business process available, the principles and concepts of account determination configuration discussed here are universally applicable. In this chapter, we have selected the most common business processes that showcase various aspects of account determination, ensuring a thorough grasp of materials management in SAP S/4HANA.

We commence our exploration with the primary purchasing business process: normal purchasing. This process not only sets the foundation for understanding materials management but also exemplifies the interconnected nature of business processes and account determination within SAP S/4HANA.

1.1 Normal Purchasing

In this section, we focus on the normal purchasing process, a fundamental activity for any organization involved in acquiring raw materials or trading items from local or international suppliers. This process is not only pivotal for ensuring the smooth operation of business activities but also plays a significant role in the financial management of an enterprise.

The normal purchasing process encompasses several key steps: placing orders with suppliers, followed by the receipt of these items into storage, then recording the supplier's invoice in the system, then payment to the supplier, and finally clearing the invoice. This sequence of actions represents the backbone of materials procurement in organizations and requires meticulous management.

We will thoroughly explore each step of this process, highlighting the critical points where financial transactions occur and accounting entries are generated. Our exploration will also extend to the examination of general ledger accounts affected at each stage, emphasizing the importance of precise accounting entries in the SAP system.

Further, we will delve into the configuration aspect of account determination in SAP. Understanding how to configure account determination for the normal purchasing process is crucial for ensuring that financial postings accurately reflect the actual business transactions, thereby maintaining the integrity of an organization's financial statements.

This section aims to provide you with a deep understanding of the normal purchasing process in SAP, from both operational and financial viewpoints. Now, let's start with an overview of the normal purchasing process.

1.1.1 Business Process Overview

The normal purchasing process flow is shown in Figure 1.1. The steps that cause accounting entries are highlighted.

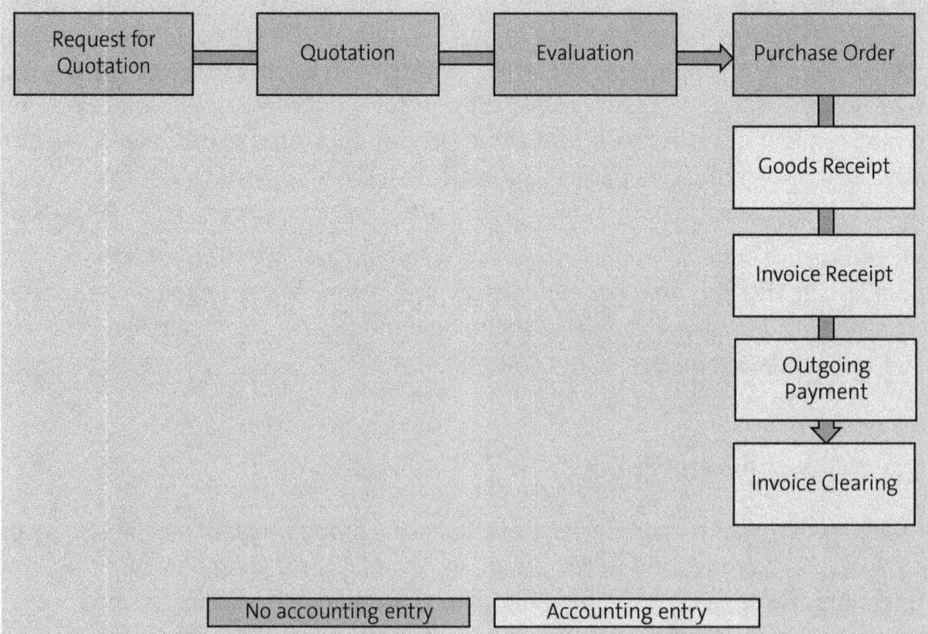

Figure 1.1 Normal Purchasing Process Flow

Let's take a quick look at each of these steps in the following sections. We will put more focus on the steps that cause accounting entries.

Request for Quotation

The process starts with the company sending requests for quotations to different suppliers: asking if an item is available, how much it would cost, and the different conditions related to purchasing and delivery. This step has no accounting impact.

Quotation

The suppliers respond with quotations that give all the details requested. This step has no accounting impact.

Evaluation

The company evaluates the different quotations and chooses a supplier. This step has no accounting impact.

Purchase Order

The company creates a purchase order to buy the items. This order serves as a contract between the company and the supplier that includes all the details related to purchasing, delivery, invoicing, and payment. This step has no accounting impact.

Goods Receipt

The company receives and accepts ownership of the purchased items. This step results in an accounting entry.

The goods receipt has an impact on inventory management, as it increases the stock quantities, and has an impact on accounting when the accounting entry is posted. It's important to understand the difference between these two points of view:

- From the inventory management point of view, the goods receipt is posted to confirm that a company has received some items that it needs to track, even if these items are not owned by the company yet.
- From an accounting point of view, the goods receipt accounting entry is posted only when the company accepts ownership of the purchased items, meaning that if anything happens to these items, then that is the responsibility of the purchasing company. The point of acceptance of ownership depends on legal requirements and agreements with the suppliers.

To understand these two different points of view, consider the following common business cases:

- **Case 1: The company doesn't accept ownership of the items received until they pass a quality check**

  - When the company receives the items into storage before the quality check, it posts a goods receipt in inventory management to quality blocked stock without an accounting entry.
  - After the items pass the quality check, the company posts another inventory movement to release the items to normal stock, and that results in posting the accounting entry.
- **Case 2: The company accepts ownership of the items before the quality check**
  - When the company receives items into storage before the quality check, it posts a goods receipt in inventory management for quality-blocked stock—but in this case, with an accounting entry.
  - After the items pass the quality check, the company posts another inventory movement to release the items to normal stock without an accounting entry.
- **Case 3: The agreed-upon incoterms in the contract state that the company takes ownership of the items at the doorstep of the supplier**
  - At the doorstep of the supplier, the company posts the goods receipt in inventory management with an accounting entry, even before shipping the items into its storage.

The account determination configuration in SAP S/4HANA is very flexible and can satisfy all these cases. We'll look at the configuration in detail in the next sections.

Table 1.1 shows an example of a goods receipt accounting entry that applies to all these cases. This example includes the main accounts that should be configured in the accounting determination for goods receipt in normal purchasing.

Debit	Credit	Debit Amount ($)	Credit Amount ($)
Inventory		1,000	
Price variance		100	
Freight costs		50	
	Inventory GR/IR		1,100
	Freight GR/IR		50

Table 1.1 Accounting Entry of Goods Receipt into Inventory

Now let's take a quick look at the meaning and use of each of these accounts. In later sections of this chapter, you'll learn how to configure account determination for each of these accounts in SAP S/4HANA:

- **Inventory**
  This is a balance sheet account. For moving average costing, the debit amount equals the number of units multiplied by the purchase price. For standard costing, the debit amount equals number of units multiplied by the standard cost. The difference between the purchase price and the standard cost goes to the price variance account.

- **Price variance**
  This is a profit and loss (P&L) account. For standard costing, if there's a difference between the item's standard cost and the purchase price, the difference is posted to this account. The difference value can be either a debit (loss) or a credit (gain), depending on whether the standard cost is higher or lower than the purchase price. You can use the same account for gain and loss, or configure two separate accounts.
- **Freight costs**
  This is a P&L account. For any purchasing condition (e.g., freight, customs, duties, handling), you can either have separate expense accounts, or you can post to the inventory account. This depends on the business requirements. You can also define a separate account for every condition, one account for all of them, or a combination of both. The value posted to this account equals the value of the purchasing condition(s) in the purchase order.
- **Inventory GR/IR**
  The GR/IR is a balance sheet account. It's an intermediate account that is cleared later when the invoice is posted. In real life, the goods receipt and invoice receipt are two separate events, so you can't post directly to the vendor account (accounts payable) at the time of goods receipt and instead use this intermediary account. The value posted to this account is always equal to the total value to be paid to the vendor equals the number of units multiplied by the purchase price.
- **Freight GR/IR**
  This is a balance sheet account. For any purchasing condition (e.g., freight, customs, duties, handling), you can configure separate GR/IR accounts, and you can also post one condition or all of them to the same GR/IR as inventory; in this case, there will be no freight GR/IR. The value posted to this account equals the value of the purchasing condition(s) in the purchase order.

Note

According to the financial standards, the valuation of inventory should be at the landed cost, meaning that the full cost to buy, handle, and deliver the stock until it's ready to be used should be posted to the inventory account. That's why the standard configuration is to post the purchasing conditions to the inventory account. You also have the option to post to separate expense accounts if requested.

The next step in the process after the goods receipt is supplier invoice receipt.

Invoice Receipt

The company receives the supplier invoice. This step results in an accounting entry. It's very common that the value in the supplier invoice is different from the value agreed upon in the purchase order. This can be due to additional costs, changes in prices, and so on. You only post the invoice receipt after the invoice has been accepted by the accounting department. In some cases, the invoice receipt can happen before the goods receipt. This has no impact on the account determination. An example of the accounting entry for invoice receipt is shown in Table 1.2.

Debit	Credit	Debit Amount ($)	Credit Amount ($)
Inventory GR/IR		1,100	
Freight GR/IR		50	
Price variance / inventory		50	
Unplanned delivery costs		100	
Input VAT		130	
	Accounts payable		1,430

Table 1.2 Accounting Entry of Supplier Invoice Receipt

There can also be additional lines in the accounting entry for withholding taxes if the legal requirement is to post it at the time of invoice receipt. This will be explained in detail in Chapter 4, Section 4.6. Now let's take a quick look at the meaning and use of each of these accounts. In later sections of this chapter, you'll learn how to configure the account determination for each of these accounts in SAP S/4HANA:

- **Inventory GR/IR**
  This is the same account used in the goods receipt accounting entry. For GR/IR, the debit value per unit is always the same as the credit value posted at goods receipt. For example, if at goods receipt we receive 100 units with a credit to GR/IR of $1,100, then the GR/IR per unit is 1100 divided by 100 equals $11. At invoice receipt, if the invoice is only for 50 units, then the GR/IR value in the invoice receipt will be $11 multiplied by 50 equals $550. The balance remaining in the GR/IR account will be cleared when the rest of the 100 units are invoiced.
- **Freight GR/IR**
  This is the same account used in the goods receipt accounting entry. The value is calculated the same way as for the inventory GR/IR account.
- **Price variance / inventory**
  The value posted here is equal to the difference between the purchasing value agreed on in the purchase order and the purchasing value received in the invoice. This value can be posted to either an inventory account or a price variance account based on (1) whether the items being purchased are valuated using moving average or standard costing and (2) whether the stock quantity available is less than the quantity being invoiced:
  - If the items being purchased are valuated on standard costing, then the variance will be posted to a price variance account regardless of the stock quantity.

  - If the items are (1) evaluated on moving average and (2) the stock quantity available of the item is equal or more than the quantity being invoiced, then the variance will be posted to an inventory account.
  - If the items are (1) evaluated on moving average and (2) the stock quantity available of the item is less than the quantity being invoiced, then this variance will be split between both the inventory account and the price variance account with the same ratio as invoice quantity to available quantity. For example, if the available quantity is 30 units and the invoiced quantity is 120 units, then 25% of the variance will go to the inventory account (30 units / 120 units) and the rest will go to the price variance account.

  Both the inventory and the price variance accounts are the same as those for the goods receipt posting.
- **Unplanned delivery costs**
  When posting the supplier invoice, you also have the option to post additional expenses related to the purchasing process to P&L expense accounts. You can configure an account to be automatically determined for unplanned delivery costs and you can also manually insert other expense accounts while posting the invoice. The values posted are inserted manually based on the invoice received.
- **Input VAT**
  This is a balance sheet account. The value is calculated automatically based on the tax codes inserted in the invoice to follow the legal tax requirements. More details on the VAT process and the related account determination configuration are in Chapter 4, Section 4.3.
- **Withholding tax**
  This is a balance sheet account. The withholding taxes can be posted at invoice receipt or at time of outgoing payment, depending on the legal requirements. More details on the withholding tax process and the related account determination configuration are in Chapter 4, Section 4.6.
- **Accounts payable**
  This is a balance sheet account. The value posted is the total value owed to the supplier according to the invoice, including VAT and withholding taxes. This is also called the supplier reconciliation account. More details on the business concepts behind and account determination of reconciliation accounts are explained in Chapter 4.

The next step in the normal purchasing process is the outgoing payment and invoice clearing.

Outgoing Payment and Invoice Clearing

The company pays the supplier and marks the invoices as cleared. This step results in an accounting entry. There are many ways the outgoing payments can happen. This will be covered in detail in Chapter 4, Section 4.4 and Section 4.5. For the sake of completing the normal purchasing process, let's consider a simple outgoing payment scenario in which you pay the supplier directly through a bank transfer. The accounting entry is shown in Table 1.3.

Debit	Credit	Debit Amount ($)	Credit Amount ($)
Accounts payable		1,430	
	Withholding tax		29
	Bank		1,300
	Cash discount		100
	Rounding differences		1

Table 1.3 Accounting Entry of Supplier Outgoing Payment

The outgoing payments accounting entry and the related account determination configuration will be explained in detail in Chapter 4, Section 4.4 and Section 4.5.

Now let's look at the account determination configuration of each of the accounts used in the goods receipt and invoice receipt accounting entries of the normal purchasing process.

1.1.2 Inventory Accounts (Transaction Key Technique)

When you post the goods receipt in SAP S/4HANA, there are some mandatory fields that must be inserted in the transaction, and these fields are essential for the account determination. Figure 1.2 shows an overview of how the inventory account is determined based on three input fields: **Plant**, **Material**, and **Movement Type**.

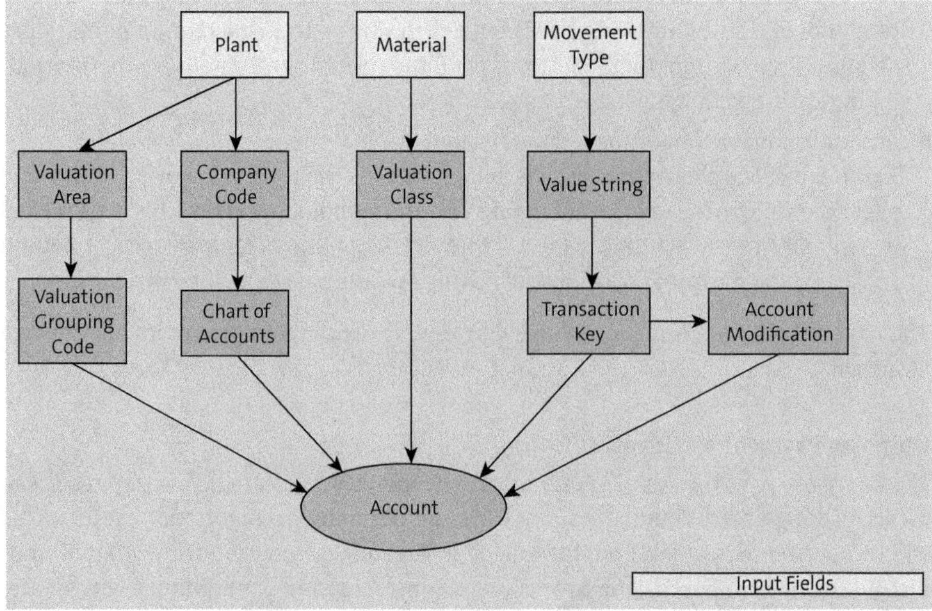

Figure 1.2 Overview of Determination of Inventory Account

The same account determination technique is used to determine different accounts in the inventory management area. We call this the *transaction key technique*.

In the following sections, we'll take a detailed look at how this transaction key technique can be configured. All of the configuration activities related to the inventory account determination can be found in the SAP configuration menu at **Materials Management • Valuation and Account Assignment • Account Determination • Account Determination without Wizard**.

The objective of this configuration is to get to the last configuration screen, where you assign the general ledger account number to a combination of chart of accounts, transaction key, account modification, valuation grouping code, and valuation class. Figure 1.3 shows the assignment of one of the general ledger accounts as an example.

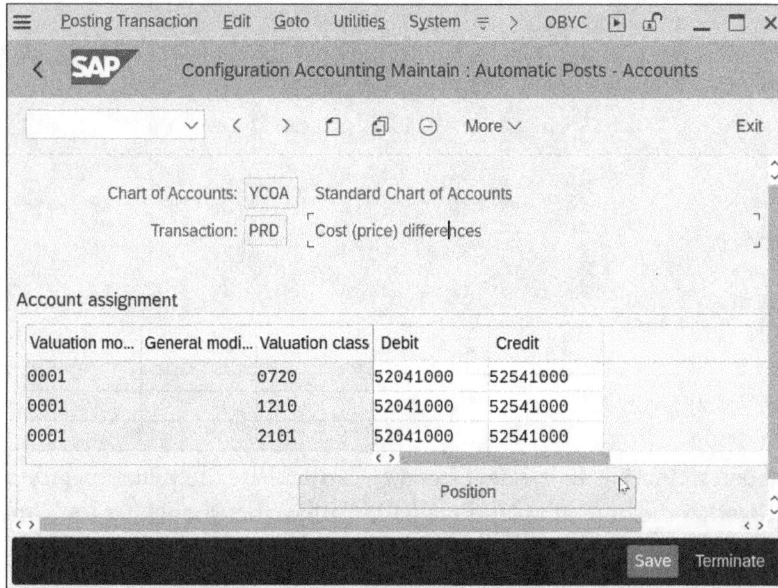

Figure 1.3 Assign General Ledger Account in Inventory Management: Example

In the following sections, we'll look at the meaning and configuration of each of the fields in Figure 1.3.

Define Valuation Level

The first step is to choose whether the valuation is on the level of the company code or the plant. This choice is made once when you perform the initial system configuration and can't be changed in the future.

This choice impacts two things:

- The material valuation: When calculating the cost of inventory, should it be calculated on the plant level or company code level?

- The account determination: Should it be configured on the plant level or on company code level?

SAP's recommendation is to choose the plant for the valuation level.

If you choose to set the valuation level to the plant, then the term *valuation area* in the different account determination screens will refer to the plant. If we choose to set the level to the company code, then this term will refer to the company code.

To choose the valuation level, go to SAP configuration menu path **Enterprise Structure • Definition • Logistics—General • Define Valuation Level (OX14).** The configuration screen is shown in Figure 1.4.

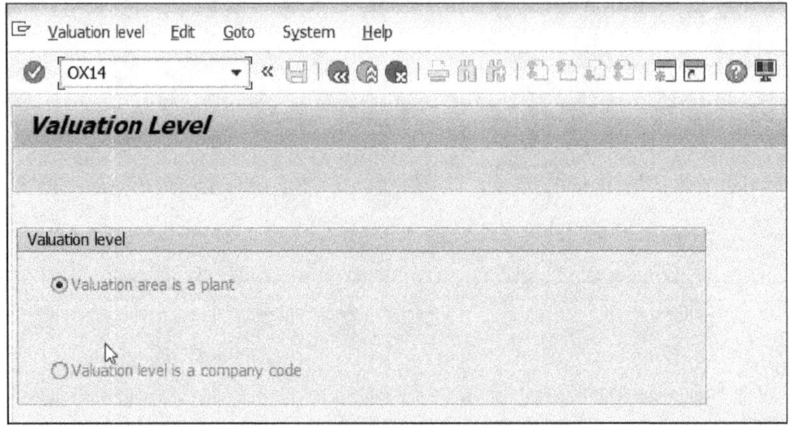

Figure 1.4 Define Valuation Level

For the account determination, in the next step you can activate the valuation grouping and group together the different plants (valuation areas) that should use the same configuration.

Group Together Valuation Areas

This can be used to group together different valuation areas so that they will use the same account determination setup. This option is always used in SAP S/4HANA in the following way:

1. Choose a four-digit code for the standard valuation group (usually 0001).
2. Maintain the inventory management account determination configuration on the level of the group.
3. When you define a new plant, you assign it to the group.
4. If you have a specific plant with different needs, you can assign it to a separate group or keep it unassigned.

To configure the valuation grouping, first go to SAP configuration menu path **Materials Management • Valuation and Account Assignment • Account Determination •**

Account Determination without Wizard • Define Valuation Control (OMWM), then activate the valuation grouping code. The configuration screen is shown in Figure 1.5.

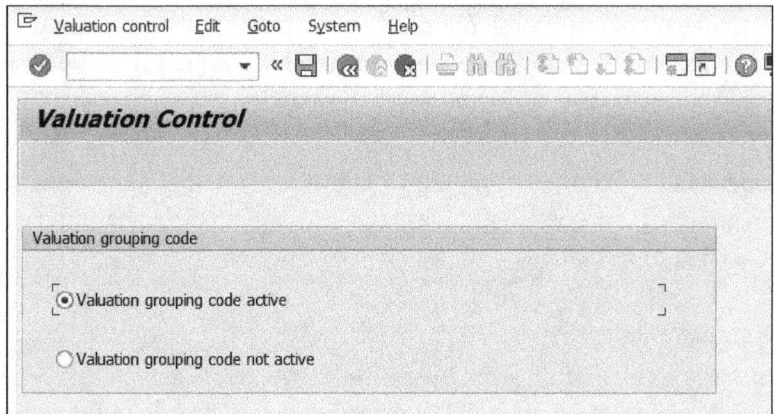

Figure 1.5 Activate Valuation Grouping Code

To assign the valuation areas to valuation groups, go to SAP configuration menu path **Materials Management • Valuation and Account Assignment • Account Determination • Account Determination without Wizard • Group Together Valuation Areas (OMWD)**, then enter a four-digit valuation group code for the plants, as in Figure 1.6.

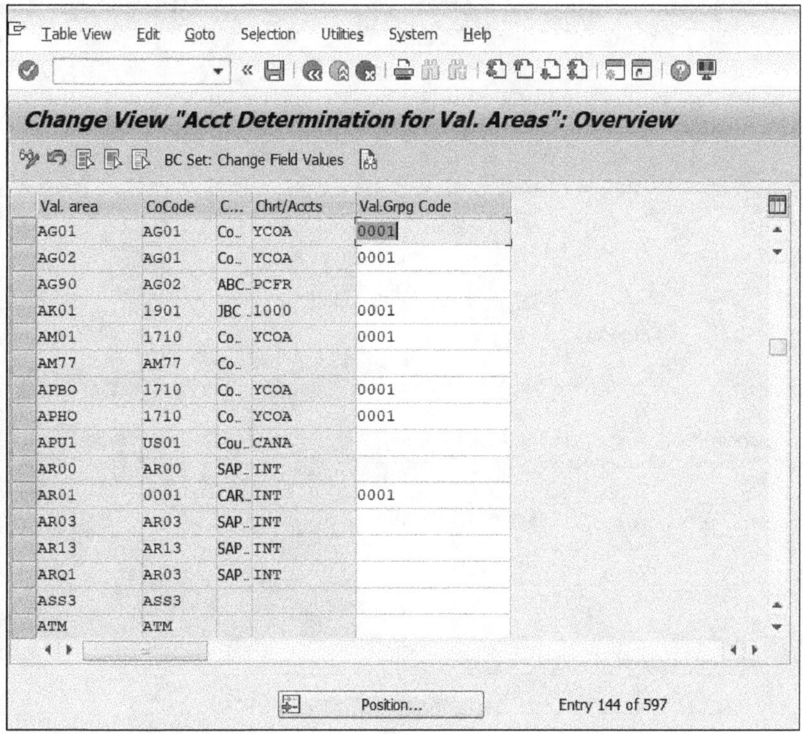

Val. area	CoCode	C...	Chrt/Accts	Val.Grpg Code
AG01	AG01	Co..	YCOA	0001
AG02	AG01	Co..	YCOA	0001
AG90	AG02	ABC..	PCFR	
AK01	1901	JBC ..	1000	0001
AM01	1710	Co..	YCOA	0001
AM77	AM77	Co..		
APBO	1710	Co..	YCOA	0001
APHO	1710	Co..	YCOA	0001
APU1	US01	Cou..	CANA	
AR00	AR00	SAP..	INT	
AR01	0001	CAR..	INT	0001
AR03	AR03	SAP..	INT	
AR13	AR13	SAP..	INT	
ARQ1	AR03	SAP..	INT	
ASS3	ASS3			
ATM	ATM			

Figure 1.6 Group Together Valuation Areas

Here, the system shows all the plants defined and assigned to company codes. You can't add additional lines here; you can only assign the existing plants (valuation areas) to valuation groups. You'll use the group codes in the next account determination configuration steps. If any valuation area is not assigned to a group, then you need to maintain a separate account determination configuration for this area. In this example, we've assigned valuation areas **AG01** and **AG02** to valuation grouping code **0001**. The configuration screen is shown in Figure 1.6.

Now that you understand the configuration links among the plant, the valuation area, and the valuation grouping code, next let's look at the links among the plant, the company code, and the chart of accounts.

Assign Plant to Company Code

This is one of the basic configuration steps in SAP S/4HANA. Before any plant can be used, it must be assigned to a company code. This step is not specific to account determination, so we won't go into the details, but you'll often want to check which plants are assigned to which company code. To do this, you can check table T001K using Transaction SE16H (SAP Data Browser). For example, company code **AG00** has only one plant (valuation area) assigned, which is plant **AG01**, as shown in Figure 1.7.

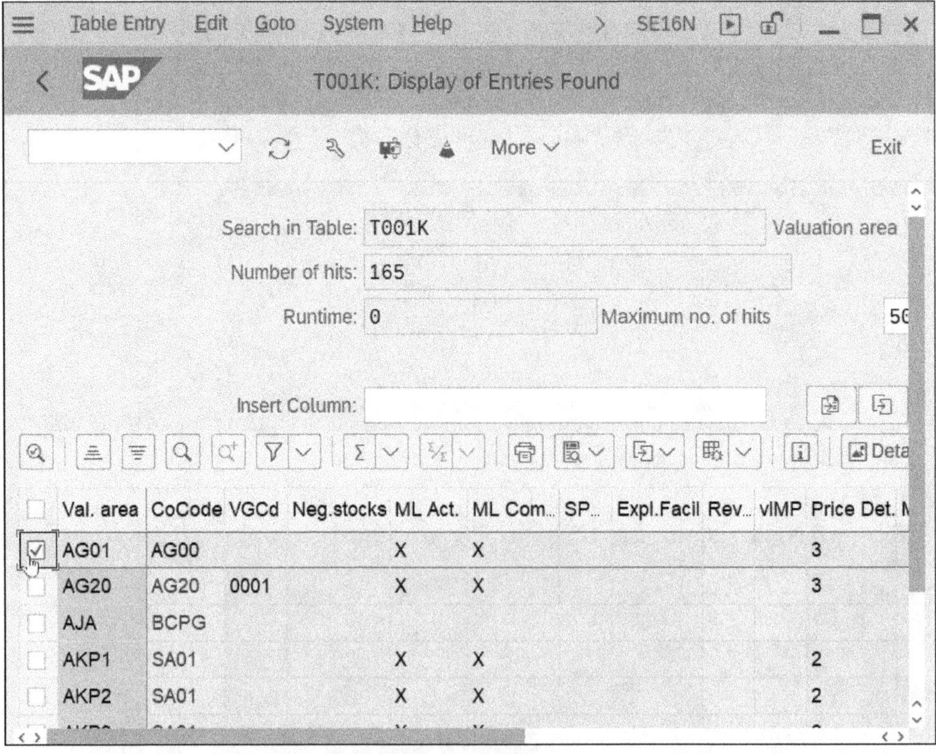

Figure 1.7 Display Plant to Company Code Assignment in Table T001K

Assign Chart of Accounts to Company Code

The chart of accounts is a registry for all the accounts that can be used in any organization. The general ledger accounts master data is created first on the level of a chart of accounts, and then can be extended to any company code assigned to this chart of accounts. Any company code in SAP S/4HANA must be assigned to an operational chart of accounts. The account determination is configured on the chart of accounts level.

Assigning a chart of accounts to a company code is one of the mandatory basic steps in SAP configuration; this is not specific to account determination, so we won't get into the details. You'll often want to check which chart of accounts is assigned to each company code. To do so, you can check table T001 using one of the data browser transactions, such as Transaction SE16H. For example, company code AG00 is assigned to the operational chart of accounts (**Chrt/Accts**) **AG00**, as shown in Figure 1.8.

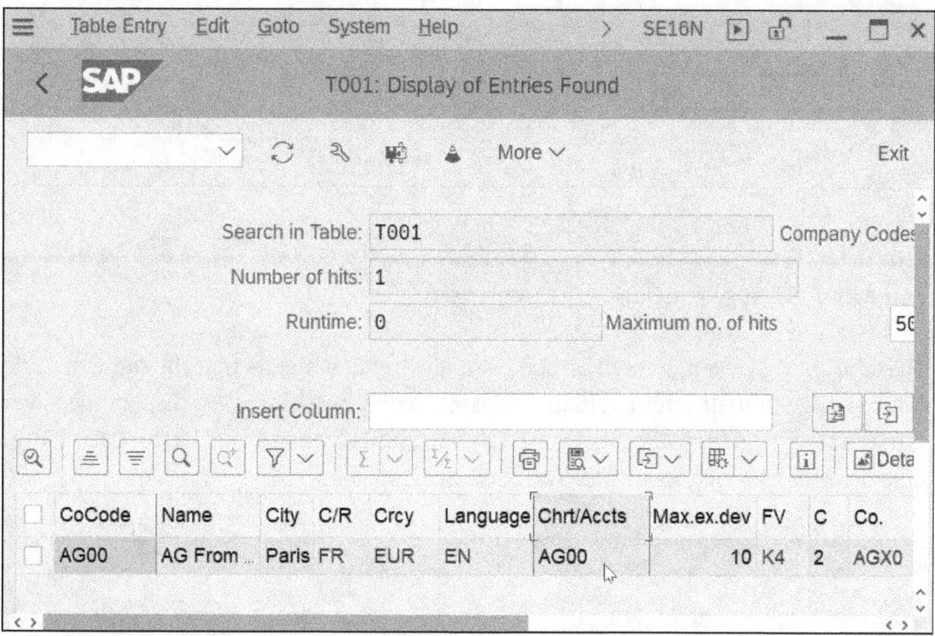

Figure 1.8 Display Chart of Account Assignment to Company Code in Table T001

Now that you understand how SAP S/4HANA uses the plant inserted in the goods receipt transaction to derive the chart of accounts and the valuation grouping code, next let's look at the configuration steps related to the material code.

Define Valuation Classes

You assign a valuation class to every material in the **Accounting 1** view of the material master data. You can see this in the material master data by using Transaction MM03, as in Figure 1.9.

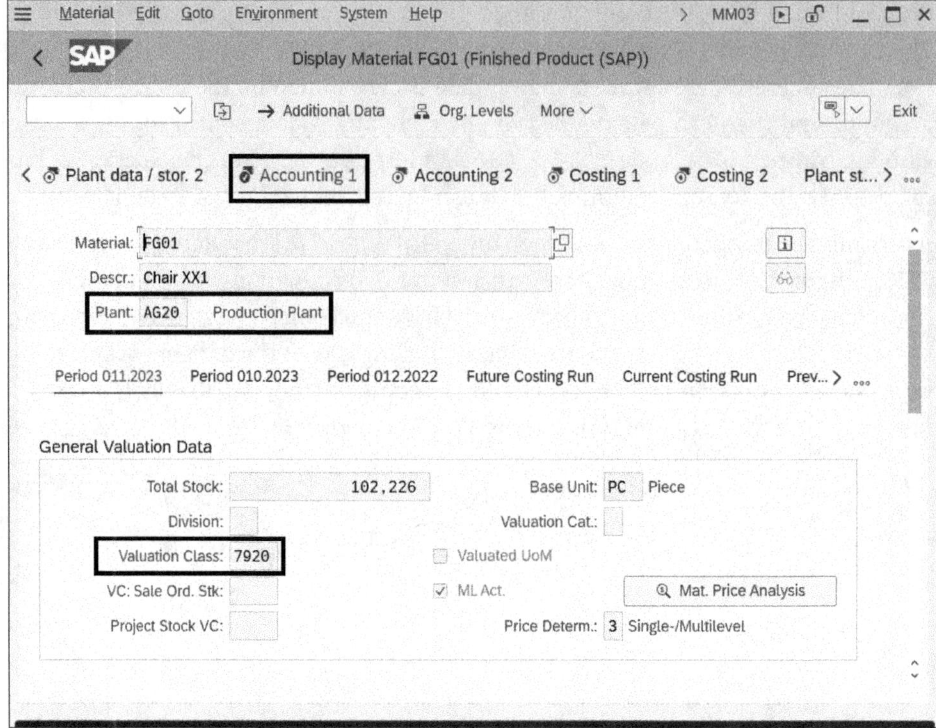

Figure 1.9 Assigning Valuation Class to Material

The accounting view is created on the plant level, which means that the same material can be assigned to different valuation classes in different plants. To display the valuation classes assigned to different materials and plants, you can check table MBEW using a data browser transaction such as Transaction SE16N.

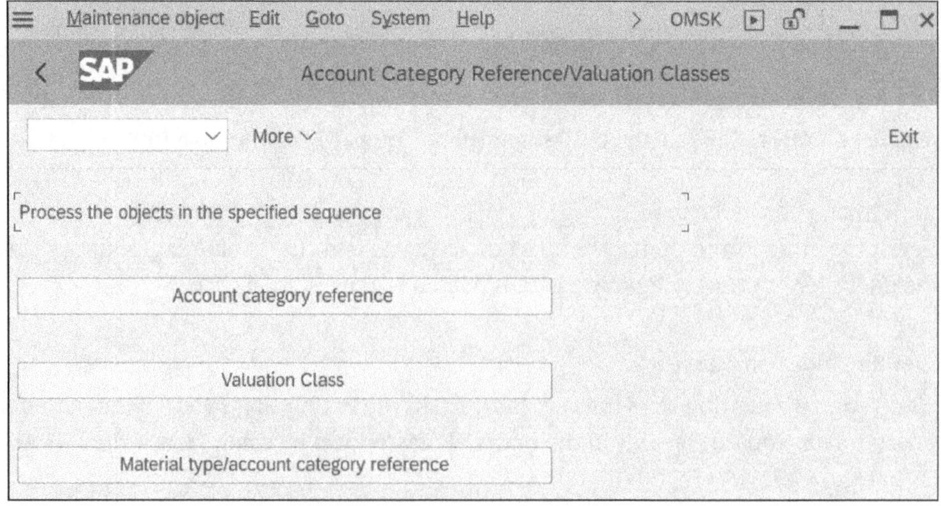

Figure 1.10 Define Valuation Classes: Main Screen

To be able to assign a valuation class in the material master data, you have to perform three configuration steps under SAP configuration menu path **Materials Management • Valuation and Account Assignment • Account Determination • Account Determination without Wizard • Define Valuation Classes (OMSK)**, as shown in Figure 1.10.

Whenever you create a material, you have to select a material type. To limit mistakes in assigning a wrong valuation class to a material, only the valuation classes assigned to the material type in the configuration can be assigned to the material. To configure this, you define four-digit account category references in the first step. Then you define four-digit valuation classes for every account category reference in the second step. The last step is to assign every material type to one account category reference. This configuration is shown in Figure 1.11.

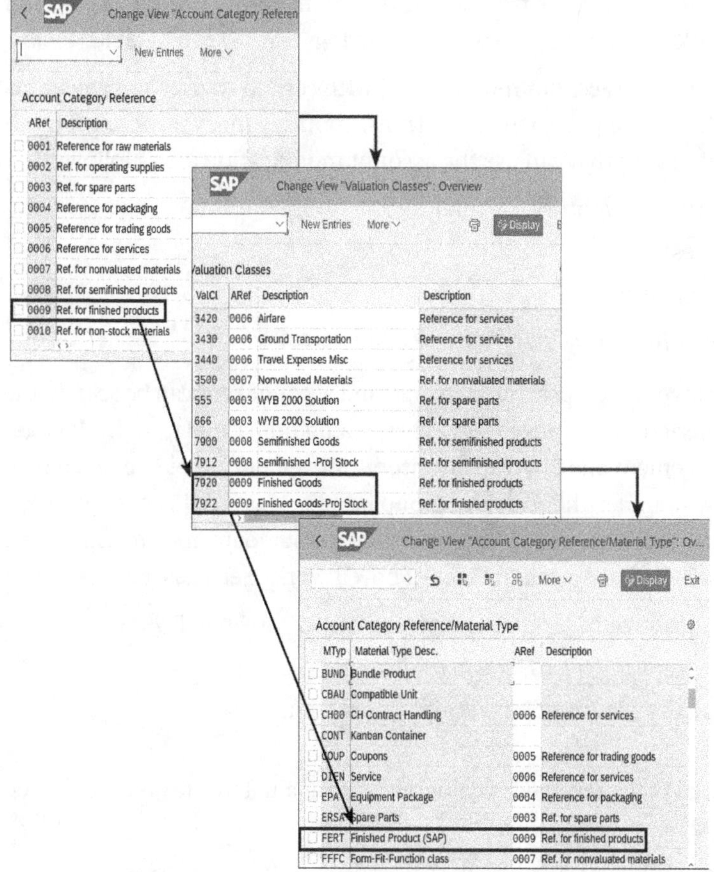

Figure 1.11 Define Valuation Classes: Steps

For example, in Figure 1.11 we've defined account category reference **0009**; then defined two valuation classes, **7920** and **7921**, and assigned them to **0009**; then assigned material type **FERT** to **0009**. Now whenever you create a material of type **FERT**, you can only choose between two valuation classes 7920 and 7921.

The material types of created materials can be displayed in table MARA.

Now that you understand how SAP is able to derive the valuation class from the material number inserted in the goods receipt transaction, next let's look at the movement type.

Define Account Grouping for Movement Types

Movement types are one of the most important configuration objects in inventory management. You usually don't change the standard configuration done by SAP; even in the few cases in which you need a new movement type, you copy one of the standard movement types.

The movement types are assigned to transaction keys through value strings. This configuration is hardcoded and predefined in SAP. You assign the transaction keys to the general ledger accounts. For example, the transaction key for inventory posting is BSX, and the one for GR/IR is WRX. The definition of the value strings can be seen in table T156W.

SAP also has the option to use account modification with certain transaction keys. This allows assigning a different account to the same transaction key for every account modification. The transaction keys that can use the account modification are as follows:

- GBB (offsetting entry for inventory posting)
- PRD (price differences)
- KON (consignment liabilities)
- KDM (exchange rate differences in the case of open items)

The links among the movement types, value strings, and account keys can be seen in the SAP configuration transaction for movement types, via menu path **Materials Management • Inventory Management and Physical Inventory • Movement Types • Copy, Change Movement Types (OMJJ)**, under the **Account Grouping** tab. Most of this configuration can't be modified, as shown in Figure 1.12. These settings are standard and are usually left untouched, but it's useful to have a general idea of how it works. Let's take a quick look.

The value string is determined based on multiple inputs, as shown in Figure 1.12:

- **Movement type**
  The movement type used in the goods receipt transaction.
- **Value updating**
  Whether the material being received is valuated or nonvaluated. This is determined by the material type.
- **Quantity updating**
  Whether the material is managed on the basis of quantity. This is determined by the material type.
- **Special Stock**
  The special stock type. For example, consignment or subcontracting. For the standard purchasing process, this is blank.

- **Movement ind.**
  The type of document that is used as a reference for the inventory movement. For example, for the standard purchasing process, this indicator is **B: Purchase Order**.
- **Consumption**
  Whether there's a consumption that will be posted—for example, to an asset or a cost object. For the standard purchasing process, this is blank.

Based on these inputs, you can see the value string assigned. The value string includes all the transaction keys that can be triggered during the material movement. For example, in Figure 1.12, you can see that for the standard purchasing process, value string WE01 is used. This value string includes the different transaction keys in the figure. You can see inventory posting (BSX), GR/IR (WRX), price difference (PRD), and others.

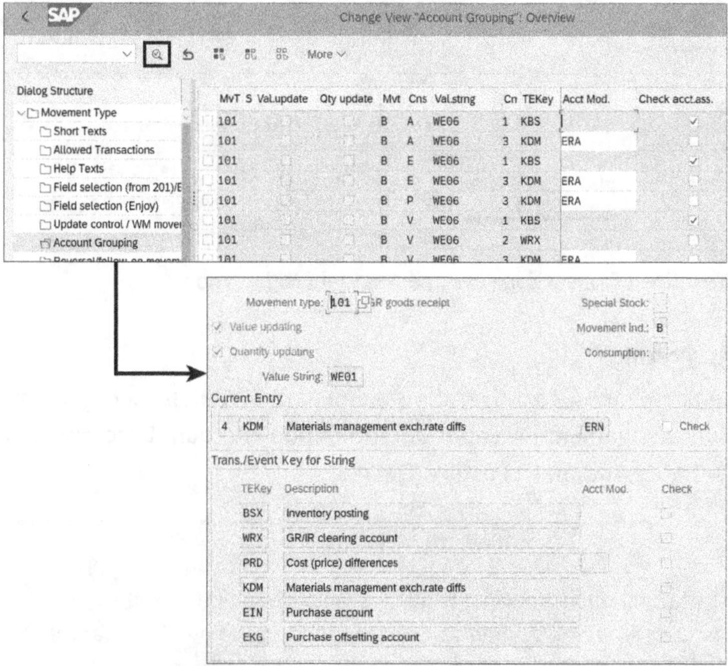

Figure 1.12 Movement Type Account Grouping

A list of all the predefined transaction keys and account modifications and when they are triggered is available via the help documentation of the configuration transaction (under **Configure Automatic Posting**). To display the documentation, go to SAP configuration menu path **Materials Management • Valuation and Account Assignment • Account Determination • Account Determination without Wizard • Configure Automatic Posting** and click the **Display Documentation** icon next to the transaction, as shown in Figure 1.13.

The transaction key for normal inventory posting is BSX.

Now you understand how all the fields needed for the account determination are determined through the three main input fields: plant, material, and movement type. The last step is to assign the accounts in the **Configure Automatic Postings** configuration transaction.

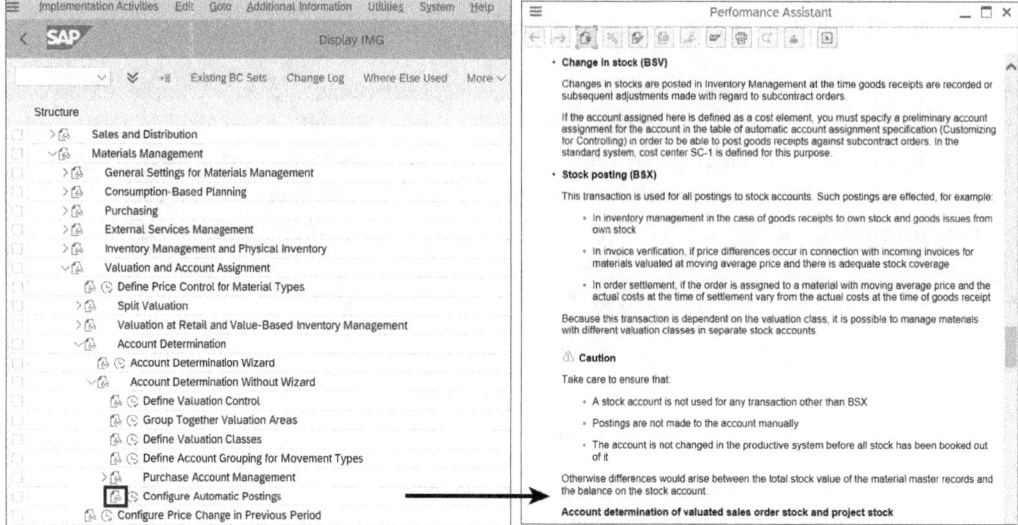

Figure 1.13 Display Predefined Transaction Keys and Account Modification

Configure Automatic Posting

To assign the accounts, follow SAP configuration menu path **Materials Management • Valuation and Account Assignment • Account Determination • Account Determination without Wizard • Configure Automatic Posting (OMWB).**

> **Note**
>
> When opening the transaction, it's normal to see a popup warning message showing the valuation areas that are not assigned to any valuation grouping. Click **Cancel** to proceed with the configuration.

This transaction includes three different screens, as shown in Figure 1.14:

- **Account Assignment**
  Here you can assign the general ledger accounts.
- **Simulation**
  Here you can simulate the account determination by giving different inputs. This is very useful to confirm that the configuration is correct and complete.
- **G/L Accounts**
  Here you can insert a company code or a valuation area and generate a report of all the accounts used in account determination in materials management. For each

general ledger account, the list displays the valuation classes, the transaction keys, and, where appropriate, the account modifications.

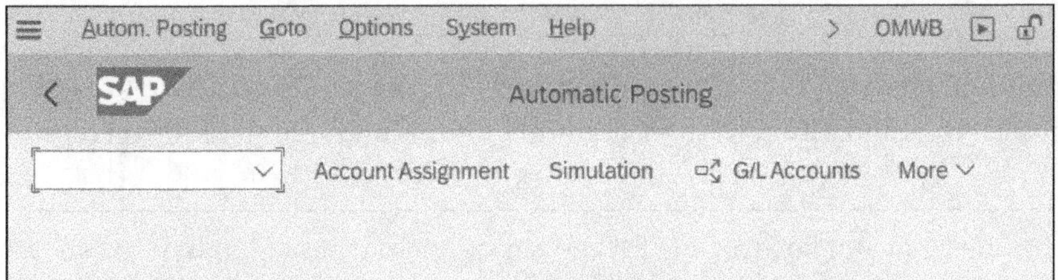

Figure 1.14 Different Screens in Automatic Posting Transaction (OMWB)

Let's start with the first screen, **Account Assignment**, to assign the general ledger accounts. When you click **Account Assignment**, you will see a screen that shows all the available transaction keys and their descriptions. Because you want to assign the inventory account, double-click the **BSX** key, then insert the chart of accounts for which you're maintaining this configuration. In the next screen, you can insert the general ledger account, but first you should check the account assignment rules and posting keys as in Figure 1.15.

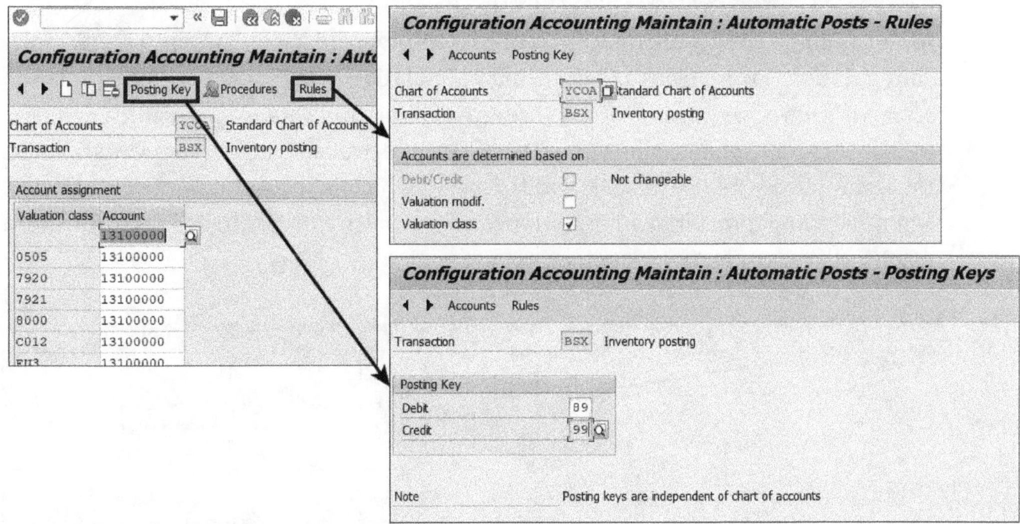

Figure 1.15 Assign Accounts, Rules, and Posting Keys

If you click **Rules**, you go to a screen where you can choose the rules you want for the account assignment for this transaction key. Every key has its own rules, and there are three rules available:

- **Debit/Credit**
  Check this box to assign different accounts in case of debit and credit.
- **Valuation modif.**
  Check this box to assign the accounts on the level of the valuation grouping codes that

you defined in the group together valuation areas step, shown earlier in Figure 1.6. This means that for every valuation grouping, you can have a different account assignment.

- **Valuation class**
  Check this box to assign the accounts on the level of the valuation classes. Use this, for example, to assign different stock accounts for raw materials and finished products.

Some of the rules may not be available for certain standard transaction keys. As shown in Figure 1.15, the **Debit/Credit** rule is grayed out (not allowed) for transaction key BSX. The main inventory account will always be the same in case of debit (goods receipt) and credit (goods issue).

Now go to the next screen by clicking **Posting Key**, as in Figure 1.15. Here you can select the posting keys that will be used for the debit and credit postings of the assigned general ledger accounts.

The last step in the account assignment screen is to assign the general ledger accounts based on the rules selected. In the example in Figure 1.15, the accounts are assigned based on the valuation classes only.

After assigning the general ledger accounts, you can run a simulation to confirm the configuration is working as expected. Click **Simulation** as in Figure 1.14. The simulation screen is shown in Figure 1.16. First, insert the simulation data—a plant and a material, then insert a movement type and select one of the available movement variations. In this example, we've selected **GR goods receipt**, which is the normal movement used when receiving a material with reference to a purchase order. To see the simulation results, click **Account Assignments**.

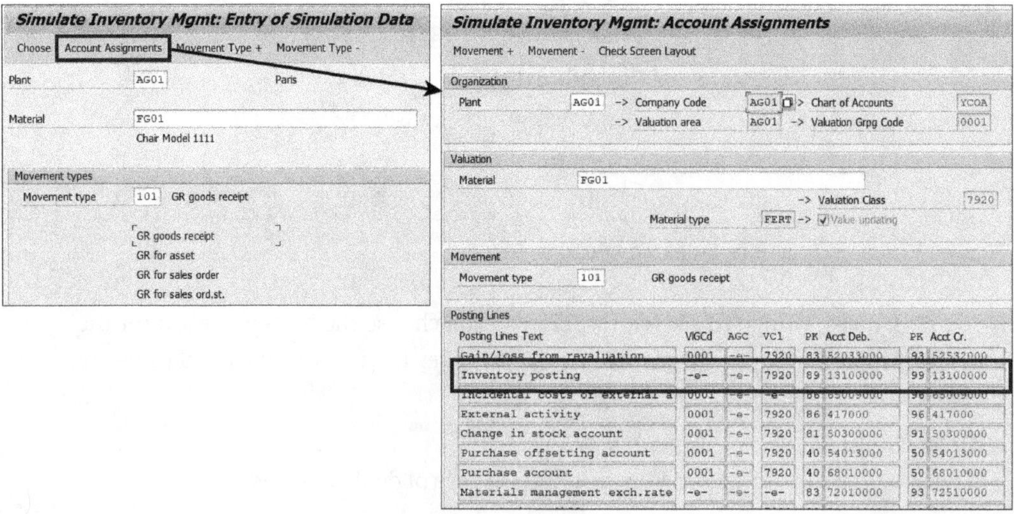

Figure 1.16 Account Assignment Simulation

In the **Account Assignments** screen, you can see how SAP was able to determine the chart of accounts, the valuation area, and the valuation grouping code from the plant. From the material code, SAP was able to determine the valuation class. In the posting lines, you can see the accounts and posting keys assigned to the inventory posting, which is as configured in Figure 1.15.

For fields that show the value **-e-**, the field is not applicable because the rule is not activated. If any general ledger account assignment is missing, you will see **--Missing--** in the **Accounts** column.

This concludes the discussion of the transaction key technique to determine inventory accounts through transaction key BSX. We'll refer to this section later whenever the same technique is used.

Another very important point when configuring the account determination in SAP is to know which general ledger account to assign. There are some fields in the master data of the general ledger account that must be maintained properly for the account determination to work without errors. You should always compare the general ledger account definitions with those of the INT or YCOA standard chart of accounts. One of the most important fields that can cause errors is **Field status group**, maintained in the general ledger account master data in Transaction FS00 under the **Create/bank/interest** tab, as shown in Figure 1.17.

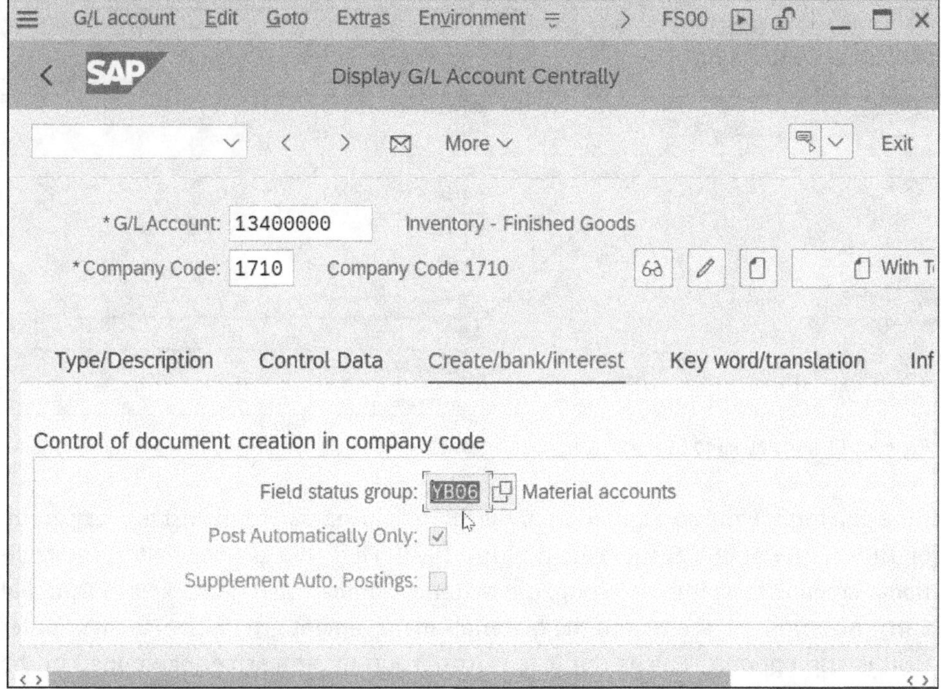

Figure 1.17 Field Status Group in General Ledger Account Master Data

The field status group controls which fields are mandatory, optional, or hidden when the general ledger account is used in any accounting entry. This is critical as, for example, for inventory accounts, SAP always passes the material code to the accounting entry, and if this field is hidden, then the system will throw an error when posting the accounting entry and will stop the process. This can happen in very critical moments such as issuing items from stock to a customer or receiving items from a vendor. This is just an example; there are many critical fields available in the field status group. This is why it's always recommended to check the general ledger account definition against a similar general ledger account in the standard chart of accounts delivered by SAP. To display the different fields available in the field status group, and the configuration of the fields (hidden, optional, mandatory), double-click the field status group in the general ledger account master data, as shown in Figure 1.17. This will take you to the field status group configuration screen shown in Figure 1.18.

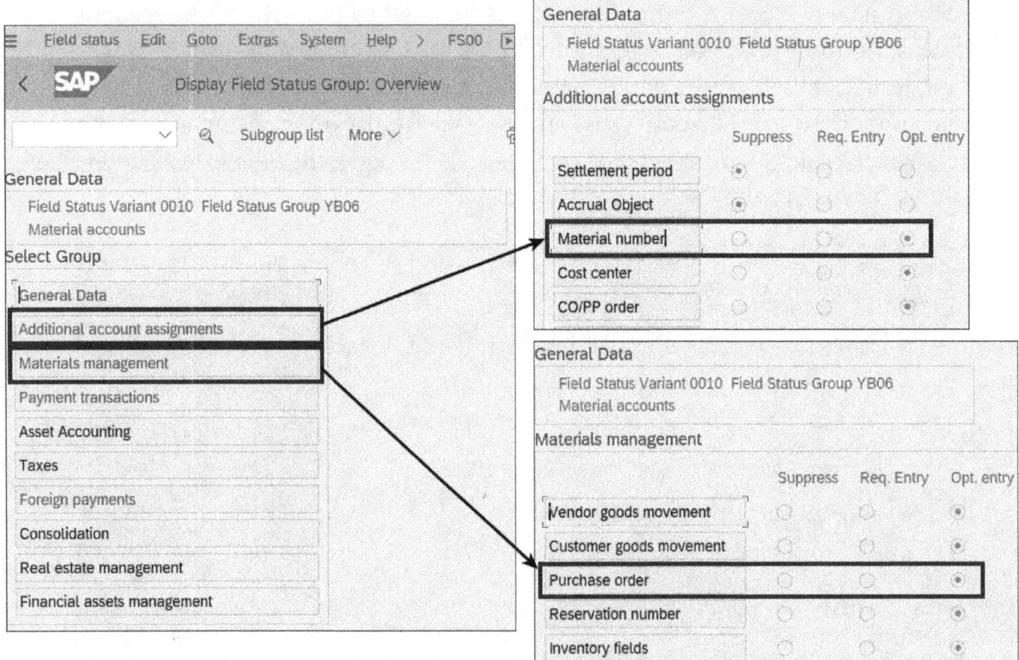

Figure 1.18 Display Field Status Group Configuration

Here you can see the configuration of the **YB06** field status group. Inside every group on the left in Figure 1.18, there are multiple fields. Here you can see that in the **Additional account assignments** group, the **Material number** field is marked as **Optional Entry**. Also, you can see that in the **Materials management** group, the **Purchase order** field is also optional. This is the field status group that must be maintained in the inventory account that you assign in the inventory account determination in Figure 1.15. This example shows the importance of comparing the general ledger accounts you assign in the account determination against the similar accounts in the standard chart of accounts provided by SAP. The same concept applies to all the accounts used in the account determination configuration.

Now that you understand the account determination of the inventory account, let's move to the next account in Table 1.1, which is the price variance or price difference account.

1.1.3 Price Variance Account

This account can also be called the *purchase price variance* (PPV) account. The PPV is a P&L account that can be used in both the goods receipt and the invoice receipt accounting entries. It's used in the following cases:

- When using standard costing, if there's a variance between the inventory standard price and the purchase price during the goods receipt or the invoice receipt. This variance can either be a gain (if the purchase price is lower than our cost) or a loss (if the purchase price is higher than our cost).
- When using moving average costing, if during the invoice receipt there's a price variance, and the stock available of the item is lower than the purchased quantity, then this variance will be split between the inventory account and PPV based on the available quantity, as explained in the invoice receipt accounting entry in Section 1.1.1.

Now let's look into the configuration steps needed to configure the account determination of the PPV accounts.

Configure Automatic Posting

The account determination of PPV follows the same steps as those in Section 1.1.2, with some differences, as shown in Figure 1.19.

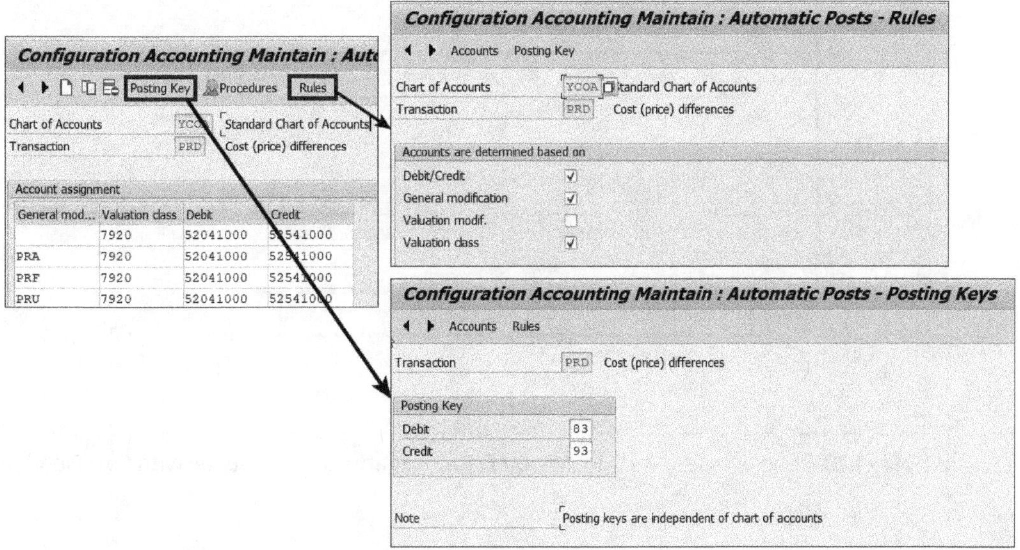

Figure 1.19 Price Variance Accounts, Rules, and Posting Keys

Note the following PPV account determination details shown in Figure 1.19:

- The predefined transaction key for PPV is **PRD**.
- You can assign different debit (PPV loss) and credit (PPV gain) accounts if needed.
- General account modification is enabled for PRD, which is used in other business cases. For purchasing price variance, no account modification is needed.

The general ledger account that you assign as the PPV account must be created as a P&L account. You can create this account either with or without a cost element, based on your business requirements. Let's look into how to use each of these options.

Create the PPV Account with a Cost Element

You create the PPV account with a cost element if you want to track the PPV in controlling reports (managerial accounting) with reference to a specific cost object. The cost object can be a cost center set by default in configuration or a profitability segment if you have margin analysis activated. The general ledger account should be created as general ledger account type P (primary costs or revenue), and the cost element should be in category 1 (primary costs/cost-reducing revenues). This can be maintained and displayed in Transaction FS00, as shown in Figure 1.20.

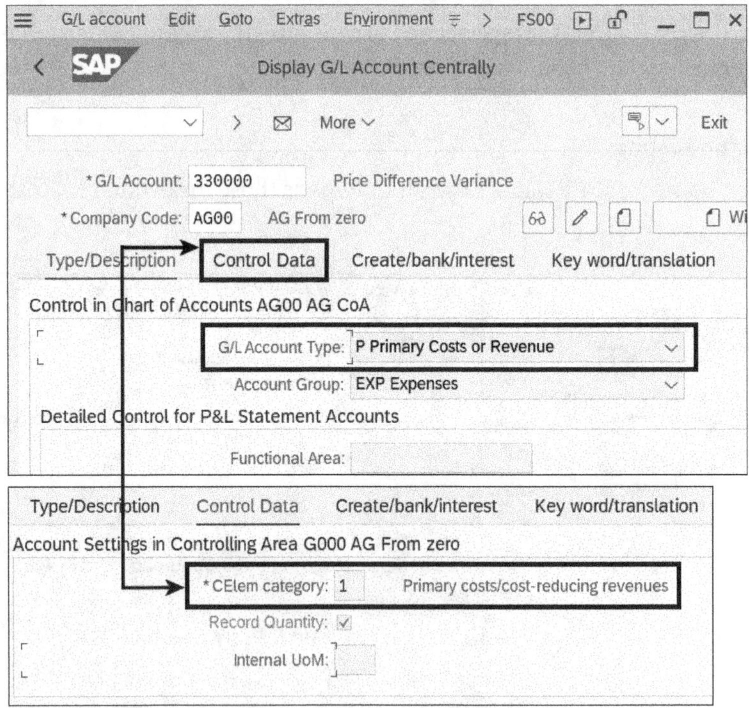

Figure 1.20 General Ledger Account Master Data: Primary Cost or Revenue with Cost Element

If you choose to create the PPV account as a cost element, then you must choose a default cost object assignment in configuration. This can be done in configuration Transaction OKB9, as shown in Figure 1.21.

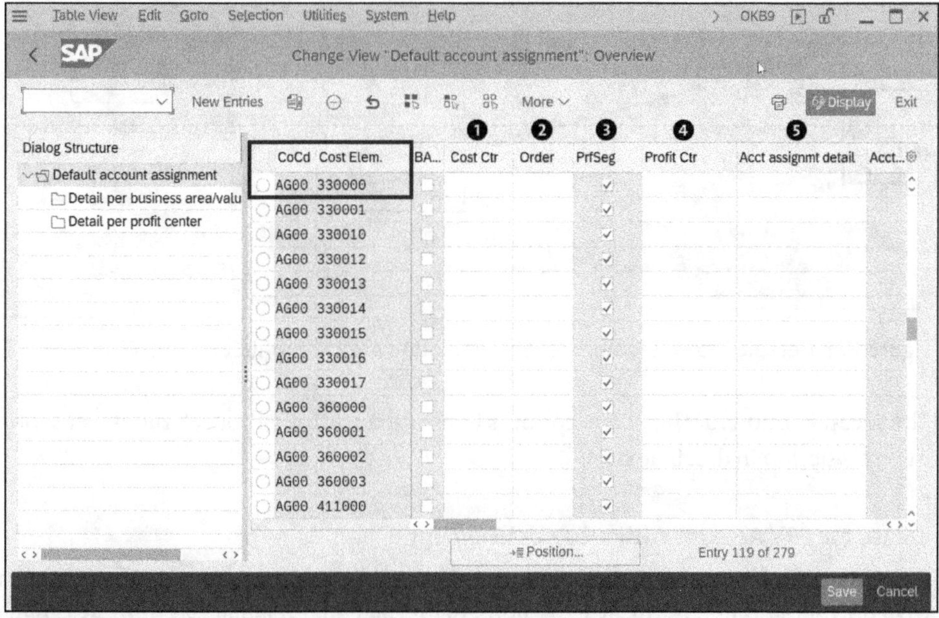

Figure 1.21 Default Cost Assignment Options in Transaction OKB9

Here, for every company code and cost element, you can choose one of the following default cost objects:

❶ **Cost Ctr** (Center)

❷ **Order**

❸ **PrfSeg** (Profitability Segment)
If you check this, then SAP will derive the profitability segment from the transaction details.

❹ **Profit Ctr**
Use this only for revenue accounts.

You can also maintain the default cost object on a more detailed level. In the **Acct assignment detail** field in Figure 1.21 ❺, you can choose different options, as shown in Figure 1.22.

First, select a value in the **Acct assignment detail** field ❶, then switch to the corresponding sub dialog ❷ and maintain the account assignment on the chosen level of detail ❸ ❹. As shown in Figure 1.22, you can maintain the account assignment on the level of the business area, valuation area (plant), or profit center.

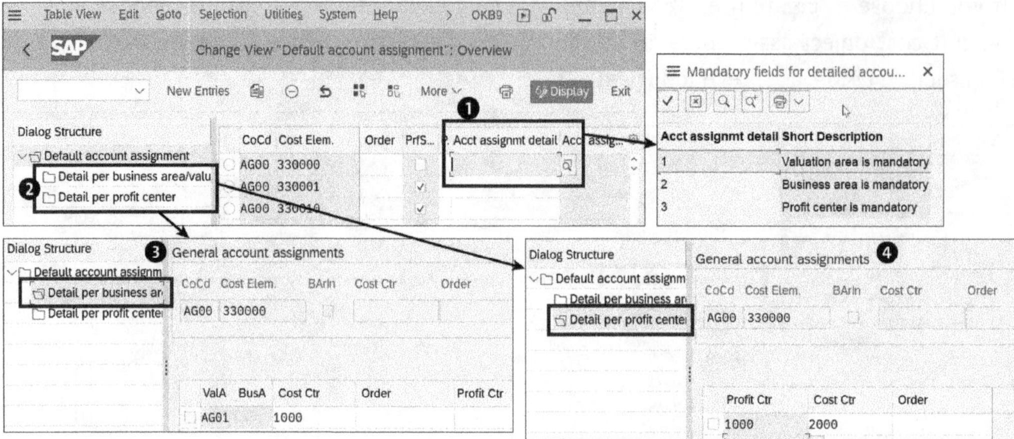

Figure 1.22 Default Account Assignment on Valuation Area or Profit Center Level

The second option for the PPV account is to create it without a cost element. Let's look into this option in detail next.

Create the PPV Account without a Cost Element

You create the PPV account without a cost element if you don't want to report on it in the controlling reports. In this case, you create the general ledger account as type N (non-operating expense or revenue), and no cost element will be created. These numbers won't be reflected at all in cost center and margin analysis reports. Because there's no cost element, no cost object assignment is needed.

Now that you understand the account determination of the purchasing price variance account and the specifics of the general ledger accounts that you must assign, let's move to the next account explained in Table 1.1, which is the inventory GR/IR account.

1.1.4 Inventory GR/IR Accounts

The inventory GR/IR account is usually called just the GR/IR account. In the context of this book, we added the word *inventory* to differentiate between this account and the freight clearing or freight GR/IR accounts that are explained in Section 1.1.5.

The inventory GR/IR account is determined following the steps explained in Section 1.1.2. For this account type, use predefined transaction key WRX. Figure 1.23 shows an overview of the standard configuration for transaction key WRX. You can choose to have different **Debit/Credit** GR/IR accounts and to use **Account Modification** for more complex scenarios.

The account you assign as the GR/IR account must be a balance sheet account. You should copy the account from the standard SAP chart of accounts to be sure you maintain all the critical fields correctly.

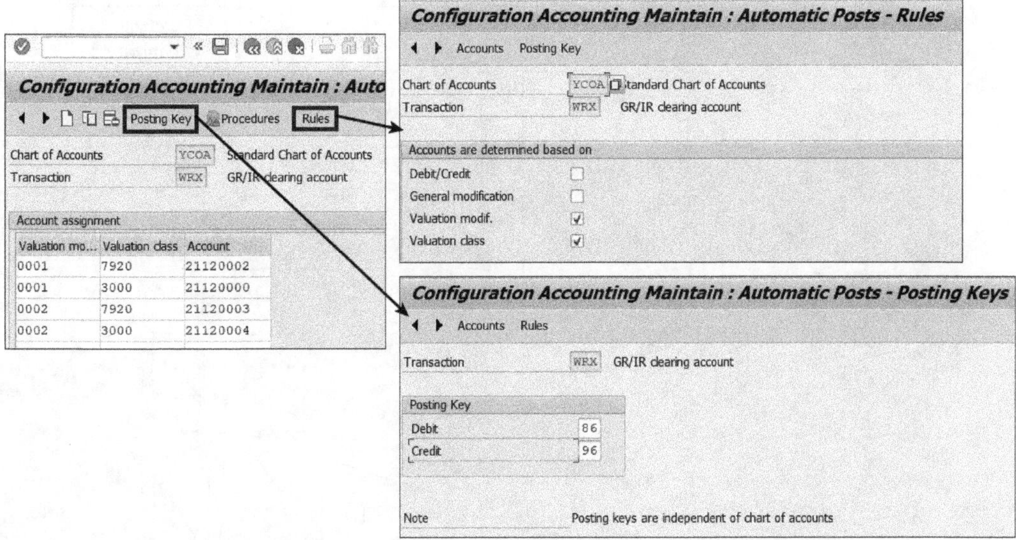

Figure 1.23 Assign Accounts, Rules, and Posting Keys for Inventory GR/IR Accounts

Next let's look at how to configure the account determination for freight expense and freight clearing (or freight GR/IR) accounts.

1.1.5 Condition-Related Accounts

Freight costs and freight GR/IR accounts are condition-related. That is, these accounts are determined based on the purchasing conditions added in the purchase order. Freight is just one example of an additional cost added to the purchasing price through purchasing conditions. There are many other common examples: customs, duties, landing fees, handling fees, and so on.

The account determination of condition-related accounts follows the same transaction keys technique, with the difference that the transaction keys are not assigned to the material movement types but assigned to the conditions in the purchasing pricing procedure.

These conditions can post to the inventory and inventory GR/IR accounts or to separate expense and GR/IR accounts based on the account determination configuration. Figure 1.24 shows an overview of the account determination for condition-related accounts.

The account is determined based on the user inputs saved in the purchase order. The fields determined based on the plant and material are the same as for the inventory account. The difference here is the account key (transaction key), which is determined based on the purchasing pricing procedure configuration.

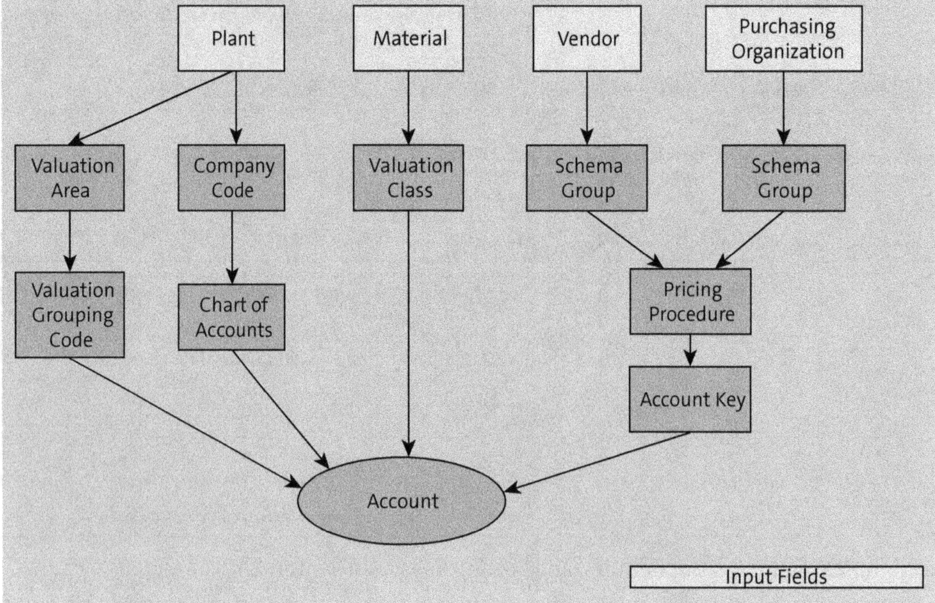

Figure 1.24 Determination of Condition-Related Accounts

The pricing procedure and pricing conditions are important configuration objects in the purchasing process that are not specific to account determination, but in order to control the account determination you must understand where to display the configuration and some important fields that impact the account determination. Let's start with the condition type definition, then consider the account key creation, and then how to assign the account key to the condition type in the pricing procedure. Then we'll take a quick look at how to identify the pricing procedure used in any purchase order so that you can analyze any account determination issues.

Define Purchasing Condition Types

To display the condition configuration, follow SAP configuration menu path **Materials Management • Purchasing • Conditions • Price Determination Process • Define Condition Types (M/06)**.

To post the condition value to separate expense and GR/IR accounts, you need to use two purchasing conditions. The first is standard freight condition FRB1, or you can copy that condition and use the same settings as shown in Figure 1.25.

The fields highlighted in the figure are the ones that impact the account determination. Checking the **Accruals** box means the value of this condition should go to a separate account. The value in the **Plus/Minus** field determines whether this account will be the expense or the GR/IR account. The value here is **A** (**Positive**), so this condition will post to the freight GR/IR account.

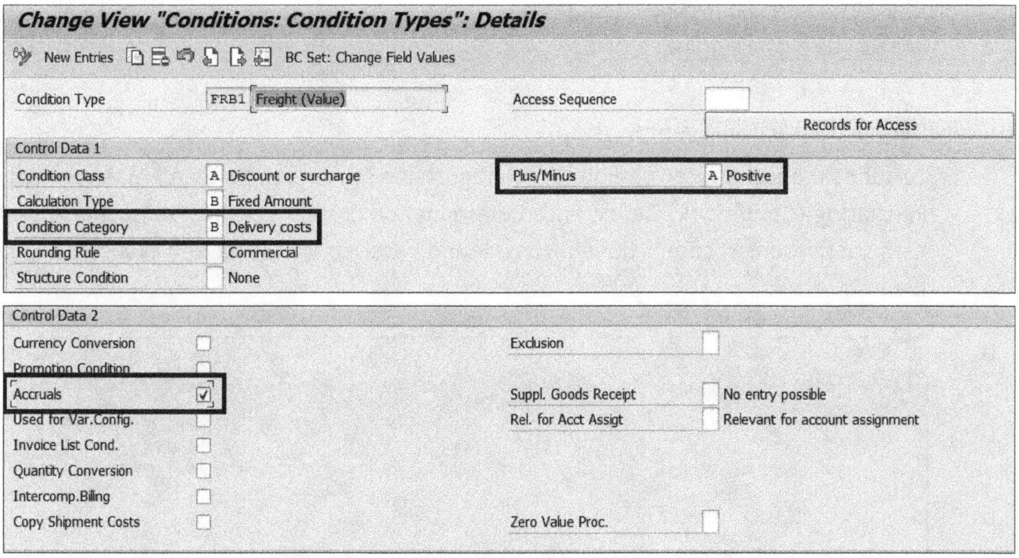

Figure 1.25 Purchasing Condition for Freight GR/IR Posting

The second condition is a new one that you can also copy from FRB1, but you have to change some fields as shown in Figure 1.26.

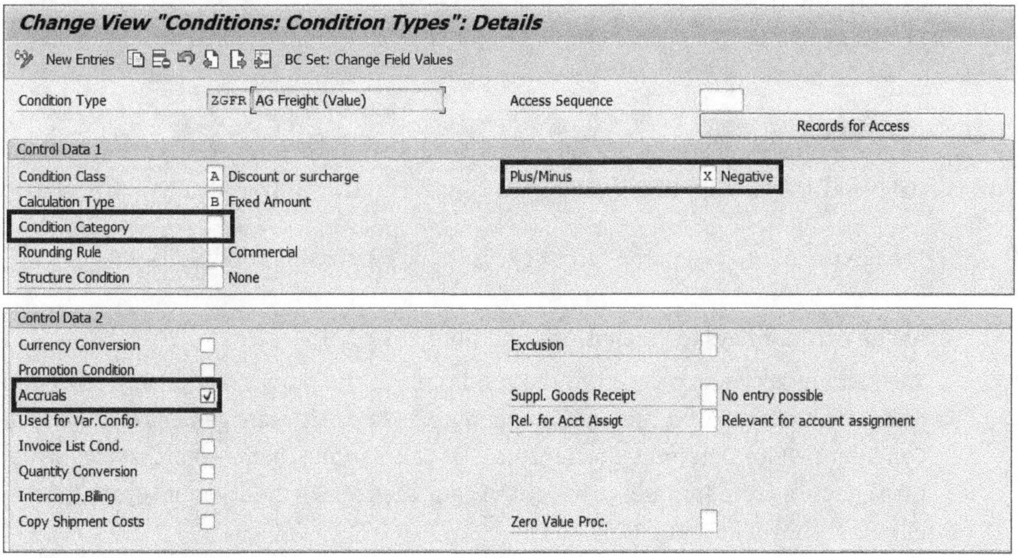

Figure 1.26 Purchasing Condition for Freight Expense Posting

Keep the **Accruals** box checked so that this condition also posts to a separate account. But change the **Plus/Minus** value to **X** (**Negative**). Leave the **Condition Category** field empty. This condition will post to the freight expense account.

Now that you've defined the conditions, the next step is to define the account keys that will be assigned to them.

Define Transaction/Event Keys

To define the account keys, follow SAP configuration menu path **Materials Management • Purchasing • Conditions • Define Price Determination Process • Define Transaction/Event Keys.** On this screen, enter a three-digit code and a description for the key, as shown in Figure 1.27.

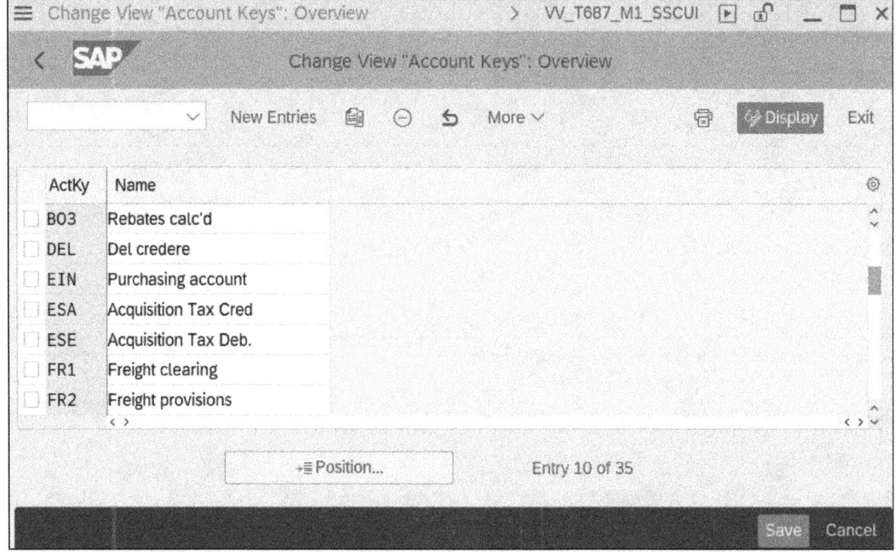

Figure 1.27 Define Transaction/Event Keys

The next step is to assign the account keys to the purchasing conditions.

Assign Account Keys to Purchasing Conditions

The account keys are assigned to the conditions in the purchasing pricing procedure definition. Follow SAP configuration menu path **Materials Management • Purchasing • Conditions • Price Determination Process • Set Calculation Schema—Purchasing (M/08)**, then choose a procedure and go to **Control data**, then assign the account keys to the conditions as shown in Figure 1.28.

Figure 1.28 shows that account key **GFR** is assigned to condition type **ZGFR**, and **FR1** is assigned to **FRB1** in the **Accruals** column.

The last step is to assign the general ledger accounts to the account keys.

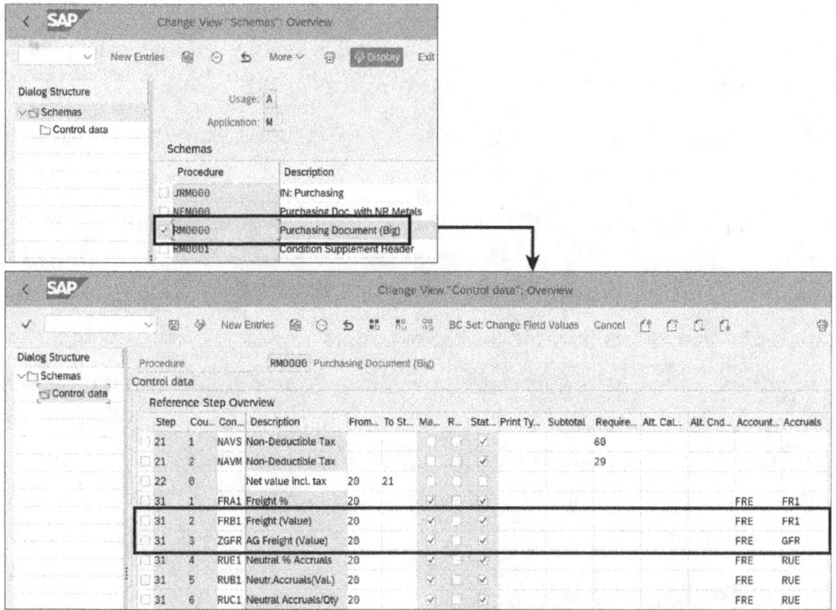

Figure 1.28 Pricing Procedure: Assign Account Keys to Condition Types

Configure Automatic Posting

The general ledger accounts are assigned in the same way as in Section 1.1.2 for inventory accounts. In Transaction OMWB, under **Account Assignment**, you will find two account keys: **FR1** and **GFR**. You can assign posting key **40** for debit postings and **50** for credit.

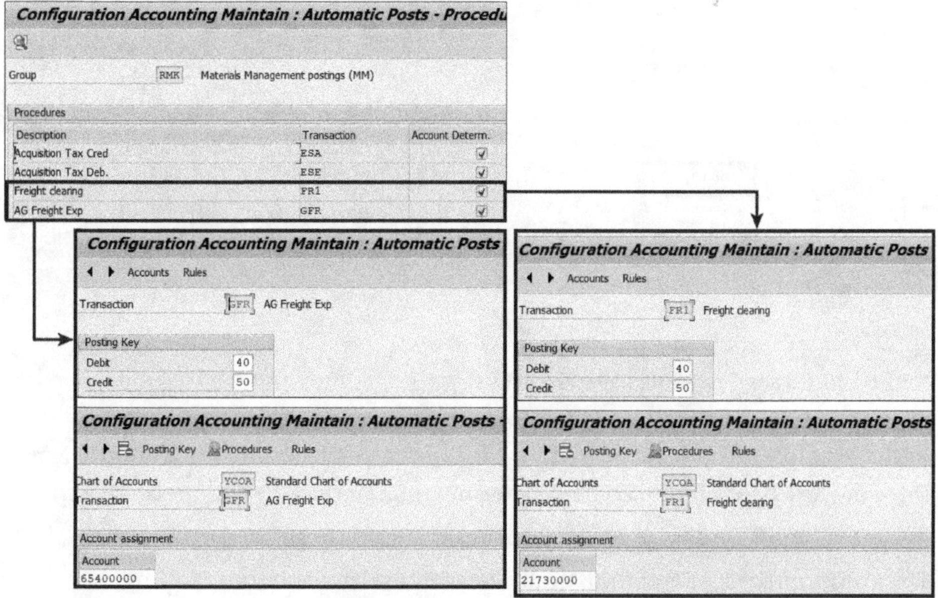

Figure 1.29 Account Assignment: Freight Expense and Freight Clearing (GR/IR)

The account assignment rules can be selected based on the business requirements. In the example in Figure 1.29, we haven't selected any rules as we need only one general ledger account for expenses and another for clearing (or GR/IR).

Now that all the configuration is done, the last step is to be sure these conditions will be filled by the user creating the purchase order.

In the next section, we'll explain briefly how the pricing procedure is connected to the purchase order.

Determination of Pricing Procedures in Purchase Orders

This is a configuration step that's not related specifically to account determination, but it's important to understand how this determination happens to be able to troubleshoot account determination issues. In any purchase order, you can identify the pricing procedure being used by checking the pricing analysis. Figure 1.30 shows Transaction ME23N (Display Purchase Order).

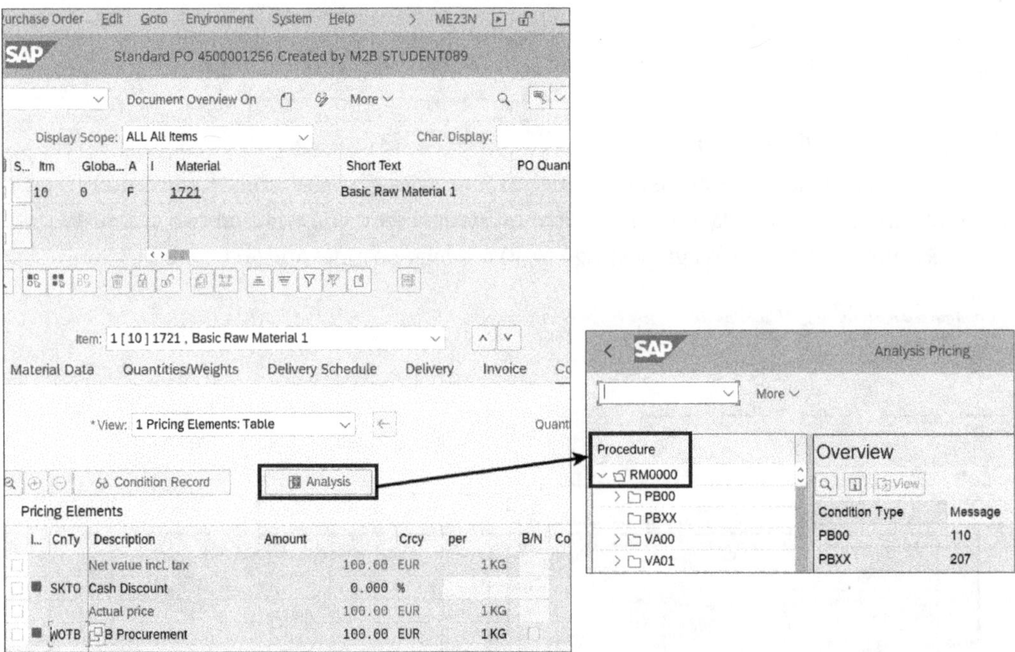

Figure 1.30 Pricing Procedure in Purchase Order

In the example, the pricing procedure used in the purchase order is **RM0000**. Now if there are any issues in the account determination of the condition-related accounts, you know where to check.

This pricing procedure is automatically determined based on the following:

- The schema group assigned to the purchasing organization in configuration Transaction OMFP
- The schema group assigned in the supplier master data, under **Purchasing Data • Additional Purchasing** data in Transaction BP

This link is maintained in configuration Transaction OMFO. Figure 1.31 shows a summary of these steps.

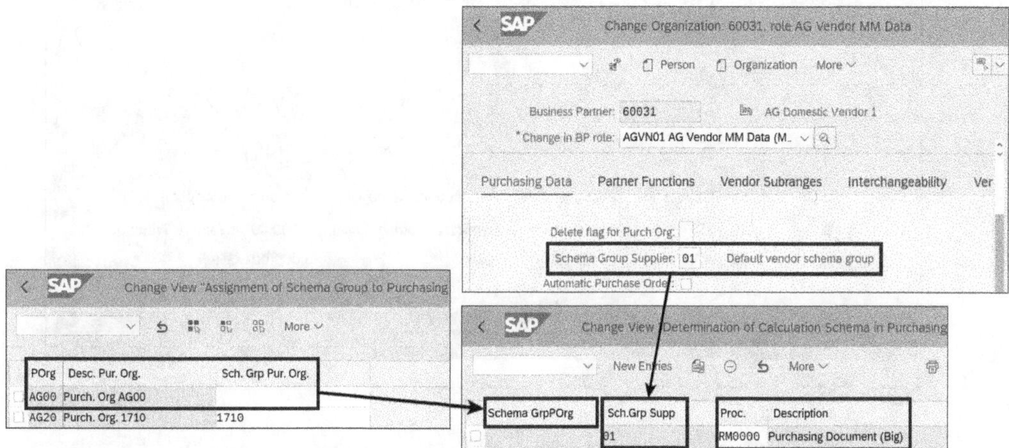

Figure 1.31 Automatic Determination of Pricing Procedure in Purchase Order

Now you understand how the condition-related accounts are determined in the purchasing process, as demonstrated in Figure 1.24.

With this we've covered all the accounts included in Table 1.1. We started by discussing how to determine the inventory, price difference, and GR/IR accounts using the transaction key technique, and then detailed how to determine the freight expenses and GR/IR accounts using the condition technique. Both these techniques are used frequently in different areas in SAP, so we will be referring to this section of the book frequently in following chapters.

Before we move on to the next accounting entry, let's quickly examine how to analyze the goods receipt accounting entry posted in SAP to be able to solve any wrong account determination.

Analyze the GR Accounting Entry

Sometimes you'll find that the wrong accounts are determined in the goods receipt accounting entry, and here it's crucial to understand how to analyze the accounting entry and the configuration to find the error.

The first step is to display the accounting entry using Transaction FB03, and then change the document layout to show the **Transaction** field, as shown in Figure 1.32.

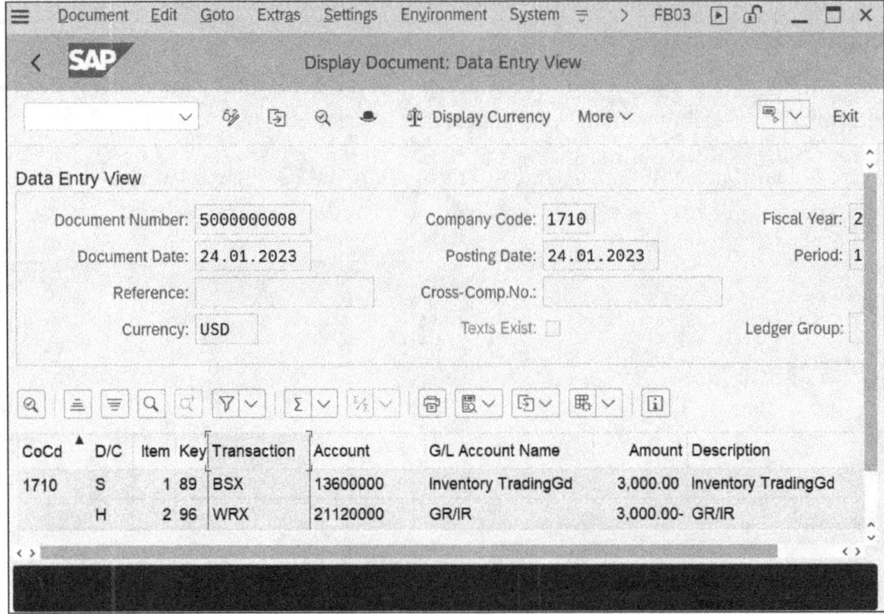

Figure 1.32 Display Accounting Entry: Transaction Key

Here you can see that the first line item is posted to account 13600000, which is determined through transaction key BSX, while the account in the second line item is determined with transaction key WRX. If there's a wrong account in the entry, then you determine the source of the error by following these steps:

1. Confirm the system is using the correct transaction key.
2. Confirm the correct general ledger account is linked to the transaction key.

You then review the configuration as explained in Section 1.1.2, based on the source of the error.

Another thing you can do to analyze inventory management account determination errors is to display all the accounts assigned in configuration Transaction OBYC by displaying table T030. SAP tables can be displayed using Transaction SE16H. Figure 1.33 shows an extract of table T030.

You can then filter on the wrong account that you found in the accounting entry to see where this account is assigned and track the configuration back from there.

With this, we've covered all the accounts included in Table 1.1. You now know how to configure the account determination of these accounts and how to analyze and solve account determination errors.

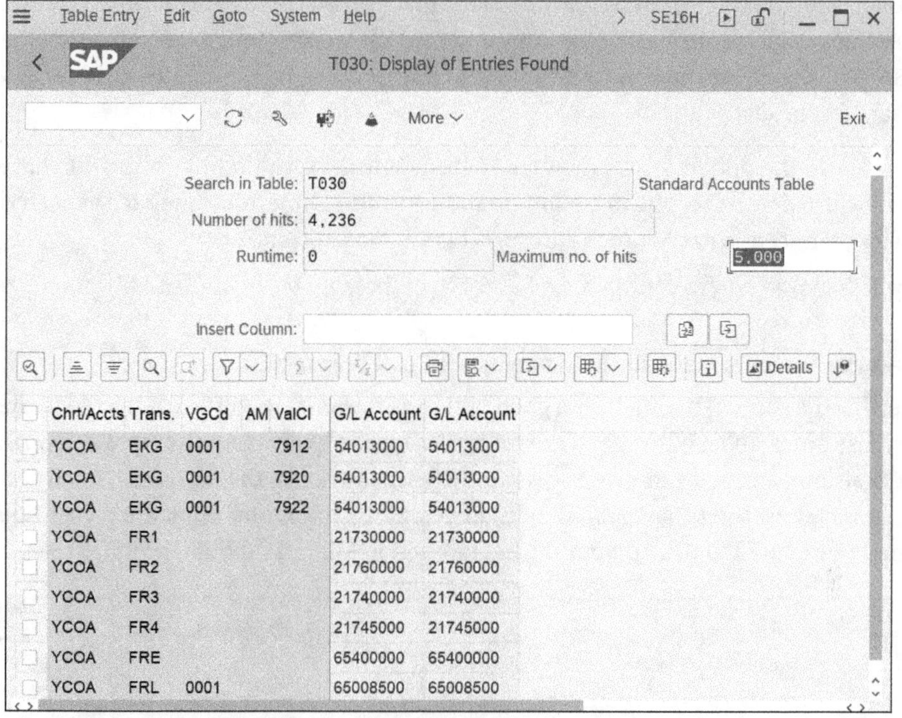

Figure 1.33 Inventory Management Account Assignment: Table T030

Now let's move to the accounts included in the next accounting entry in the normal purchasing process, shown earlier in Table 1.2. The first account in this entry is the vendor reconciliation account.

1.1.6 Vendor Reconciliation Accounts

The vendor reconciliation account is the general ledger AP account that's assigned to the vendor master data. This will be explained in detail in Chapter 4, Section 4.1.

Let's move on to the next account shown in Table 1.2, the unplanned delivery costs account.

1.1.7 Unplanned Delivery Costs Accounts

Unplanned delivery costs are delivery costs that weren't agreed upon in the purchase order. These are extra costs that you may find in the supplier invoice. These costs can be accounted for in several ways:

- Insert the accounts manually in the invoice receipt transaction (Transaction MIRO).
- Determine one account automatically in Transaction MIRO.

- Distribute the costs automatically to all the invoiced line items in Transaction MIRO. The user will only insert the amount in a field for an unplanned delivery cost, and SAP will distribute the cost to the different line items using the line item values as a base for the distribution.

There's also the option to use manual financial entries with no link to the purchase order, which doesn't require any configuration and also is not recommended as there's no clear connection to the purchasing process and no audit trail.

Now let's look at how to configure and use each of these three recommended options.

Insert the Accounts Manually in the Invoice Receipt Transaction

Inserting the account manually is a good option as the posting will be connected to the purchase order, and it offers good flexibility because you can insert multiple accounts. But it also means you can give access to the user to choose the accounts to post to, which increases the probability of human errors. To insert the unplanned cost line items, we can use the **G/L Account** tab in Transaction MIRO, as shown in Figure 1.34.

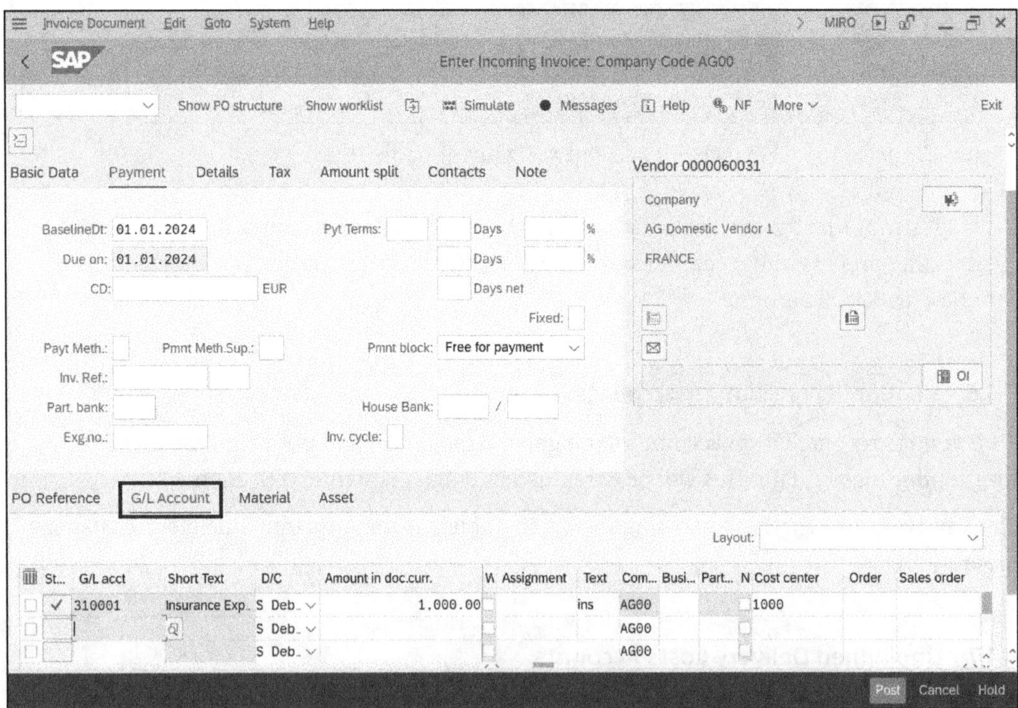

Figure 1.34 Insert Unplanned Cost Accounts Manually in Transaction MIRO

To enable this option, follow SAP configuration menu path **Materials Management • Logistics Invoice Verification • Incoming Invoice • Activate Direct Posting to G/L & Material Accounts & Assets** and check the **Dir. Posting to G/L Acct = Active** box as shown in Figure 1.35.

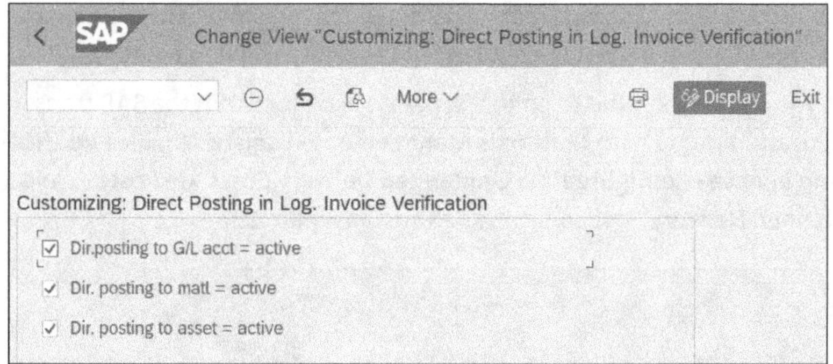

Figure 1.35 Activate Posting to General Ledger Accounts in Transaction MIRO

Next, let's look into the second option for posting unplanned delivery costs.

Determine One Account Automatically in the Invoice Receipt Transaction

Some companies prefer to post automatically to an unplanned delivery cost account and reclassify the entries later if needed. During invoice receipt, the user will only insert the amount in a field for unplanned delivery cost; SAP will determine the account automatically based on the account determination configuration. The unplanned delivery cost amounts can be inserted as shown in Figure 1.36.

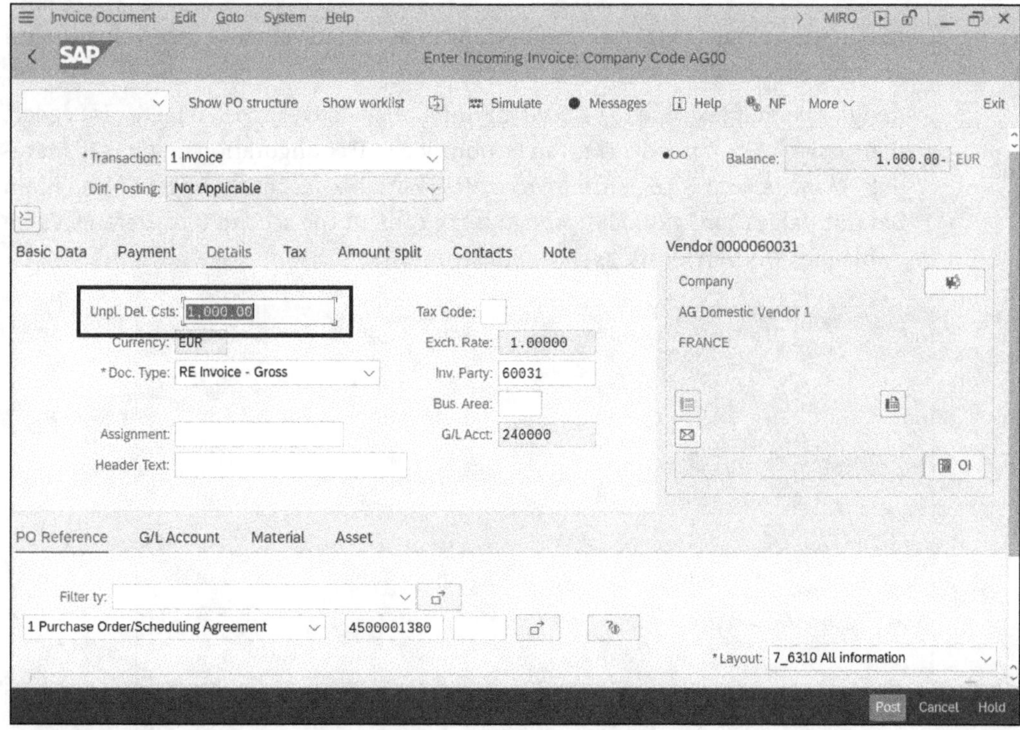

Figure 1.36 Unplanned Delivery Costs Field in Invoice Receipt

To configure the account determination of unplanned delivery costs, there are multiple steps:

1. Tell SAP that you want this value to go to a separate account, from the SAP transaction at configuration menu path **Materials Management • Logistics Invoice Verification • Incoming Invoice • Configure How Unplanned Delivery Costs Are Posted**. Select **2** in the **Unplanned Delivery Cost** column, as shown in Figure 1.37.

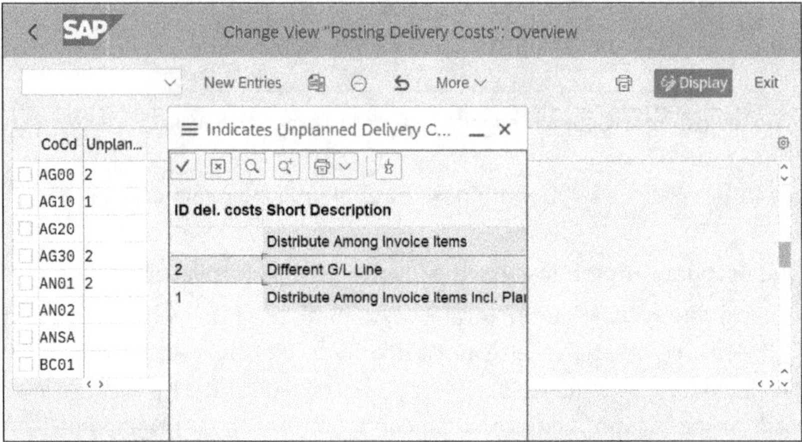

Figure 1.37 Configure How Unplanned Delivery Costs Are Posted

2. Assign the account to transaction key UPF in automatic account determination via Transaction OMWB.
3. Assign a default tax code to be used for unplanned delivery cost. If no tax is needed, then assign a 0% tax code. This can be done via SAP configuration menu path **Materials Management • Logistics Invoice Verification • Incoming Invoice • Maintain Default Values for Tax Codes.** Assign a tax code in the second box, **Default Value Unplanned Delivery Costs**, as shown in Figure 1.38.

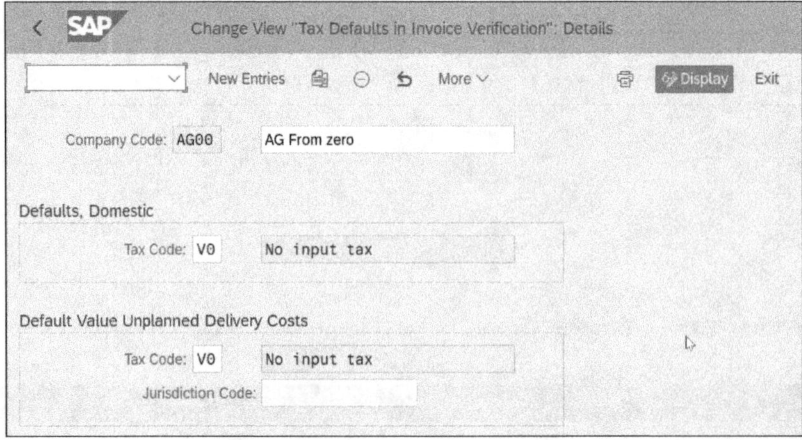

Figure 1.38 Assign Default Tax Code for Unplanned Delivery Cost Posting

4. Assign a default cost object to assign the unplanned delivery costs to. This can be done in configuration Transaction OKB9, as shown in Figure 1.39.

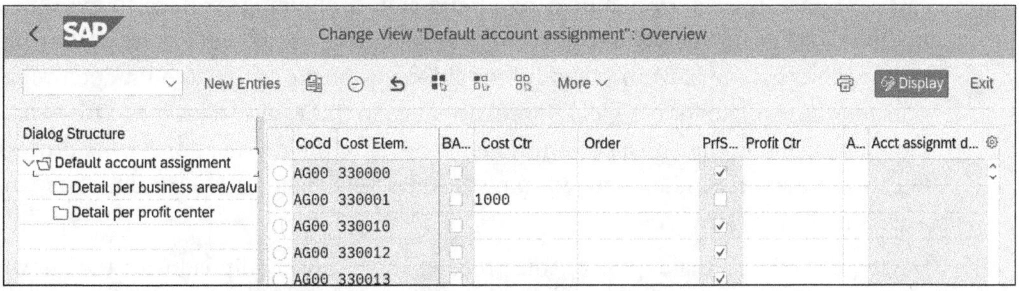

Figure 1.39 Assign Default Cost Object to Unplanned Delivery Costs Account

The last option for posting the unplanned delivery costs is to distribute the amount to all the invoiced line items.

Distribute the Costs Automatically to All the Invoiced Line Items

When you distribute the amount to the invoiced line items, it will be treated as a normal purchase price variance. The posting will follow the same treatment as explained in the Invoice Receipt subsection of Section 1.1.1. The value will be split proportionally between the invoice line items, based on either

- the value of the materials invoiced in each line item, or
- the value of the materials plus the planned delivery costs in each line item.

To choose which base you want to use, go to the same transaction as was shown in Figure 1.37, and then either choose **1** to include the planned delivery costs or leave the field empty to distribute the amount based on the material value only.

The unplanned delivery cost account is an expense account that you can define either with or without a cost element. If the account is defined with a cost element and you decide that the account is to be determined automatically without user input, then you must define a default cost assignment for this account in the same way as explained for the PPV account (Section 1.1.3).

Now that you understand how to process the unplanned delivery costs and how to configure the account determination, let's move to the next account that was shown in Table 1.2: the input tax account.

1.1.8 Input Tax Accounts

Input and output VAT account determination is configured centrally and then used in different business processes. The account determination configuration for VAT will be explained in Chapter 4, Section 4.3, so let's move to the next account in Table 1.2: the withholding tax account.

1.1.9 Withholding Tax Accounts

Withholding taxes are configured centrally and then used in the different business processes. The account determination configuration for withholding taxes will be explained in Chapter 4, Section 4.4. With this comprehensive coverage of the normal purchasing business process, we have thoroughly examined the details of the account determination and the configuration for all relevant accounts in this essential process. This exploration should not only enhance your understanding of the normal purchasing process but also lay a solid foundation for grasping the financial dynamics within materials management in SAP S/4HANA.

Having delved into the depths of this process, you're now equipped with the knowledge to navigate through the complexities of the purchasing transactions and their accounting implications in SAP. This understanding is crucial for ensuring accurate financial reporting of procurement activities.

As we conclude this section, we turn our attention to a common variation of the normal purchasing process, which involves the purchasing of consumables. This variation presents its own unique challenges and opportunities in terms of process flow and account determination, warranting a dedicated exploration. Let's examine the business process for purchasing consumables, diving into its specific requirements and account determination configuration within SAP S/4HANA.

1.2 Purchasing of Consumables

In this section, we turn our attention to the purchasing of consumables, a fundamental aspect of materials management in SAP S/4HANA. Unlike the procurement of inventory items, the process of acquiring consumable materials, such as office supplies or factory consumables, has its unique process flow and accounting implications.

Consumables are typically items that are used directly and not stored as inventory. The process of purchasing these items may seem straightforward, but it is critical to understand its impact on financial accounting and how SAP facilitates their acquisition and expense tracking.

Here, we will explore the end-to-end process of purchasing consumables, from placing the order to receiving and paying for the items. Special focus will be given to the accounting entries generated during the goods receipt of consumables, which differ from those for inventory items due to their direct consumption nature.

We also will explore the configuration aspects in SAP for accurately determining the appropriate expense accounts related to consumables. This includes setting up account determination rules that ensure correct financial postings, reflecting the nature of these transactions.

Understanding the purchasing process for consumables is crucial for businesses to maintain accurate financial records and effective control over their expenditures. This section aims to equip you with the knowledge to manage this process seamlessly within SAP S/4HANA, aligning with our book's objective of comprehending business processes and mastering account determination. Now let's look at an overview of the purchasing of consumables business process.

1.2.1 Business Process Overview

The consumables purchasing process is a variation of the normal purchasing process, and it follows exactly the same process flow outlined in Section 1.1.1. The only difference from the normal purchasing process is the accounting entry posted at goods receipt, as shown in Table 1.4.

Debit	Credit	Debit Amount ($)	Credit Amount ($)
Consumption/expense		1,000	
Price variance		100	
Freight costs		50	
	Inventory GR/IR		1,100
	Freight GR/IR		50

Table 1.4 Accounting Entry of Goods Receipt of Consumables into Inventory

Instead of posting the purchased items' value into the inventory account, you post the value directly to an expense account that you configure in the account determination. You also have to either insert a cost object when creating the purchase order or have a default cost object assigned automatically based on the configuration. All the other accounts in the accounting entry are the same as in the normal purchasing process.

Now let's take a detailed look at how to configure the account determination for this expense account.

1.2.2 Consumption/Expense Accounts

When purchasing consumables, you can either have master data created for the items you're buying, or you can buy a generic item without master data but insert a material group. The account determination configuration differs slightly based on this, as we'll cover in the following sections.

Items with Material Master Data

If the item has material master data, then it also has a valuation class maintained. SAP will use the valuation class to determine the consumption account, the same as for the determination of inventory accounts, as explained in Section 1.1.2. But it will use a different transaction key based on the account assignment category inserted in the purchase order.

When creating the purchase order, to inform SAP that this is a consumable item you insert a value in the **AcctAssgntCateg** (account assignment category) field, as shown in Figure 1.40.

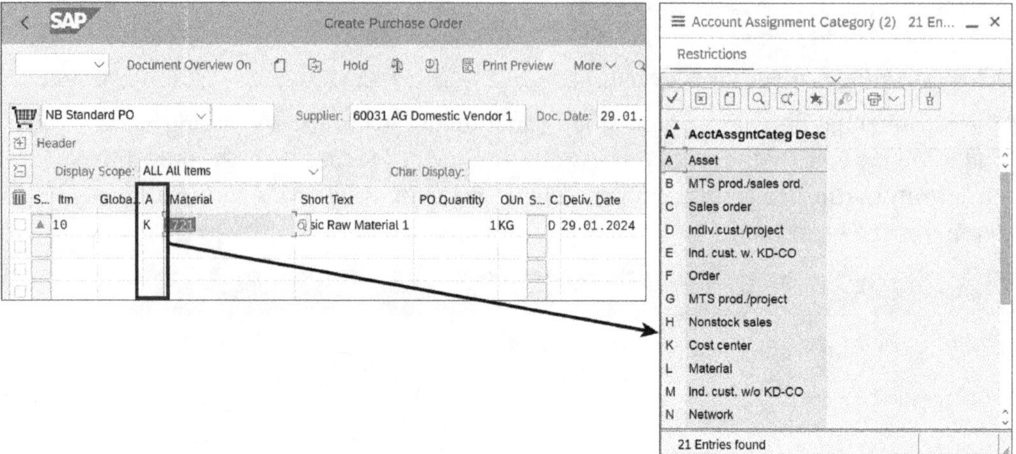

Figure 1.40 Purchase Order of Consumables with Master Data

Once you insert a value in the **AcctAssgntCateg** field, SAP will use transaction key GBB to determine the consumption account, but GBB has many account modifications that give flexibility to determine different general ledger accounts for different account assignment categories. The mapping between the account assignment categories and the GBB account modification can be seen in the SAP configuration help for Transaction OMWN (see Figure 1.10). Account assignment category K triggers transaction key GBB with account modification VBR. The consumption general ledger account can be assigned in configuration Transaction OMWN, as shown in Figure 1.41.

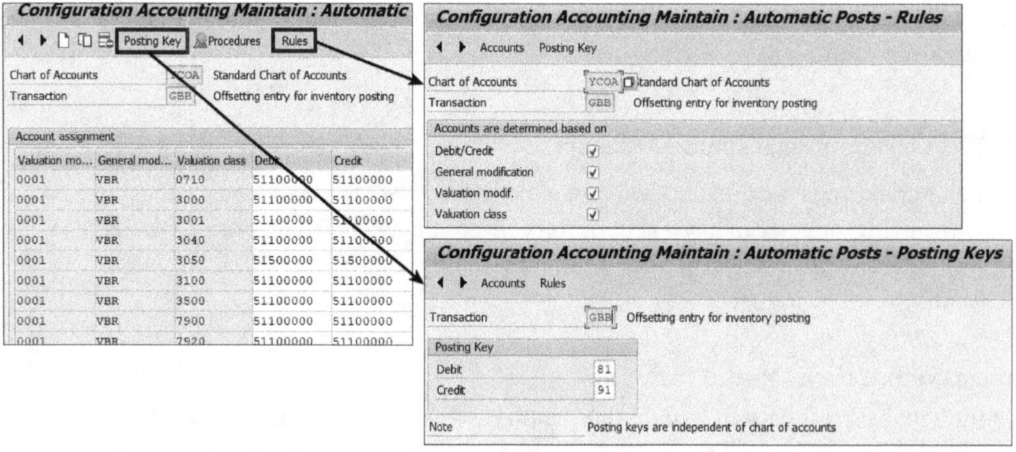

Figure 1.41 Assign Accounts, Rules, and Posting Keys for Consumption Accounts

The second option for purchasing consumables is to purchase items without material master data.

Items without Material Master Data

In the case of items without master data, there's no material code to insert in the purchase order, and it's mandatory to insert a material group. You can either allow the user to manually insert the general ledger account while creating the purchase order or you can configure the automatic account determination by assigning a valuation class to the material group in configuration.

Figure 1.42 shows how the user can manually insert the consumption general ledger account and cost center while creating the purchase order.

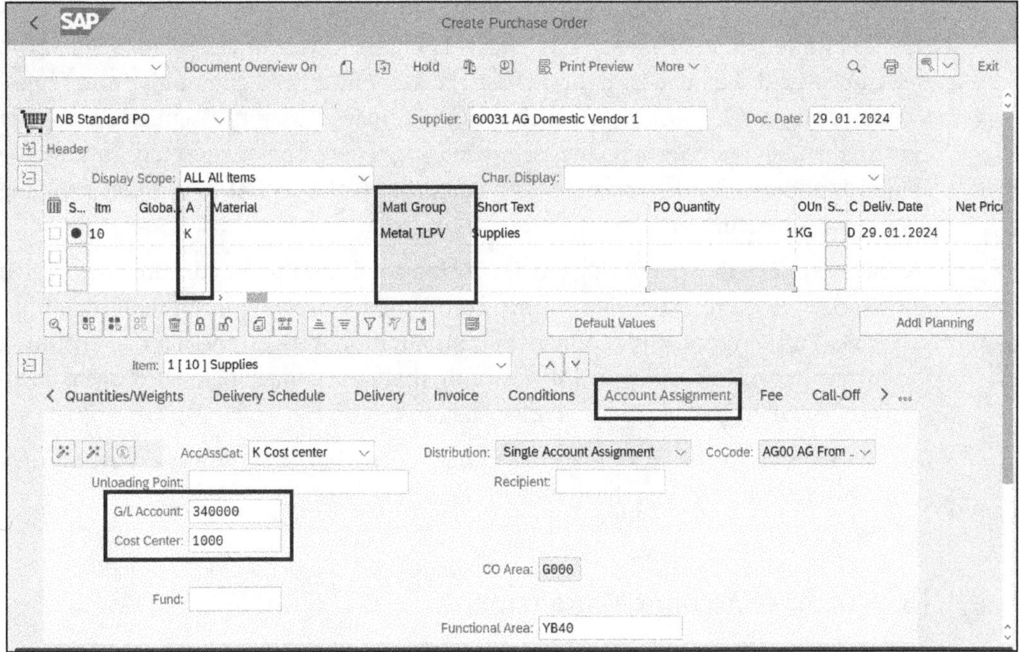

Figure 1.42 Purchase Order for Consumables without Master Data

To assign a valuation class to a material group, follow SAP configuration menu path **Materials Management • Purchasing • Material Master • Entry Aids for Items without a Material Master** and assign a valuation class as shown in Figure 1.43.

SAP will use this valuation class to determine the consumption or expense general ledger account, the same way as for items with material master data.

Now you should understand the general ledger account determination configuration needed for the purchasing of consumables process. The same concept applies to any purchasing process that should post to an expense account instead of inventory accounts—such as purchasing of services or purchase orders for travel expenses.

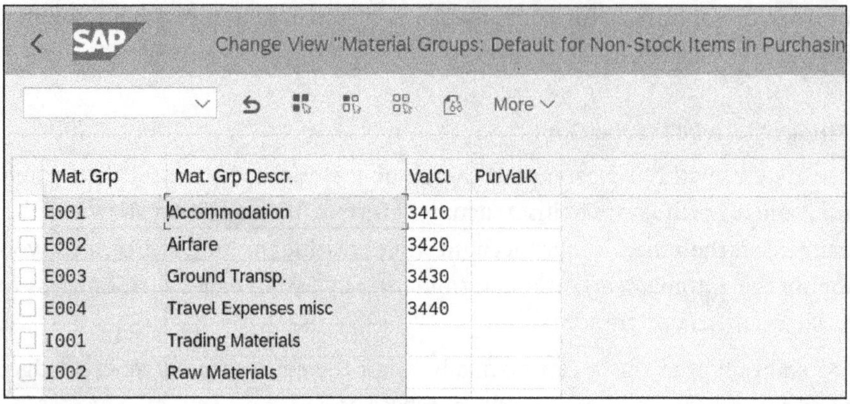

Change View "Material Groups: Default for Non-Stock Items in Purchasin

Mat. Grp	Mat. Grp Descr.	ValCl	PurValK
E001	Accommodation	3410	
E002	Airfare	3420	
E003	Ground Transp.	3430	
E004	Travel Expenses misc	3440	
I001	Trading Materials		
I002	Raw Materials		

Figure 1.43 Assign Valuation Class to Material Group

Now that we've explored the detailed process of purchasing consumables in SAP S/4HANA, we've uncovered the nuances of how these transactions differ significantly from typical inventory purchases. Consumables such as office supplies or factory essentials are directly expensed at the time of receipt, bypassing traditional inventory accounting. This unique approach underscores the versatility and adaptability of SAP S/4HANA when handling diverse procurement scenarios.

We have explored the accounting entries during goods receipt for consumables, which are directly posted to expense accounts, contrasting sharply with standard inventory processes. Furthermore, we examined the configuration steps essential for determining the correct expense accounts in SAP, ensuring that each consumable purchase is accurately reflected in the company's financial records.

Now let's move to the next material management business process that requires account determination: purchase account management.

1.3 Purchase Account Management

In this section, we'll focus on *purchase account management*, a significant function in SAP S/4HANA that's especially pertinent in countries where it is a legal requirement to track values associated with the procurement of external stock items. This process involves documenting and managing the financial values that are posted for externally procured materials, ensuring compliance with specific legal standards.

Purchase account management in SAP S/4HANA is designed to facilitate this tracking either in a single purchase account, encompassing both stock and delivery values, or in two separate accounts, depending on the legal requirements of the country. This function is crucial for organizations to maintain accurate and legally compliant records of their external procurement expenditures.

Key aspects of purchase account management we will explore include the following:

- **Activation and configuration**
  We will examine how to activate purchase account management for a company code in SAP and define the value calculation applicable to the purchase account. This includes settings in Customizing under **Materials Management** and **Invoice Verification.**
- **Updating purchase accounts**
  We will discuss the methods of updating purchase accounts, either at the receipt value or at the stock value, and how these affect postings in invoice verification and goods receipt.
- **Separate accounting documents**
  The configuration for creating separate accounting documents for purchase account postings will be addressed, highlighting its importance in financial reporting.

Throughout this section, our objective is to provide a thorough understanding of how purchase account management operates within SAP S/4HANA, emphasizing the accounting entries and the account determination configuration. Now let's start with the business process overview.

1.3.1 Business Process Overview

Purchase account management is an option you can add to purchasing processes. The process stays the same as described in Section 1.1.1, but the financial entries posted at goods receipt and invoice receipt also post to a purchasing account and a purchasing offsetting account. We'll go over the accounting entries in the following sections.

Goods Receipt with Purchase Account Management

In the example shown in Table 1.5, we see the goods receipt accounting entry when purchase account management is activated.

Debit	Credit	Debit Amount ($)	Credit Amount ($)
Inventory		1,000	
Price variance		150	
	Inventory GR/IR		1,100
	Freight GR/IR		50
Inventory purch. acct.		1,100	
Freight purch. acct.		50	
	Purch. offsetting acct.		1,150

Table 1.5 Accounting Entry of Goods Receipt into Inventory with Purchase Account Management

In this example, we purchase items for a price of $1,100 and a freight cost of $50. The inventory account is debited for $1,000, which is the standard cost. The rest of the purchase value ($1,100 + $50) goes to the price variance account, which is debited for $150. The inventory GR/IR is credited with the items' purchase price of $1,100, and the freight GR/IR is credited with the freight value of $50. Because purchase account management is activated, the purchase price is also posted to the inventory purchasing account, which is debited for $1,100, and the freight value is posted to the freight purchasing account, which is debited for $50. The purchase accounts' values are offset by posting a credit to the purchasing offsetting account of $1,150.

Invoice Receipt with Purchase Account Management

Next, let's assume we post the invoice receipt with an additional price variance of $20, as shown in Table 1.6.

Debit	Credit	Debit Amount ($)	Credit Amount ($)
Inventory GR/IR		1,100	
Freight GR/IR		50	
Price variance		20	
	Accounts payable		1,170
Freight purch. acct.		20	
	Purch. offsetting acct.		20

Table 1.6 Accounting Entry of Supplier Invoice Receipt with Purchase Account Management

The inventory GR/IR and the freight GR/IR are both debited with the same value as the one posted at goods receipt to clear the GR/IR accounts. We also have $20 posted to price variance. Because purchase accounts should reflect the full value spent to purchase the items, the $20 price variance will also be posted to the purchase and the purchase offsetting accounts.

Now let's look at how to configure the automatic account determination for each of the purchasing management accounts.

1.3.2 Inventory Purchasing Management Account

This account is determined using the same technique as outlined in Section 1.1.2. The general ledger account can be assigned in configuration Transaction OMWB to the transaction key EIN, as shown in Figure 1.44.

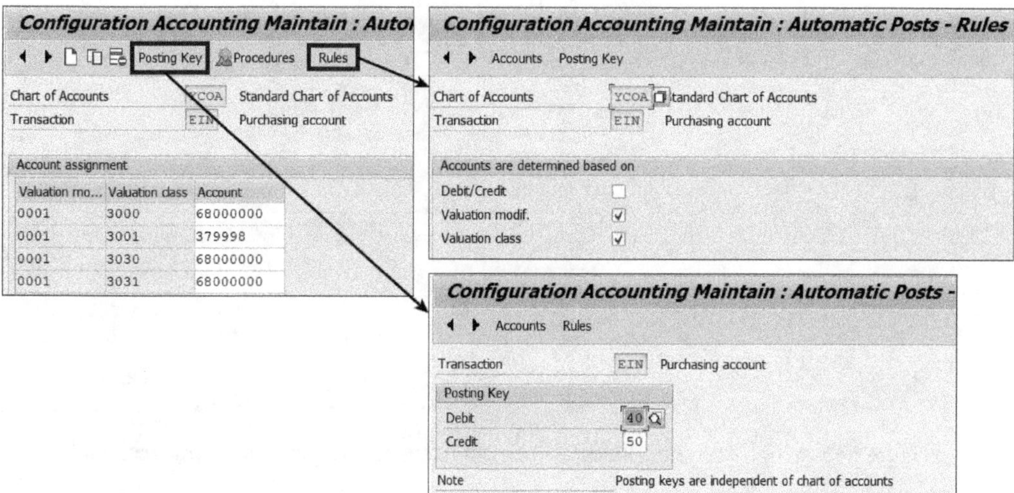

Figure 1.44 Assign Accounts, Rules, and Posting Keys for Inventory Purchasing Account

1.3.3 Freight Purchasing Management Account

The account determination for this account is configured via the same technique as outlined in Section 1.1.5. You assign the account keys to the purchasing condition types in the pricing procedures through Transaction M/08, where you assign the account keys for the purchasing account management accounts in the **Account Key** column. In Figure 1.45, you can see account key **FRE** assigned to freight condition **FRB1.**

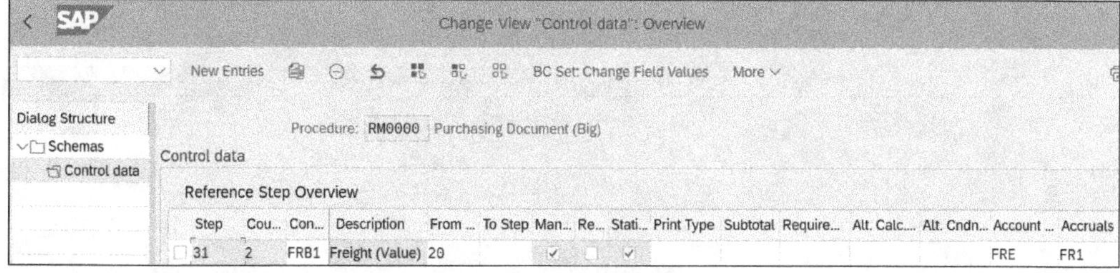

Figure 1.45 Assign Freight Purchasing Account Key in Pricing Procedure

> **Note**
>
> The account key assigned in the accruals columns is used to determine the freight GR/IR or clearing account, as explained in Section 1.1.5.

The next step is to assign the account key to a general ledger account in configuration Transaction OMWB, as shown in Figure 1.46.

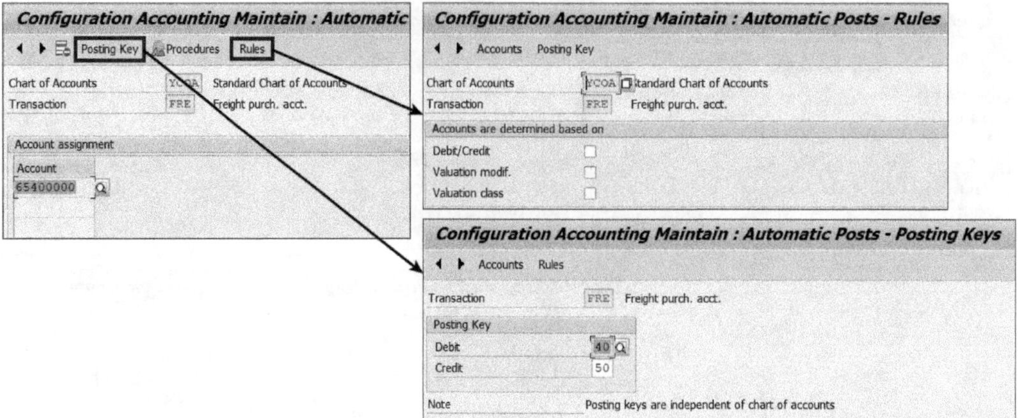

Figure 1.46 Assign Accounts, Rules, and Posting Keys for Freight Purchasing Account

1.3.4 Purchasing Management Offsetting Account

This account is determined using the same technique as outlined in Section 1.1.2. The general ledger account can be assigned in configuration Transaction OMWB to transaction key EKG, as shown in Figure 1.47.

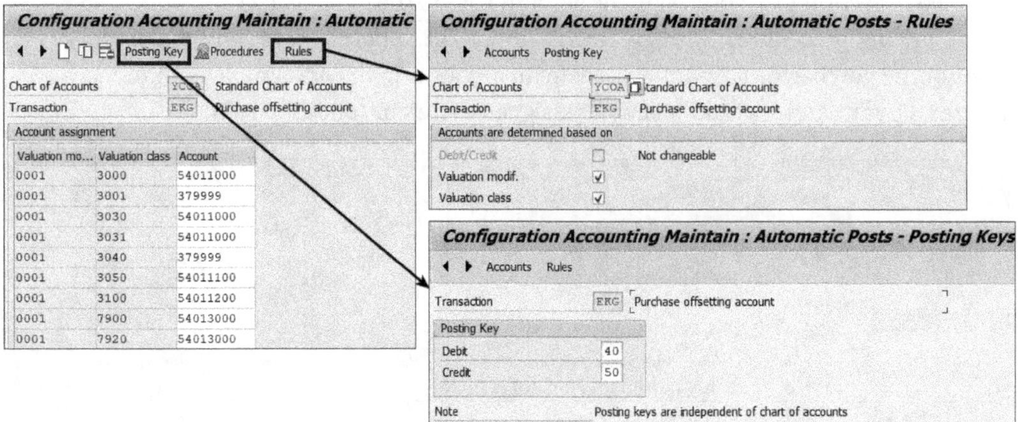

Figure 1.47 Assign Accounts, Rules, and Posting Keys for Purchasing Offsetting Account

Now you should understand how to configure the account determination for all the accounts related to purchasing account management.

Through our exploration of purchase account management in SAP S/4HANA, we've offered valuable insights into how this function allows for the tracking of values associated with external procurement. This process, crucial in countries with specific legal requirements, involves managing the financial values of externally procured materials in dedicated accounts. We've shown you the steps needed to activate and configure purchase account management and update purchase accounts, and we've discussed the implications of price variances.

You've learned how purchase account management plays a pivotal role in ensuring transparency and compliance in financial reporting, particularly in tracking both stock and delivery values. By understanding the practical application and configuration of purchase account management, you are now better prepared to ensure that your SAP systems align with legal requirements and accurately reflect the financial aspects of procurement activities. This knowledge forms a critical part of effective materials management and underscores the integrated nature of SAP S/4HANA.

Now let's proceed to the next business process: vendor consignment.

1.4 Vendor Consignment

In this section, we dive into the concept of *vendor consignment*, a distinct and strategic procurement process utilized in SAP S/4HANA. This specialized approach is where goods are shipped to the buyer, but the ownership remains with the vendor until the items are either consumed or sold. By storing the vendor's stock at the buyer's location without immediate ownership transfer, this process introduces a unique dynamic in inventory and financial management.

Vendor consignment offers substantial benefits for businesses by significantly reducing inventory holding costs and associated risks. It allows companies to operate more efficiently, as they are only financially accountable for stock that is actually used or sold. This method is particularly advantageous in managing products with unpredictable demand, high value, or limited shelf life.

In our exploration of vendor consignment, we will cover the following:

- The difference between vendor consignment and customer consignment.
- Accounting and financial impacts—that is, the financial transactions and accounting entries that occur when the consigned goods are consumed or sold. We'll also discuss how these transactions are recorded in SAP S/4HANA.
- We will also dive into the specifics of configuring the account determination for the vendor consignment process.

Vendor consignment in SAP S/4HANA not only enhances procurement efficiency but also aligns closely with an organization's financial and operational strategies. This process is used by many retail and wholesale chains, such as Costco. Now let's first look into the difference between vendor consignment and customer consignment, as this is a common point of confusion.

1.4.1 Vendor Consignment versus Customer Consignment

The consignment process exists in two forms: vendor consignment and customer consignment. To understand the difference, let's take Costco as an example. Costco is a wholesale chain selling products to final customers, and one of the product categories it sells is electronic devices. Let's assume Costco sells Samsung electronic devices using the consignment model. Vendor consignment is from the point of view of Costco (in this case, called the *consignee*) when getting the devices from Samsung (the *consignor*), while customer consignment is from the point of view of Samsung when sending the devices to Costco, as shown in the example in Figure 1.48.

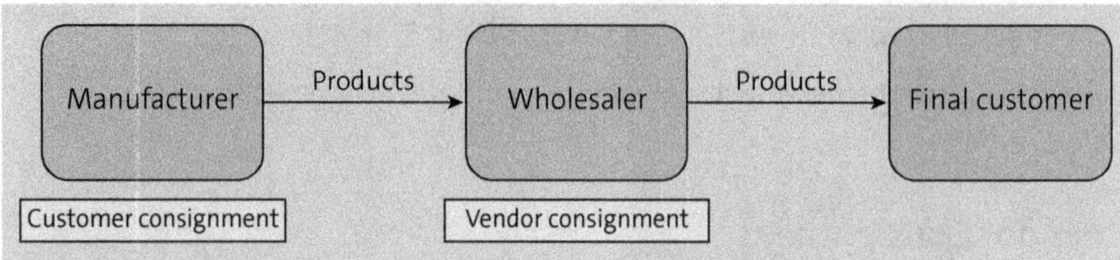

Figure 1.48 Difference between Vendor and Customer Consignment

In this section we will discuss the vendor consignment process; the customer consignment process will be discussed in Chapter 2.

Next, let's look at the vendor consignment process flow and discuss which accounting entries are posted in which steps throughout the process. Then we'll walk through how to configure the account determination for these accounting entries.

1.4.2 Business Process Overview

The vendor consignment process flow is illustrated in Figure 1.49. The steps that cause accounting entries are highlighted in green.

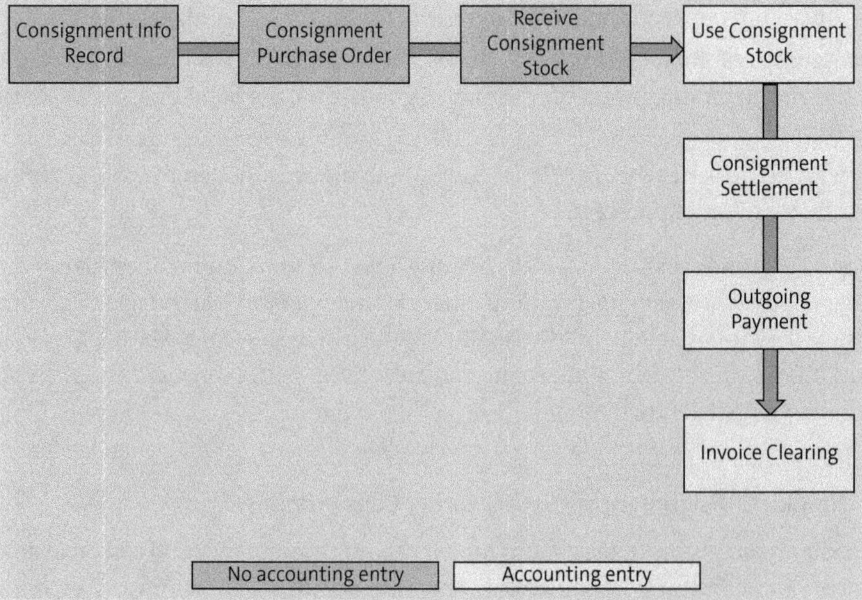

Figure 1.49 Vendor Consignment Process Flow

Let's take a quick look at each of these steps. We'll put more focus on the steps that cause accounting entries.

Consignment Info Record

The consignment info record documents the agreement you have with a supplier and includes details such as delivery instructions, planned delivery time, and (importantly) prices. The prices in the consignment info record are what will be used to valuate the consignment inventory in the following steps. This step has no accounting entry.

Consignment Purchase Order

This is a normal purchase order like the one used in the process flow in Section 1.1. To specify to the system that this is a consignment order, use item category K on the level of the purchase order item, which means that the same purchase order can have some normal line items and other consignment line items. No price is inserted for the consignment line items. The price is to be picked automatically from the consignment info record when you consume the consignment stock in the following steps. This step has no accounting entry.

Receive Consignment Stock

You receive the consignment stock from the supplier into your storage locations. The consignment stock doesn't belong to your company, so this transaction shouldn't impact your accounting records. The transaction should only impact the stock quantities without value, and you should be able to see in your stock reports that you have available consignment stock that's owned by the supplier. This step has no accounting entry.

Use Consignment Stock

You can use the consignment stock available in your custody either internally or externally. With internal usage, you consume the items for production or any other reason; with external usage, you sell the items. The stock usage can be done either in one step or in two steps, and the accounting entries will change based on which option you use, as discussed ahead.

Two-Step Process

In the two-step process, you transfer the consignment stock to your own stock and then consume it internally or externally. When transferring the consignment stock to your own stock, you post the accounting entry shown in Table 1.7.

Debit	Credit	Debit Amount ($)	Credit Amount ($)
Inventory		10,000	
Price variance		1,000	
	Consignment GR/IR		11,000

Table 1.7 Accounting Entry: Transfer from Consignment Stock to Own Stock

Also, the stock transfer will post a material document that will reduce the quantity of consignment stock and increase the quantity of own stock, and this is reflected in stock reporting. Now let's consider the meaning and use of each of the accounts in the accounting entry:

- **Inventory**
  This is the same stock balance sheet account explained in Section 1.1.2. If the item is valuated with the moving average cost, then the value posted here will be equal to the price maintained in the consignment info record multiplied by the quantity while If the item is valuated with the standard cost, then the value posted here is equal to the standard cost multiplied by the quantity. The difference between the standard cost and the purchase price maintained in the consignment info record goes to the purchasing price variance account, which is explained next.
- **Price variance**
  This is the same PPV account explained in Section 1.1.2. The PPV account will appear in this posting only if the item is valuated with the standard cost and there's a difference between the standard cost and the price maintained in the consignment info record. The value posted to this account is equal to the difference between the standard cost and the price maintained in the consignment info record multiplied by the quantity.
- **Consignment GR/IR**
  This is a similar account to the one used in the normal purchasing process, as explained in Section 1.1.1, but it's determined using a different transaction key. We'll look into the account determination configuration after the business process overview. The value posted to the consignment GR/IR account is equal to the value that you must pay to the supplier, which is the value maintained in the consignment info record.

We've already covered the concepts behind and configuration of all the accounts used in this accounting entry in previous sections, so now let's move to the second step in the consignment stock usage.

The second step is to use the stock you transferred to own stock in the first step. There are a lot of usage options available, such as selling the stock, consuming it in house, or using it in production. The accounting entry for the second step will depend on the usage transaction. All the different transactions are covered in other chapters of this book: the sales related usage is explained in Chapter 2, and the production related usage is explained in Chapter 3.

Now you can understand why this is a two-step process for the consignment stock usage: in the first step you move the consignment stock to your own stock, and in the second step you use this stock as needed. Next, let's look into the second option for consignment stock usage: the one-step process.

One-Step Process

In the one-step process for consignment stock usage, you consume directly from the consignment stock. You can consume the consignment stock any way you want—for example, in-house consumption, sales, or production. In this case, the accounting entry will be as shown in Table 1.8.

Debit	Credit	Debit Amount ($)	Credit Amount ($)
Expense account		11,000	
	Consignment GR/IR		11,000

Table 1.8 Accounting Entry: Direct Consumption of Consignment Stock

Now let's consider the meaning and use of each of the accounts in the accounting entry:

- **Expense account**
  This account depends on the consumption scenario, as explained in different parts of the book. For example, if you sell the consignment stock, then this will be the COGS account, which will be explained in Chapter 2. If you consume it in house, then it will be as explained in Section 1.2.2. The value posted to the expense account is equal to the price maintained in the consignment info record multiplied by the quantity used.
- **Consignment GR/IR**
  The consignment GR/IR account is the same account as we discussed earlier in this section. The choice between the one-step and two-step process depends on the business scenario. Now that you understand the use consignment stock step of the vendor consignment process, let's move to the next step: consignment settlement.

Consignment Settlement

Once you use the consignment stock, you're liable to the vendor for the price of this stock and should update your accounting records accordingly. Normally you do this settlement on a monthly basis: you check all the consignment stock that you used and post a supplier invoice. When you post the consignment settlement, the accounting entry is as shown in Table 1.9.

Debit	Credit	Debit Amount ($)	Credit Amount ($)
Consignment GR/IR		11,000	
	Supplier accounts payable		11,000

Table 1.9 Accounting Entry: Consignment Settlement

Now let's consider the meaning and use of each of the accounts in the accounting entry:

- **Consignment GR/IR**
  - This is the same account and the same value as in Section 1.1.1.
- **Supplier accounts payable**
  - This is the supplier reconciliation account. It's the same as that explained in the normal purchasing process in Section 1.1.1.
  - This entry may also include VAT, like the normal IR purchasing process.

Now you know that the consignment settlement step is responsible for posting the supplier invoice. The next steps in the vendor consignment process are shown in Figure 1.49: outgoing payment and invoice clearing. These steps are the same as in the other purchasing processes and have already been explained in Section 1.1.

With this, you should understand the different process steps in the vendor consignment process. We also discussed the different accounting entries posted and when they are posted, and you've seen the different process options available for the consignment stock usage.

Now, let's look at the account determination of the different accounts in the vendor consignment process. We've already explained the account determination of all the accounts, as they are also used in the previous purchasing processes, except the consignment GR/IR account. Let's look into that.

1.4.3 Consignment GR/IR Account

The consignment GR/IR account is determined using the same technique as in Section 1.1.2. The difference between the consignment GR/IR and the normal GR/IR comes from the special stock indicator you insert when you post the consignment GR/IR goods receipt, which triggers a different transaction key. For example, to post a direct consumption from consignment stock to a cost center, use movement type 201 with the special stock indicator K in Transaction MIGO, as shown in Figure 1.50.

This special stock indicator triggers transaction key KON instead of the regular transaction key for the normal GR/IR account, which is WRX. This is explained in Section 1.1.2. Figure 1.51 shows the assignment of the consignment GR/IR account.

You should now understand the business process overview of the vendor consignment process and how to configure the account determination for the different accounts used in the accounting entries throughout the process.

Throughout this section, we examined the details of managing consignment stock, from receiving and using the stock to the crucial step of consignment settlement. Each step of the process highlights the vital role of accurate accounting and inventory management in maintaining financial integrity and operational efficiency.

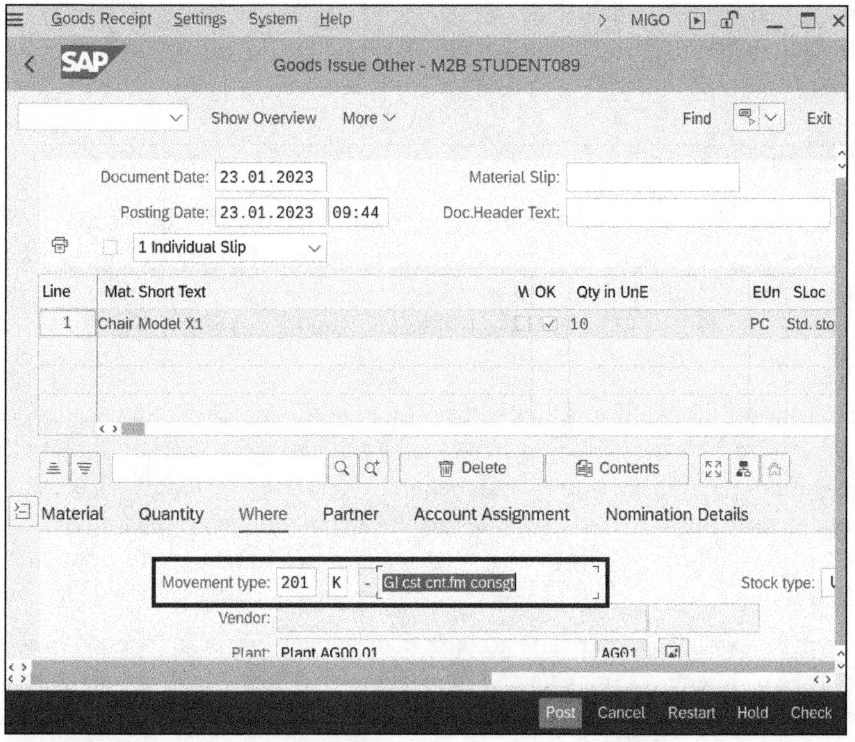

Figure 1.50 Consumption from Consignment Stock

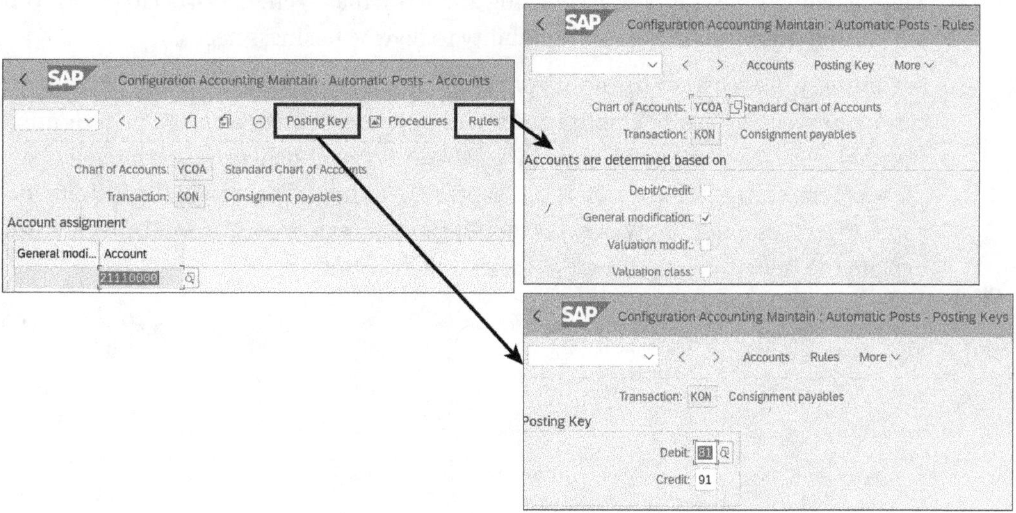

Figure 1.51 Assign Accounts, Rules, and Posting Keys: Consignment GR/IR

With a comprehensive understanding of the accounting entries associated with each step and the necessary configurations within SAP, you are now well-equipped to handle the vendor consignment process. This process not only offers a strategic advantage in inventory management but also ensures precise financial tracking and compliance within the SAP system.

1.5 Summary

As we reach the end of our comprehensive journey through the various procurement processes in SAP S/4HANA, let's reflect on the details of each process we've explored. This chapter has provided a deep dive into key procurement activities, each with its unique characteristics and importance within the broader context of materials management.

We began with the standard procedure of normal purchasing, dissecting each step, from purchase order creation to payment settlement, while paying special attention to the corresponding accounting entries and their impact on financial reporting. We then shifted our focus to the purchasing of consumables, highlighting how their direct consumption nature influences accounting entries and necessitates specific configurations in SAP.

Our exploration extended to purchase account management, a crucial function in certain regions, which requires tracking and management of financial values associated with external procurement, ensuring compliance and accurate financial reporting. Finally, we explored vendor consignment, a strategic procurement process offering significant benefits in inventory management. We examined how SAP S/4HANA adeptly handles the complex accounting and stock management associated with vendor consignment, reflecting its adaptability to diverse business needs.

In summary, this chapter has not only provided a detailed understanding of each procurement process but also emphasized the importance of accurate account determination and effective configuration in SAP S/4HANA. As we conclude this chapter, we pave the way for a seamless transition into Chapter 2. This next chapter will further enhance your understanding of how account determination works in SAP S/4HANA and will explain the key business processes in sales and distribution.

Chapter 2
Sales and Distribution

In this chapter, we'll dive into the sales and distribution module in SAP S/4HANA, a module integral to managing all business processes related to sales. In this chapter, our focus will be on those processes that have a significant accounting impact and thus are crucial for the accurate financial representation of sales activities.

The aim of this chapter is to provide a comprehensive understanding of the main business processes within the sales and distribution module, particularly emphasizing how to configure account determination for these processes. We'll begin by examining the steps in the business process flow, scrutinizing the accounting entries posted at each stage, and understanding the general ledger accounts involved. Then, we'll navigate through the detailed configuration steps in SAP S/4HANA to set up these general ledger account determinations. Furthermore, we'll explore the nature of each general ledger account and discuss the important fields in the general ledger account master data.

Note that the commonalities and shared areas between materials management and sales and distribution. One such shared area is inventory management and the related account determination, as explained in Chapter 1, Section 1.1.2, in our discussion of inventory accounts and the transaction key technique. The account determination configuration for inventory-related accounts in the materials management module is equally applicable to sales and distribution.

Let's embark on this learning journey by starting with the most common sales process in SAP S/4HANA: the standard sales from stock. This process is not only fundamental but also sets the foundation for understanding more complex sales and distribution processes.

2.1 Sales from Stock

In this section, we explore the *standard sales from stock process*, a cornerstone of the sales and distribution module in SAP S/4HANA. This process, widely recognized as the standard order-to-cash process, is pivotal in the sales operations of organizations handling either in-house manufactured items or externally procured trading items. We'll follow the complete journey of a product from the moment it is available for sale to the finalization of the transaction.

The standard sales from stock process is initiated with a customer inquiry, a critical starting point where customer needs are identified. Next, a response follows with a quotation tailored to meet these specific requirements. Upon customer agreement, the process advances to the creation of a sales order. This sales order is the formal agreement between the company and the customer, marking the commencement of the transaction. Subsequent steps involve creating an outbound delivery, which encompasses picking and packing the products. The goods issue is then posted, signifying the physical movement of goods out of the warehouse, and the customer billing document is printed. This document is an essential component for accounting, as it records the sales transaction and triggers the invoicing process. The culmination of this process is the receipt of payment from the customer, followed by the clearing of the billing document. This step marks the completion of the financial transaction and is critical for maintaining accurate financial records.

Throughout this section, we'll look closely at each step in this process flow. We'll focus on understanding the specific accounting entries that are generated at each stage, from goods issue to payment receipt. This discussion includes examining how sales revenue is recorded, the impact on inventory and cost of goods sold (COGS), and how customer payments are processed and recorded.

Additionally, we'll explore how to configure SAP S/4HANA for account determinations related to this process. Understanding how to set up these configurations is crucial for ensuring that financial postings accurately reflect the sales transactions, thereby maintaining the integrity of financial reporting and compliance with accounting standards.

As we proceed, our exploration will provide a comprehensive understanding of the standard sales from stock process, a critical component of the order-to-cash cycle in SAP S/4HANA.

Next, let's have a detailed look at this process flow.

2.1.1 Business Process Overview

The standard sales from stock process flow is shown in Figure 2.1. The steps that have an accounting impact are highlighted.

Let's quickly look at each step. Note that we'll focus more on the steps that cause accounting entries:

1. **Inquiry**
   The customer contacts us and inquires about the different conditions and the availability of one or more of our products. This step has no accounting impact.
2. **Quotation**
   We respond to the customer with the requested details. This step has no accounting impact.

3. **Sales order**
   When interested in buying products, a customer sends us a purchase order. Based on the customer's order, we create a sales order. The sales order is the main document in this process, and it includes all the details related to selling the items. We process all the following steps with reference to this sales order. This step has no accounting impact.
4. **Outbound delivery**
   Based on our agreement with the customer captured in the sales order, we create an outbound delivery specifying what items to be issued to the customer and when. The warehouse team uses this document to get ready. This step has no accounting impact.
5. **Pick and pack**
   When the time comes to issue products to the customer, we pick the products to be issued, and we pack them. This step has no accounting impact. After picking and packing the products, we post the goods issue.
6. **Goods issue**
   We issue the products to the customer. This step is also called a *post goods issue*, and it causes the accounting entry shown in Table 2.1.

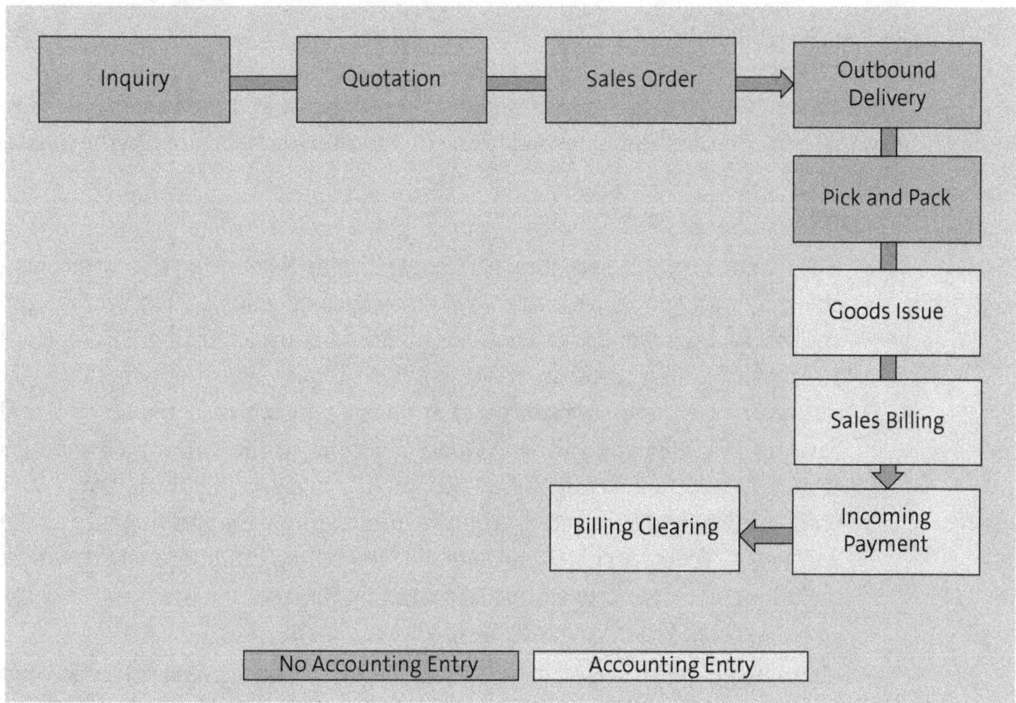

Figure 2.1 Standard Sales from Stock Process Flow

We can also have a second accounting entry if we've configured the COGS split function, which splits any posting to the COGS account to different accounts based on the cost elements included in the sold item's standard cost estimate. The second accounting entry is shown in Table 2.2.

Debit	Credit	Debit Amount ($)	Credit Amount ($)
COGS		1000	
	Inventory		1000

Table 2.1 Accounting Entry of Post Goods Issue

Debit	Credit	Debit Amount ($)	Credit Amount ($)
COGS: Raw materials		600	
COGS: Manufacturing activities		250	
COGS: Overhead		150	
	COGS split clearing		1000

Table 2.2 Accounting Entry of COGS Split

Now, let's quickly look at the meaning of these accounts and the values posted to each of them. The account determination configuration is structured in the following ways:

- **COGS**
  - A profit and loss (P&L) account. An important step is choosing the correct general ledger account type for COGS. Your choice depends on the profitability analysis type activated and whether you use sales order costing. Two options are available: First, you can create a COGS account with the general ledger account type **N: non-operating cost or revenue**. In this case, the account will have no cost element and won't post anything to the controlling module. Select this option if costing-based profitability analysis is activated or if sales order costing is not activated. When we use costing-based profitability analysis, the COGS value is posted through the sales billing conditions when we post the billing document, not directly posted by the account when we post the goods issue, and thus you don't need a cost element.
  - Alternatively, you can create the COGS account with the general ledger account type **P: primary cost or revenue** and cost element category **1: Primary costs/ cost-reducing revenues**. In this case, the account will post to the controlling module. Use this option if accounting-based profitability analysis (also called *margin analysis*) is activated or sales order costing is activated. In this case, the

COGS account is posted directly to the accounting-based profitability analysis at the time of posting the goods issue, so the cost element is needed.

- The value posted to COGS is the cost of the items issued. This cost can either be moving average cost or standard cost depending on the inventory costing method you use.

– **Inventory**
  - This account is a balance sheet account.
  - The value posted is the cost of the items issued (same as the COGS value).

– **COGS split and clearing accounts**
  - This COGS split is an optional feature that is mainly used to split COGS values based on the different cost components in accounting-based profitability analysis.
  - These accounts are P&L accounts of the same type as the standard COGS account.
  - You can have one account assigned to every cost component used in our standard cost calculation. Cost components can include, for example, raw materials; manufacturing activities (e.g., labor, machinery etc.); and manufacturing overheads.
  - The total value posted to all the COGS split accounts is equal to the value posted to COGS, which is equal to the cost of the items issued. The value is split between these accounts in the same ratio as the cost estimate. For example, if our cost estimate is $100 for one item of the product, calculated as: raw materials $60 + manufacturing activities $25 + manufacturing overheads $15, then the COGS value will be split in the same ratios.
  - The COGS split clearing account is posted with the sum of the values posted to the COGS split accounts to write them off. We can also configure the original COGS account to be used in this case instead of a clearing account.
  - Now, you should understand the accounting entry posted at the time of post goods issue and how the COGS account can be split to different accounts. The next step after the post goods issue is to issue the sales billing document.

7. **Sales billing**
This step marks the completion of the sales order. Now, we issue the sales billing document to the customer. This step causes the accounting entry shown in Table 2.3.

Debit	Credit	Debit Amount ($)	Credit Amount ($)
Accounts receivable (AR)		1350	
Sales discount		250	

Table 2.3 Accounting Entry of Sales Billing

Debit	Credit	Debit Amount ($)	Credit Amount ($)
Withholding tax (WHT)		50	
	Value-added tax (VAT)		150
	Sales revenue		1500

Table 2.3 Accounting Entry of Sales Billing (Cont.)

Now, let's quickly look at the meaning of these accounts and the values posted to each of them. The account determination configuration is structured in the following ways:

- **Accounts receivables (AR)**
  - This account is a balance sheet account.
  - The value posted is the total value to be paid by the customer according to the invoice including VAT and WHT.
  - This account is also called a *customer reconciliation account*. More details on the business concept and account determination of reconciliation accounts are provided in Chapter 4.
- **Sales discount**
  - A P&L account with general ledger account type **P: Primary cost or revenue** and cost element category **12: Cost Deductions.**
  - The value posted is the discount value or percentage inserted in the billing document.
  - You can configure the account determination of some discounts or surcharges to post to the sales revenue account instead of separate accounts for discounts or surcharges.
- **Withholding tax (WHT)**
  - This account is a balance sheet account.
  - WHT can be posted at the sales billing or at the incoming payment depending on the legal requirements.
  - More details on the WHT process and the related account determination configuration steps are provided in Chapter 4.
- **Output VAT**
  - This account is a balance sheet account.
  - The value is calculated automatically based on the tax codes inserted in the billing to follow the legal tax requirements.
  - More details on the VAT process and the related account determination configuration can be found in Chapter 4.

- **Sales revenue**
  - A P&L account with general ledger account type **P: Primary cost or revenue** and cost element category **11: Revenues.**
  - The value posted is the gross sales value before discounts or surcharges.
  - You can configure the account determination of some discounts or surcharges to post to the sales revenue account, in such case the value posted here will include these discounts.

The next step is to receive the customer payment and allocate it to the sales billing document to clear it.

8. **Incoming payment and clearing**
   When you receive a payment from the customer, you may clear one or more billing documents. This step causes an accounting entry that can differ based on the payment method (i.e., cash, bank transfer, check). This topic will be explained in detail in Chapter 4.

For the sake of completing this process, let's assume we're receiving a payment through a bank transfer. The accounting entry is shown in Table 2.4.

Debit	Credit	Debit Amount ($)	Credit Amount ($)
Incoming transfers		1000	
Cash discount		100	
WHT		50	
	AR		1150

Table 2.4 Accounting Entry of Incoming Customer Payment with Bank Transfer

With this step, we've completed our overview of the sales from stock business process, starting from customer inquiry to sales billing and then to receiving the payment and clearing the sales billing document. We also covered the various accounting entries posted in these steps and explored the general ledger accounts used for these accounting entries.

Now, let's look into the account determination configuration steps for each account used in goods issue and the sales billing accounting entries in a standard sales from stock process.

Inventory Accounts

Inventory accounts are always determined in the same way in SAP S/4HANA, as explained in Chapter 1, Section 1.1.2. Thus, in this section, we'll start with the other accounts.

2.1.2 COGS Accounts at Goods Issue

The first account we'll explore is the COGS account, to which we posted earlier, as shown in Table 2.1. When posting the COGS account at goods issue, the account is determined in the same way as described in Chapter 1, Section 1.1.2.

The standard movement type for goods issue from stock with reference to an outbound delivery is **601: GD goods issue:delvy**, and the transaction key that determines the COGS account is GBB. This transaction key includes some flexibility in the account determination configuration since it can have account modifications that can be configured.

Figure 2.2 shows the account grouping configuration screen in Transaction OMJJ for movement type **601** in SAP S/4HANA.

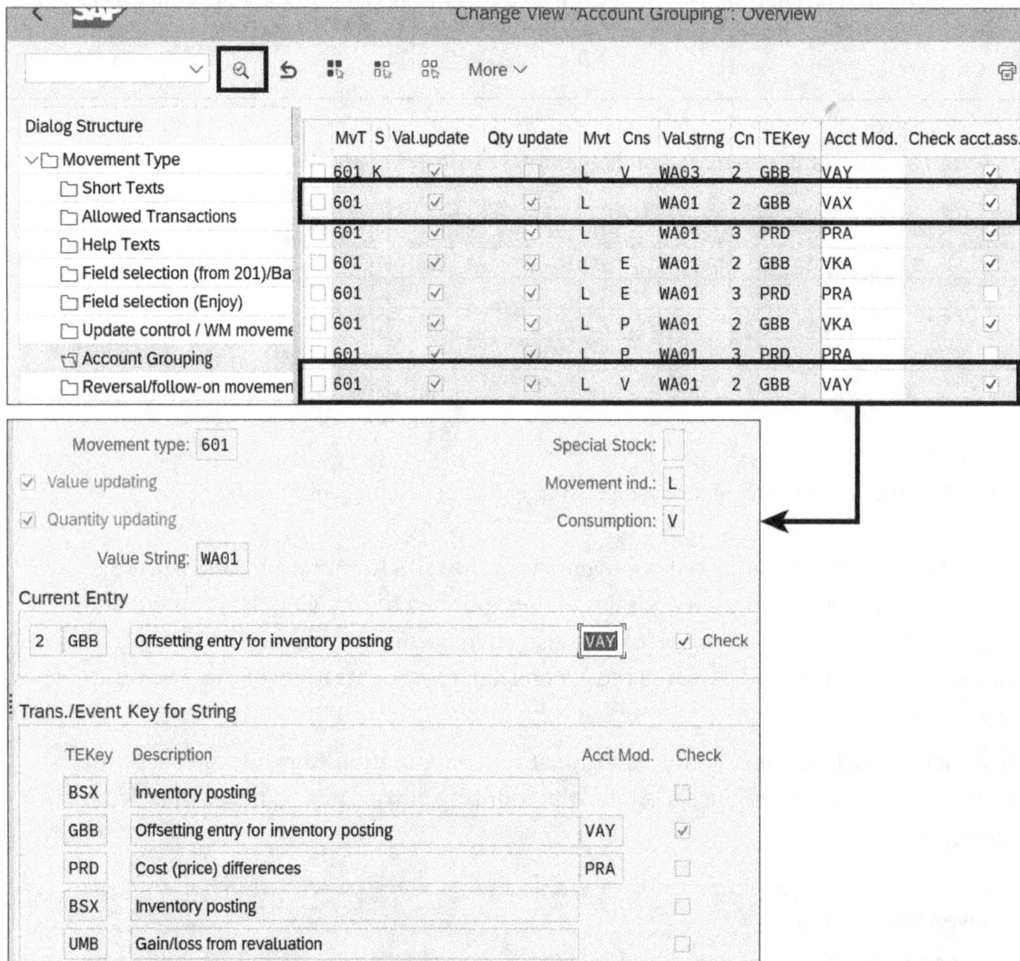

Figure 2.2 Movement Type Account Grouping for Movement 601: Goods Issue for Delivery

The two standard account modifications used to determine the COGS account under transaction key **GBB** are account modifications **VAX** and **VAY.** VAX is determined if the **Consumption** field is blank, while VAY is determined if the **Consumption** field contains a value, as shown in Figure 2.2.

When accounting-based profitability analysis is activated, SAP S/4HANA automatically determines the COGS general ledger account based on **Consumption: V.** So, in this case, the account modification VAY is used. This setup is the standard configuration, but these account modifications can be changed. You can use new account keys or change assigned keys on the screen shown in Figure 2.2.

Now that we know the transaction key (**GBB**) and account modification (**VAY**) used in the account determination, we can assign the COGS general ledger account in the SAP S/4HANA configuration via Transaction OMWB, as described in Chapter 1, Section 1.1.2.

Figure 2.3 shows the automatic account assignment configuration for the COGS general ledger account.

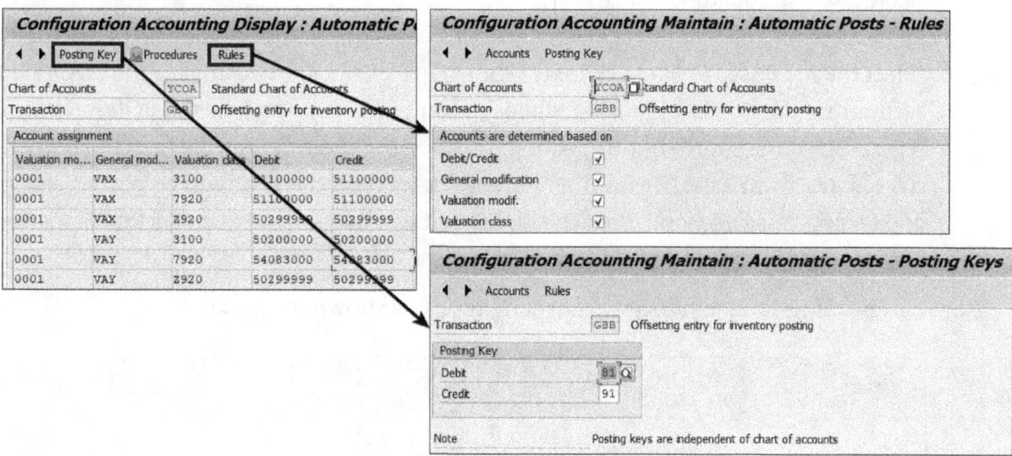

Figure 2.3 Assign Accounts, Rules, and Posting Keys for COGS Account

Now, you understand how to configure the accounting determination for the two accounts shown earlier in Table 2.1—the inventory account and the COGS account.

Now, let's turn the other accounts involved in the process.

2.1.3 COGS Split Accounts

If the function to split the COGS account is activated in SAP S/4HANA, then you'll also have a second accounting entry posted at goods issue, as shown earlier in Table 2.2. Now, let's learn how to configure the account determination for a COGS split.

All the configuration steps related to the account determination of COGS split can be found in the SAP configuration by following the menu path **Financial Accounting •**

General Ledger Accounting • Periodic Processing • Integration • Materials Management • Define Accounts for Splitting the Cost of Goods Sold. Our first step is to define a cost splitting profile, as shown in Figure 2.4.

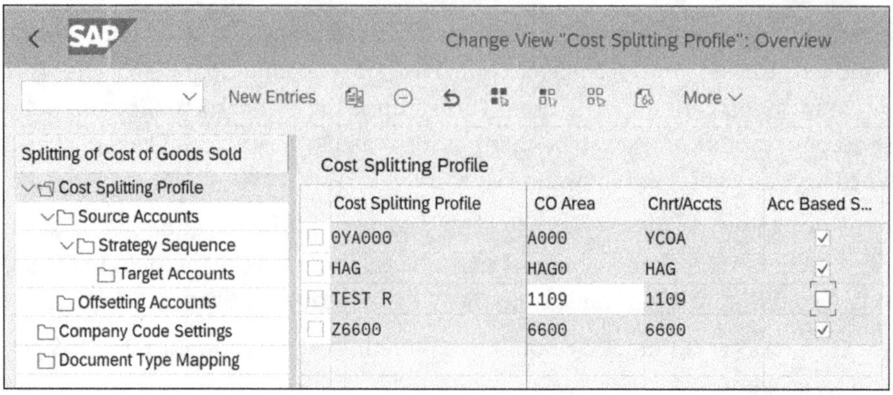

Change View "Cost Splitting Profile": Overview

New Entries More

Splitting of Cost of Goods Sold
Cost Splitting Profile
Source Accounts
Strategy Sequence
Target Accounts
Offsetting Accounts
Company Code Settings
Document Type Mapping

Cost Splitting Profile

Cost Splitting Profile	CO Area	Chrt/Accts	Acc Based S...
0YA000	A000	YCOA	✓
HAG	HAG0	HAG	✓
TEST R	1109	1109	☐
Z6600	6600	6600	✓

Figure 2.4 COGS Splitting Profile

Select the **Acc Based Split** checkbox, as shown in Figure 2.4, which means the COGS split accounting entry will be posted whenever an entry is posted to the COGS account, regardless of the source of this posting. This posting can originate from manual financial entries, from sales, or from any other source, and the split will be posted. If you don't select this checkbox, then the COGS split will happen only when a COGS posting occurs with reference to a sales order.

This profile is later assigned to company codes, as shown in Figure 2.5.

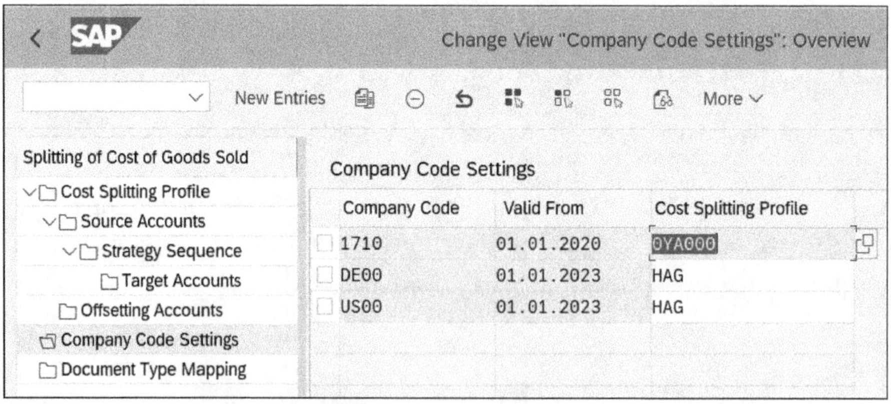

Change View "Company Code Settings": Overview

New Entries More

Splitting of Cost of Goods Sold
Cost Splitting Profile
Source Accounts
Strategy Sequence
Target Accounts
Offsetting Accounts
Company Code Settings
Document Type Mapping

Company Code Settings

Company Code	Valid From	Cost Splitting Profile
1710	01.01.2020	0YA000
DE00	01.01.2023	HAG
US00	01.01.2023	HAG

Figure 2.5 COGS Splitting: Assign Splitting Profile to Company Codes

After creating the splitting profile, the next step is to assign the source COGS accounts that should trigger the COGS splitting posting, as shown in Figure 2.6.

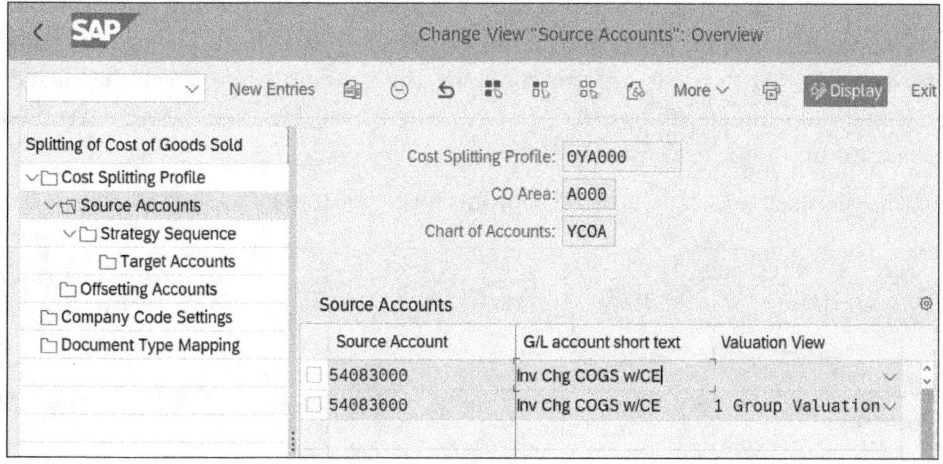

Figure 2.6 COGS Splitting: Assign Source Accounts

On the level of every source account and valuation view, you can define different COGS splitting accounts. The valuation view can be legal, group, profit center, or segmentation valuation. If any valuation view is not specified at this point, then it will be split using the rule in the line with no valuation view. For example, as shown in Figure 2.6, the source COGS account **54083000** will be split on the group valuation view using a different structure than on the other valuation views.

Next, assign strategy sequences to every source account and valuation view, as shown in Figure 2.7.

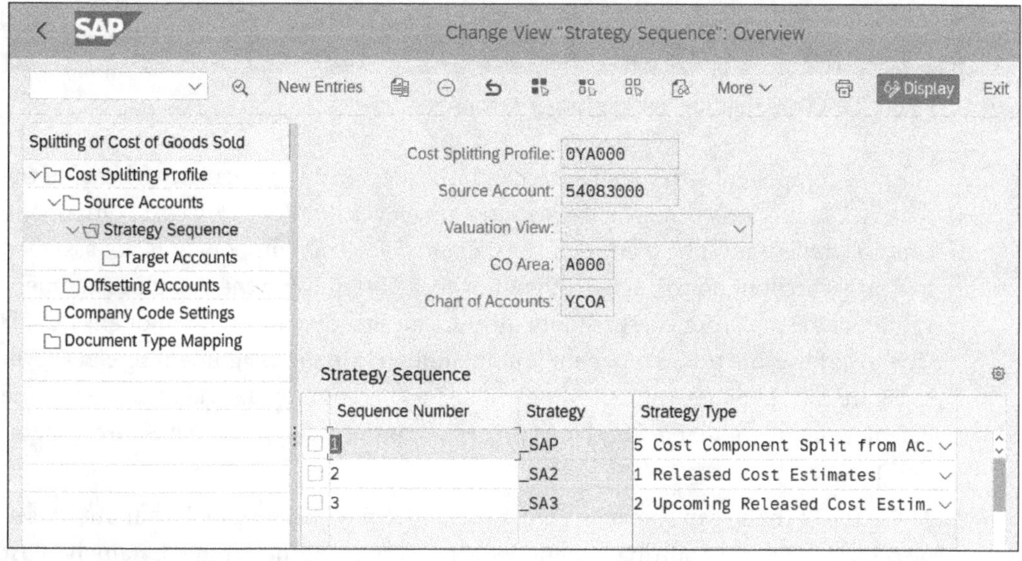

Figure 2.7 COGS Splitting: Strategy Sequences

How the COGS is split to different accounts is based on the cost components used in the cost estimate of the material. You can specify a cost estimate to use as a reference for the split. For example, we can configure the system to first check the current released cost estimate. If no current cost estimate is available, then the system should check the upcoming released cost estimate.

Then, we'll need to assign the target accounts for every strategy, as shown in Figure 2.8.

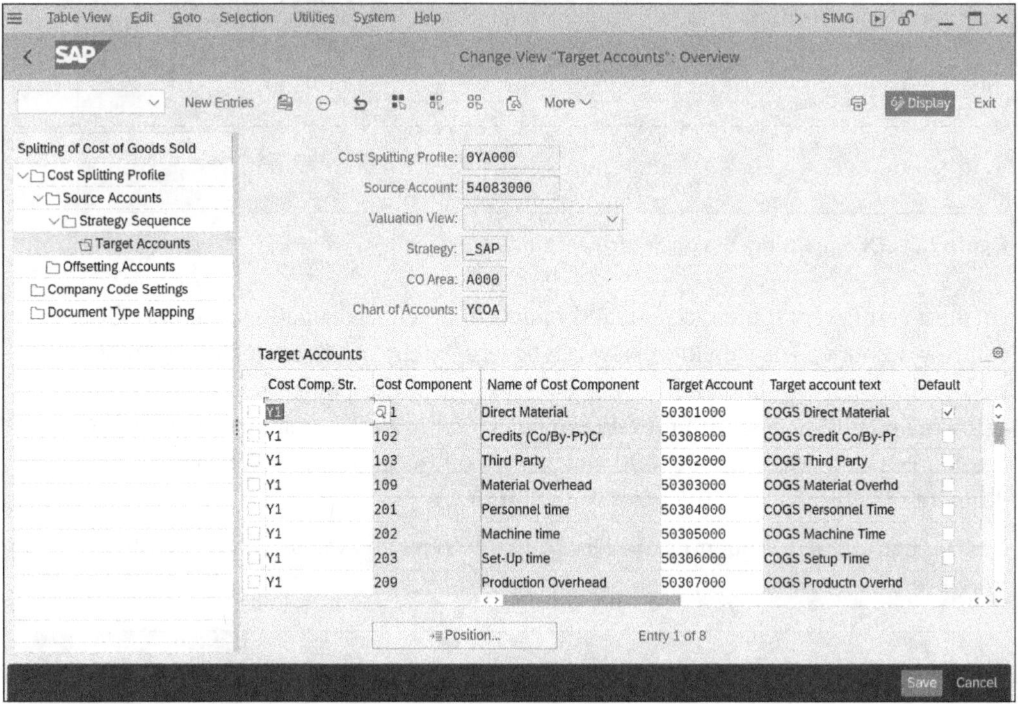

Cost Comp. Str.	Cost Component	Name of Cost Component	Target Account	Target account text	Default
Y1	1	Direct Material	50301000	COGS Direct Material	✓
Y1	102	Credits (Co/By-Pr)Cr	50308000	COGS Credit Co/By-Pr	
Y1	103	Third Party	50302000	COGS Third Party	
Y1	109	Material Overhead	50303000	COGS Material Overhd	
Y1	201	Personnel time	50304000	COGS Personnel Time	
Y1	202	Machine time	50305000	COGS Machine Time	
Y1	203	Set-Up time	50306000	COGS Setup Time	
Y1	209	Production Overhead	50307000	COGS Productn Overhd	

Figure 2.8 COGS Splitting: Assign Target Accounts

The cost component structure is used in the cost estimate to calculate the cost of the material. In this structure, you can assign a target account for every cost component. One assignment must be marked as the default. If any cost component is not assigned to a target account on this screen, then it will be posted using the default assignment value. For example, our cost estimate includes a value on the cost component 110, for labor overhead, but this component is not maintained in the assignment screen shown in Figure 2.8. Therefore, the COGS split value for this component will be posted to the default assignment marked in the screen, which is the assignment **COGS Direct Material**.

At the top of the screen shown in Figure 2.8, notice the final result of the inputs based on which the COGS splitting general ledger accounts are determined, namely, **Cost Splitting Profile**, **Source Account**, **Valuation View**, and **Strategy**.

Now, assign the offsetting accounts to the source accounts on the splitting profile level, as shown in Figure 2.9.

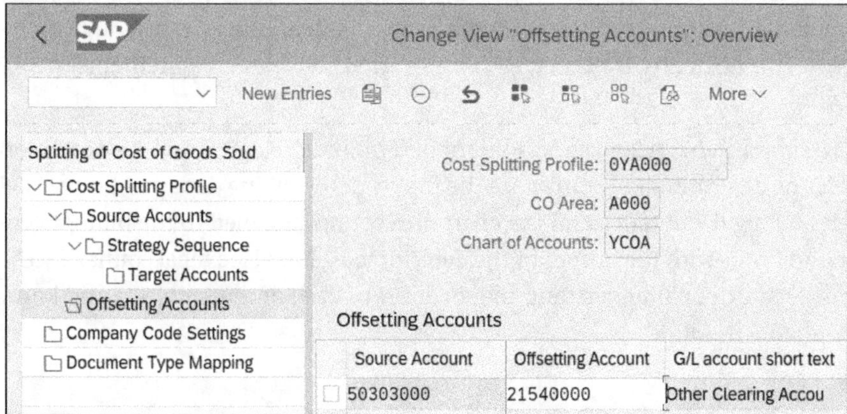

Figure 2.9 Splitting: Clearing Accounts

If any source account is not assigned to an offsetting account in this screen, then this source account will be used also as an offsetting account in the COGS splitting accounting entry. Thus, the source account is debited in the original entry and then credited with the same value in the splitting entry, so the net value posted to the source account is zero, and the only value we have in our P&L is for the COGS splitting target accounts.

At this point, you know how to configure the COGS splitting feature in SAP S/4HANA. This feature that wasn't available in the older versions of SAP. This new feature is essential for the reporting of accounting-based profitability analysis (also called margin analysis) as it allows us to report the details of the COGS account per cost element after the split.

With this step, you now know how to determine the general ledger accounts shown earlier in Table 2.2. You can handle all the account determinations required for the post goods issue step: inventory, COGS, and COGS split.

Now, let's move to the account determination of the accounts shown earlier in Table 2.3, starting with the sales revenue account.

2.1.4 Sales Revenue, Discounts, and Surcharges Accounts (The Condition Technique)

The account determination process for sales revenue accounts and for sales discounts and surcharges follows the same technique. In this book, we refer to this technique as the *condition technique* for account determination. From now on, whenever we use the term "sales revenue account determination," we are also referring to sales discounts and surcharges.

The condition technique offers great flexibility in account determination because you can assign general ledger accounts based on any combination of some of the fields available in the sales order. For example, we can determine the revenue account based on the material, customer, country, distribution channel, order reason, and many other fields. This flexibility is needed to satisfy various business requirements in SAP S/4HANA.

Figure 2.10 shows an overview of the account determination logic using the condition technique. The determination starts with the inputs of the billing document type used in the sales billing documents, and the chart of accounts assigned to the billing company code and ends with the output of the determined general ledger accounts. By the end of this section, you'll understand the meaning of each of these objects and know how to configure this link.

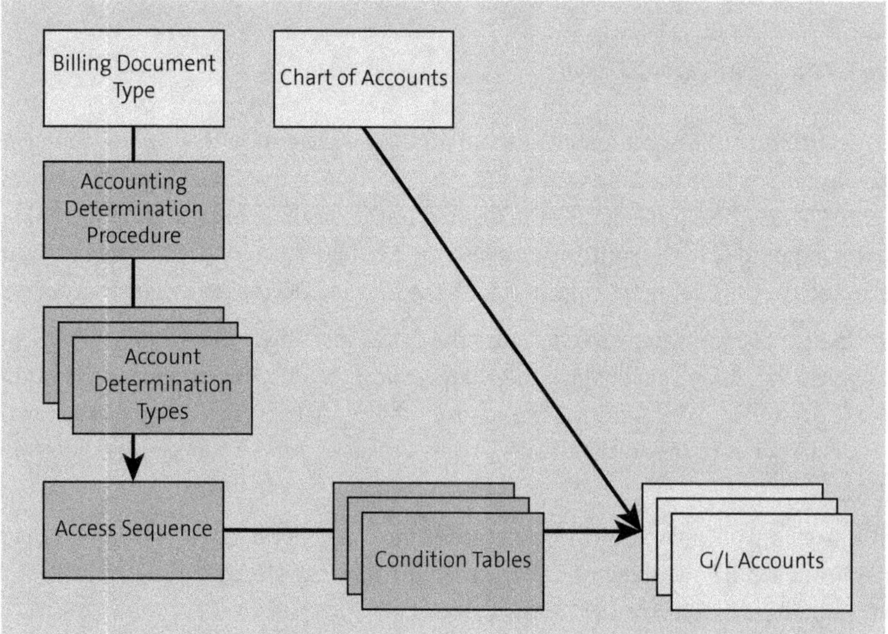

Figure 2.10 Sales Revenue Account Determination Logic (The Condition Technique)

All the configuration steps for the sales revenue account determination can be performed in the SAP configuration by following the menu path **Sales and Distribution • Basic Functions • Account Assignment/Costing • Revenue Account Determination**. The various configuration steps are shown in Figure 2.11.

Now, let's take a detailed look at the configuration steps for the account determination of the sales revenue, discounts, and surcharges.

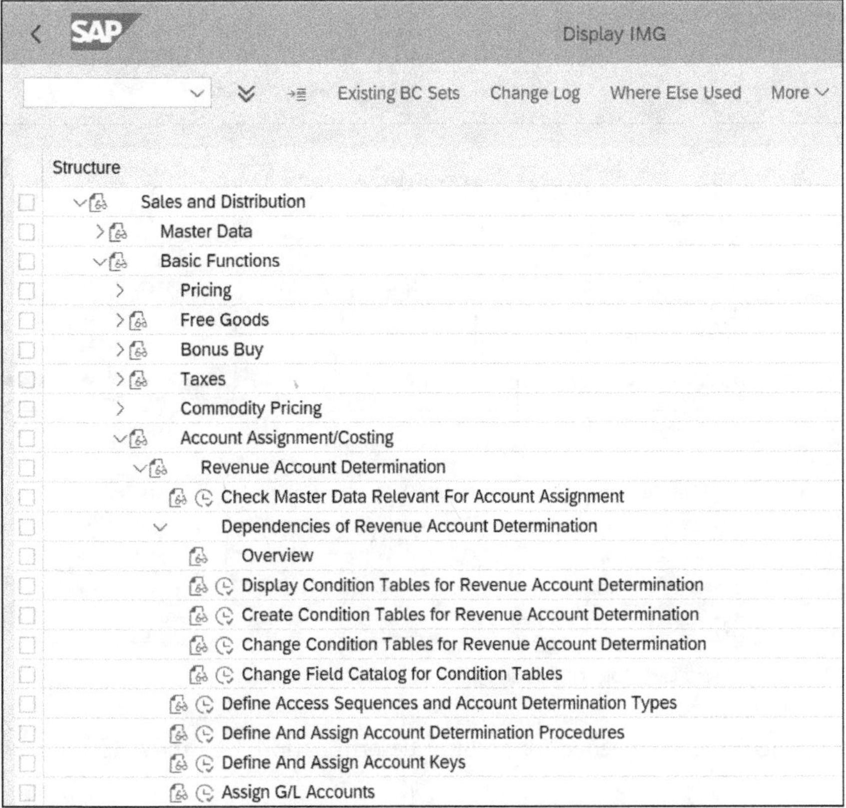

Figure 2.11 Sales Revenue Account Determination: SAP Configuration Menu

Change Field Catalog for Condition Tables

First, start by activating the different fields you want to use in account determination. A field catalog provided by SAP includes all the fields that you can choose from. As shown in Figure 2.12, you can activate a new field from the allowed available fields when you check the available field values.

You can also deactivate a field by deleting its line from the screen. Then, the field can no longer be used in account determination.

> **Note**
>
> Only the fields that are not already assigned to condition tables can be removed. We'll cover the assignment of fields to condition tables in the next configuration step.

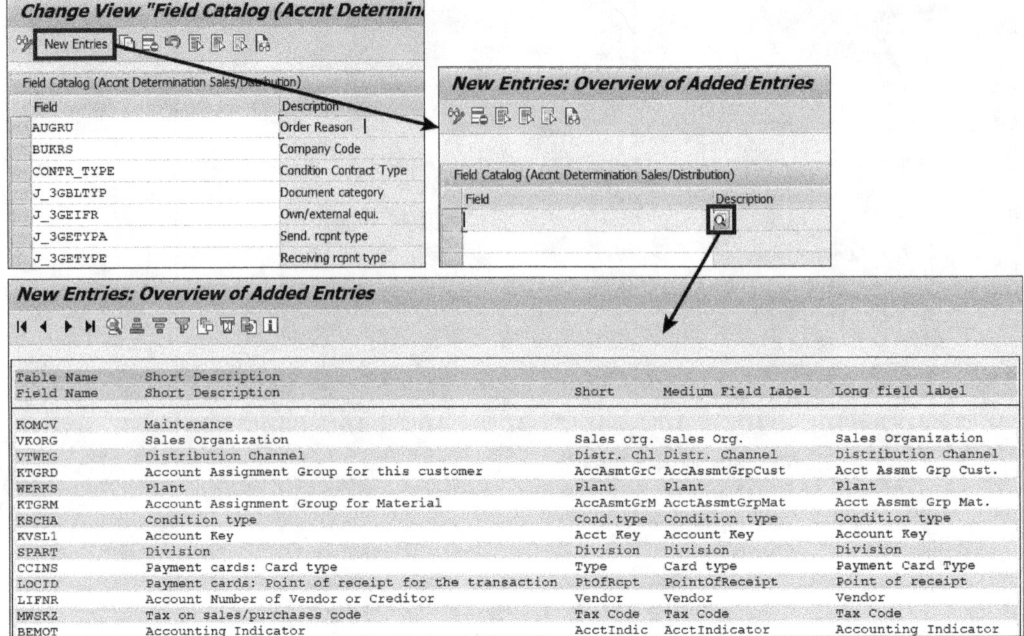

Figure 2.12 Change Field Catalog for Revenue Account Determination

After activating the fields needed for account determination, let's move to the next step.

Create Condition Tables for Revenue Account Determination

The condition tables combine the fields you want to use for account determination. For example, as shown in Figure 2.13, condition table 001 combines the fields on the left: **Sales Organization**, **Acct Assmt Grp Cust.** (account assignment group customer), **Acct Assmt Grp Mat.** (account assignment group material), and **Account Key**. On the right, notice the available fields that we activated in the previous section. Each table has a code of 3 digits between 001 and 999. The numbers from 001 to 500 are reserved for SAP standard tables. To define new condition tables, you must use numbers from 501 to 999.

To add one of the available fields to the table, double-click on the field, and it will move to the left, which means it has been added to the table. Then, save and generate the table by clicking on the icon.

> **Note**
>
> Once a table is generated, you can no longer add or remove the fields. To insert or remove a field, you must create a new condition table.

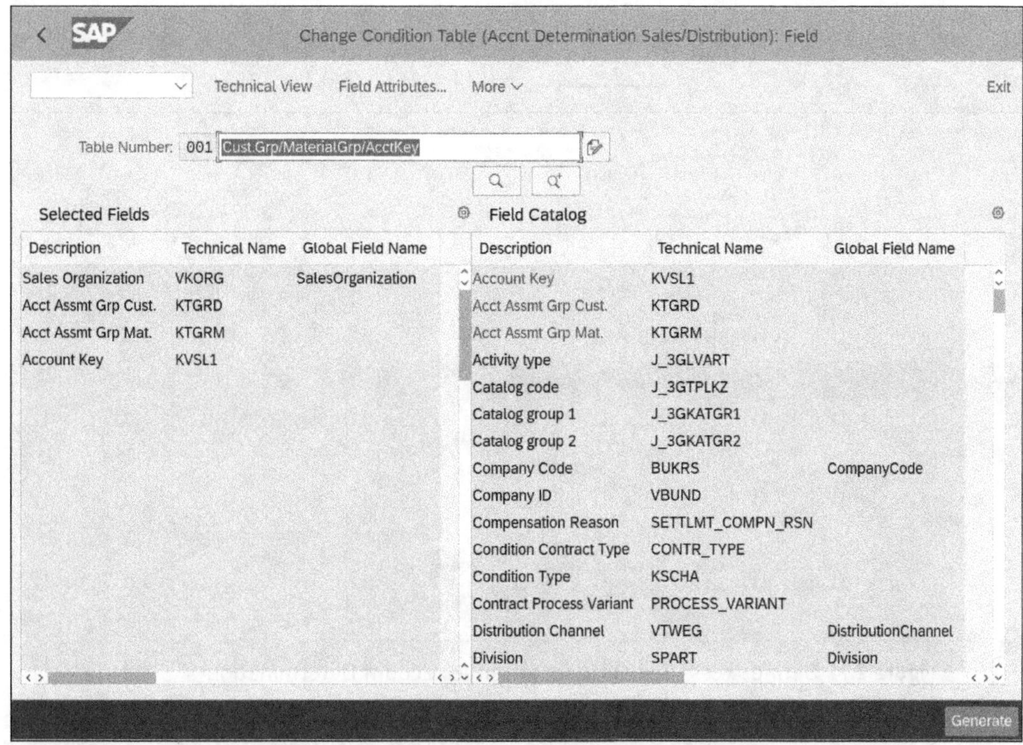

Figure 2.13 Change Condition Tables for Revenue Account Determination

You can define multiple condition tables, each table with different fields. The next step is to configure the sequence in which SAP should check these tables.

Define Access Sequences and Account Determination Types

First, let's assign our condition tables to access sequences. Then, we can assign each access sequence to one or more account determination types.

Figure 2.14 shows the screens used to define access sequences.

Usually, you can use the standard access sequence **KOFI** for revenue account determination, but if needed, you can create a new access sequence, as shown in Figure 1.14 ❶. Then, as shown in ❷, you can assign the different condition tables we created in the previous step.

For every condition table, we'll assign an access number. SAP accesses the condition tables in the sequence of access numbers (from lower to higher). For example, as shown in Figure 2.14, SAP will access the condition table **017**, then **001**, then **002**, and so on, based on the access numbers and *not* the condition table number.

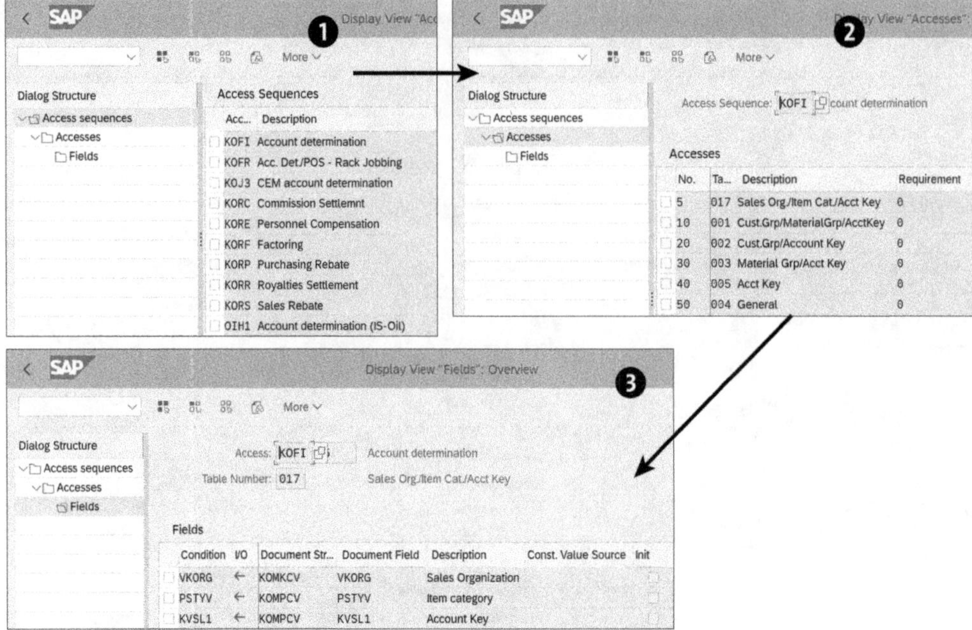

Figure 2.14 Define Access Sequence for Revenue Account Determination

> **Project Note**
>
> Usually, we create the condition tables with the same access sequence numbers we want when we start a configuration project, but in many cases, you may need to create additional condition tables later with numbers that don't match the access sequence and the account determination analysis gets confusing. Always remember that SAP accesses the condition tables based on the access sequence numbers, not the condition table numbers.

As shown in Figure 2.14 ❸, you can check the fields assigned to any condition table, as we configured in the previous section. You can also see where the values in the condition table fields will come from. For example, the **Sales Organization VKORG** field in our condition table **508** will copy the value from the field **VKORG** in the document structure **KOMKCV**. Notice also see two available configuration fields:

- **Const. Value Source**: Allows you to configure any field in the condition table to always have a constant value in this access sequence. When this field is maintained, the system uses this constant in the account determination regardless of the sales document.
- **Init**: Allows the field to have an initial value. For example, if we select this option for the **Account Assignment Grp Cust.**, then even if this field is empty in the sales document, SAP will use this condition table for the account determination and will look to a general ledger assigned to this field with the initial value (empty).

We usually configure the access sequence to access the tables that have more field requirements first. For example, as shown in Figure 2.15, we want the general ledger to be determined based on a combination of fields A, B, C, and D. However, if no general ledger is maintained on this level, then we want to determine the general ledger based on A, B, and C, and then again based on A.

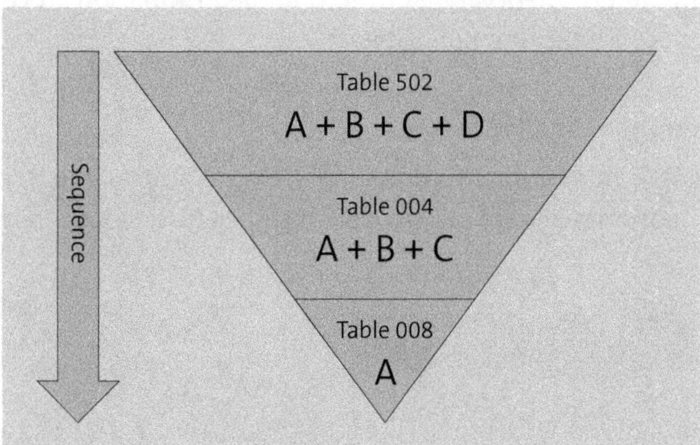

Figure 2.15 Access Sequence Logic

SAP Table

All the details of the access sequences, the assigned tables, fields, and constants can be seen in SAP table T682Z.

After creating the access sequence, the next step is to assign it to account determination types, which are also called *condition types* (**CTyp**), as shown in Figure 2.16.

Change View "Conditions: Types": Overview

New Entries More

Condition Types

Con...	Name	Acc...	Description
J3GK	CEM Cross-Co. Code	KOJ3	CEM account determination
KOFI	Acct determination	KOFI	Account determination
KOFK	Acct Determ.with CO	KOFI	Account determination
KOJ3	CEM Acc. Determntn	KOJ3	CEM account determination
KORC	Commission Settlemnt	KORC	Commission Settlemnt
KORE	Personnel Compens.	KORE	Personnel Compensation
KORP	Purchasing Rebate	KORP	Purchasing Rebate
KORR	Royalties Settlement	KORR	Royalties Settlement
KORS	Sales Rebate	KORS	Sales Rebate

Figure 2.16 Assign Access Sequences to Account Determination Types

SAP Table

The assignment of account determination types and access sequences can be displayed in SAP table T685.

Next, we need to assign account determination *types* to account determination *procedures*. Then, we can assign these procedures to billing types.

Define and Assign Account Determination Procedures

First, let's define our account determination procedures. In their definitions, you can specify which account determination types are assigned to the procedure, as shown in Figure 2.17.

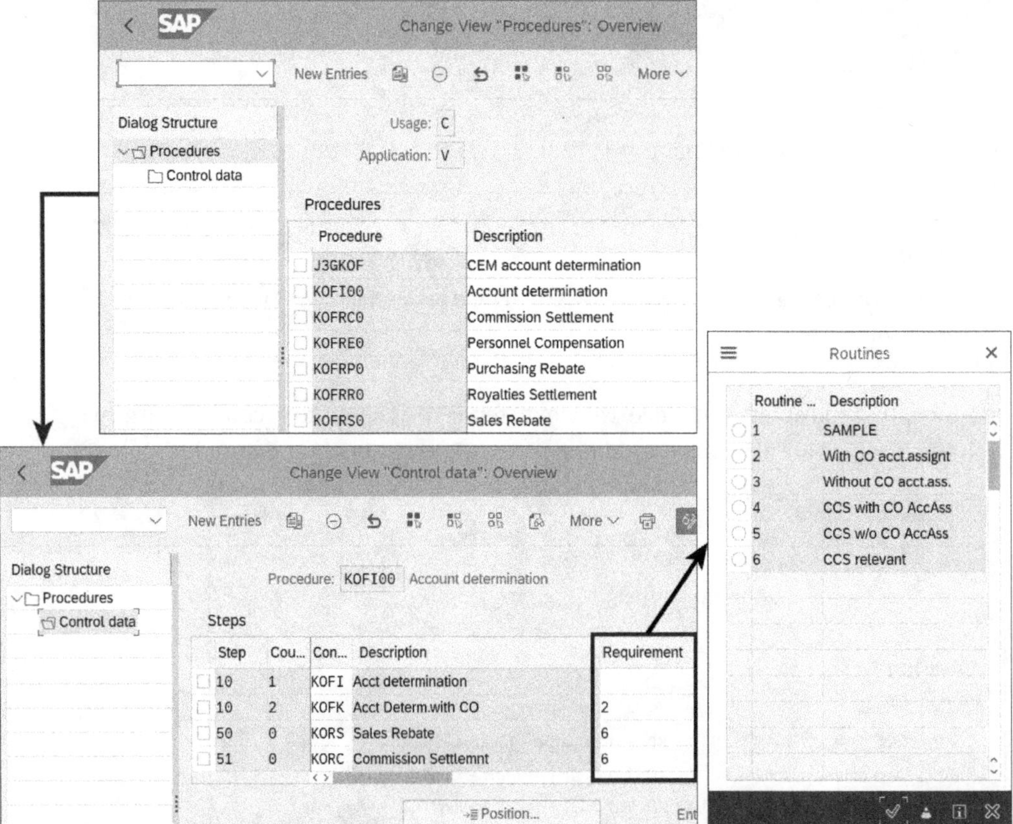

Figure 2.17 Define Account Determination Procedure

According to this configuration, if the sales document has a cost object assignment, then the condition type KOFK will be triggered. If no cost object exists, then condition type KOFI will be triggered.

SAP Table

The assignment of account determination procedures and account determination types can be displayed in SAP table T683S. The usage is "C," and application is "V."

The next step is to assign account determination procedures to billing types, as shown in Figure 2.18.

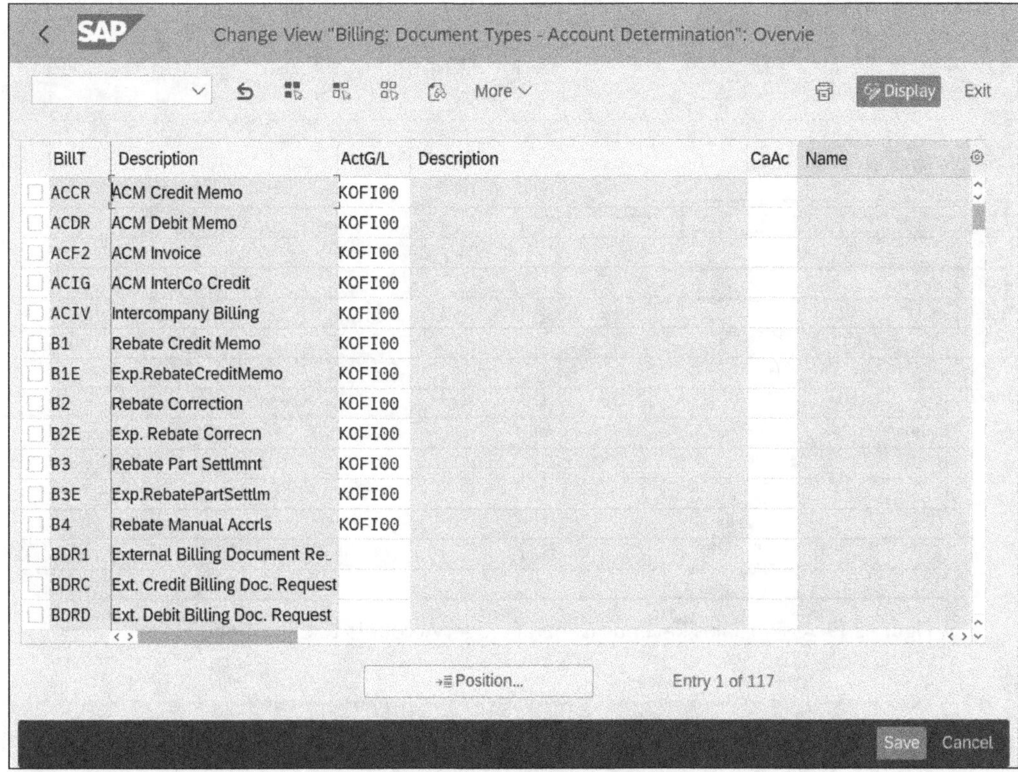

Figure 2.18 Assign Account Determination Procedures to Billing Types

SAP Table

The assignment of account determination procedures and billing types can be displayed in SAP table TVFK.

A billing type can be automatically determined when creating a sales billing document through the configuration maintained in the sales and distribution module. In this step, you'll maintain the complete link from the billing document to the condition table, as shown in Figure 2.10.

Now, let's learn how to assign general ledger accounts to condition tables.

Assign General Ledger Accounts

To access this configuration transaction, as shown in Figure 2.11, follow the menu path **Sales and Distribution • Basic Functions • Account Assignment/Costing • Revenue Account Determination**. Select the **Assign G/L Accounts** option. Alternatively, you can use Transaction VKOA.

Let's review the configuration screens for assigning general ledger accounts. As shown in Figure 2.19 ❶, you can see all the condition tables created regardless of the access sequences used.

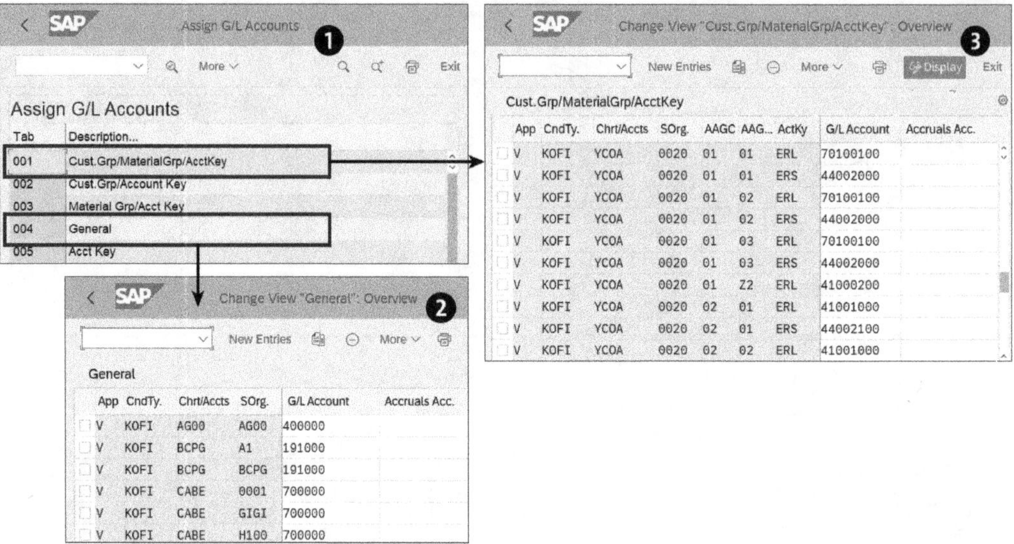

Figure 2.19 Assign Sales Revenue General Ledger Accounts to Condition Tables

> **Note**
>
> The condition tables on this screen are sorted based on the table number, which is not related to the actual access sequence. Instead, the actual access sequence is based on the **Access Number** in each **Access Sequence**, as described in the previous step.

Now, as shown in ❷, you can maintain the general ledger accounts for table 004. In the definition of this condition table, we only have one field: sales organization (**SOrg.**). However, in the general ledger assignment lines, we can see additional fields: application (**App**), condition type (**CndTy.**), and chart of accounts (**Chrt/Accts**). These three fields are mandatory for account determination and are added to all condition tables. If you maintain the general ledger accounts only in table 004, all sales revenues, discounts, and surcharges will be posted to the one general ledger account assigned to the sales organization.

To determine separate general ledger accounts, you must assign general ledger accounts to another table that has more fields. For example, as shown in Figure 2.19 ❸, you can assign general ledger accounts to table 001. This table allows us to assign a separate general ledger account for every combination of the following elements:

- Customer account assignment group (**AAGC**): Allows different general ledger determinations based on the customer billed.
- Material account assignment group (**AAGM**): Allows different general ledger determinations based on the material in the billing line item.
- Account key (**ActKy**): Allows different general ledger determinations based on the sales condition, for example, sales revenue, customer discount, material discount, etc.

To avoid errors in account determination that can stop the sales process and cause business loss, we normally maintain the general ledger accounts on the detailed level (e.g., table 001) and the general level (e.g., table 004). Then, we'll configure the access sequence to check the detailed table first to satisfy the business requirement. If no general ledger account is assigned there, then the general table should be check.

Now, let's learn how to configure and use the three fields mentioned earlier: customer account assignment group, material account assignment group, and account key.

Define and Assign Customer Account Assignment Groups

The customer account assignment groups (AAGC) are first defined in the configuration and then assigned in the customer business partner master data. The material account assignment groups (AAGM) are also defined in the configuration then assigned in the material master data.

To define the customer and material account assignment groups, as shown earlier in Figure 2.11, follow the menu path **Sales and Distribution • Basic Functions • Account Assignment/Costing • Revenue Account Determination**. Then, select the **Check Master Data Relevant For Account Assignment** option. Figure 2.20 shows the configuration steps for defining account assignment groups to customers and materials.

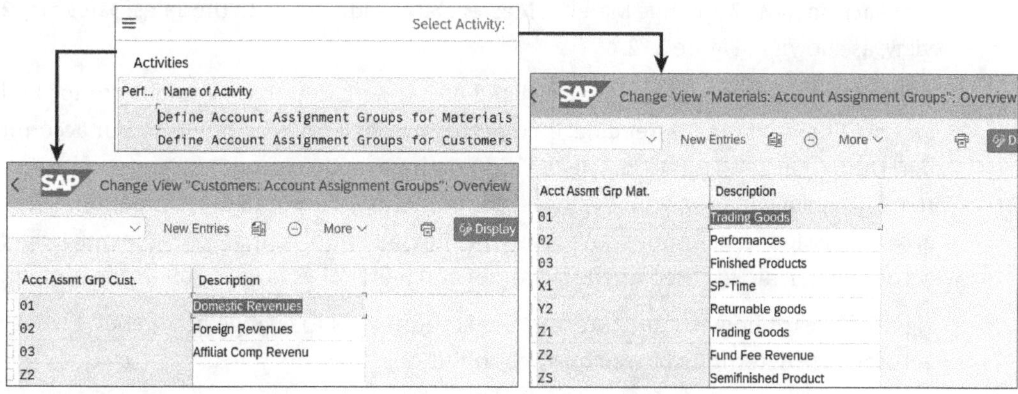

Figure 2.20 Define Customer and Material Account Assignment Groups

First, let's define the account assignment groups for materials. The definition is simply a code and a description. Then, define the account assignment groups of customers. The next step is to assign the materials account assignment groups in the materials master data and assign the customer account assignment groups in the business partner master data, let's explore more on this step.

We can assign the customer account assignment groups in the customer business partner master data. Run Transaction BP and then choose the role that includes the sales billing tab, which is **FLCU01** in standard SAP configuration. The field is highlighted on the screen shown in Figure 2.21.

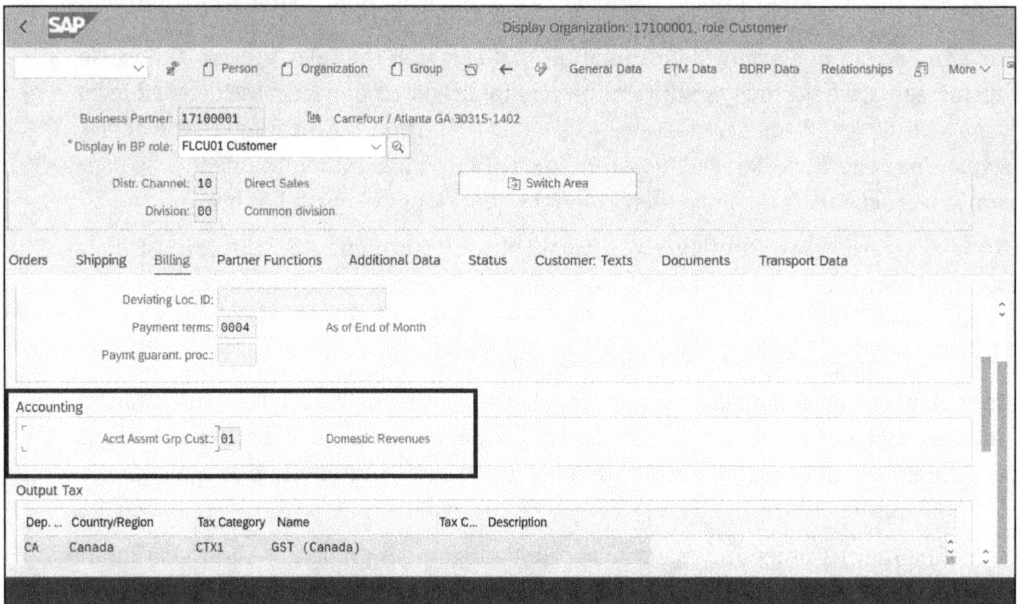

Figure 2.21 Assign Customer Account Assignment Groups

Material account assignment groups can be assigned in the material master data. Run Transaction MM02 (Change Material Master Data) and then go to the **Sales: sales org. 2** view, as shown in Figure 2.22.

With this step, we've defined material and customer account assignment groups and assigned them in the master data, linking the materials and customers to our account determination configuration. The decision on how many account assignment groups to maintain depends on your business requirements for account determination. These account assignment groups give you the flexibility to change the general ledger account determined based on the customer and material used in the billing document.

Next let's learn how to configure and use account keys to change how general ledger accounts are determined based on sales conditions.

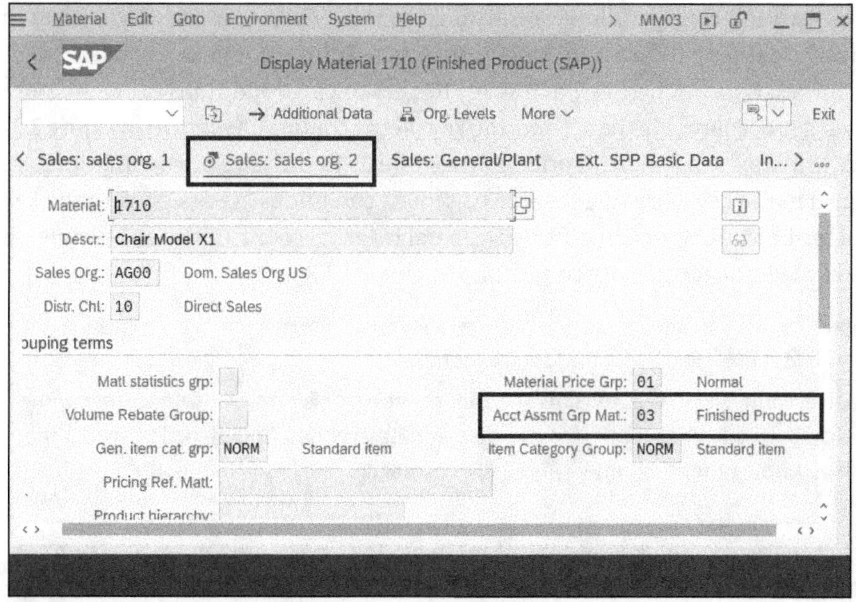

Figure 2.22 Assign Material Account Assignment Groups

Define and Assign Account Keys

To define account keys in the SAP configuration, as shown earlier in Figure 2.11, follow the menu path **Sales and Distribution • Basic Functions • Account Assignment/Costing • Revenue Account Determination.** Select the **Define And Assign Account Keys** option. The configuration steps are shown in Figure 2.23.

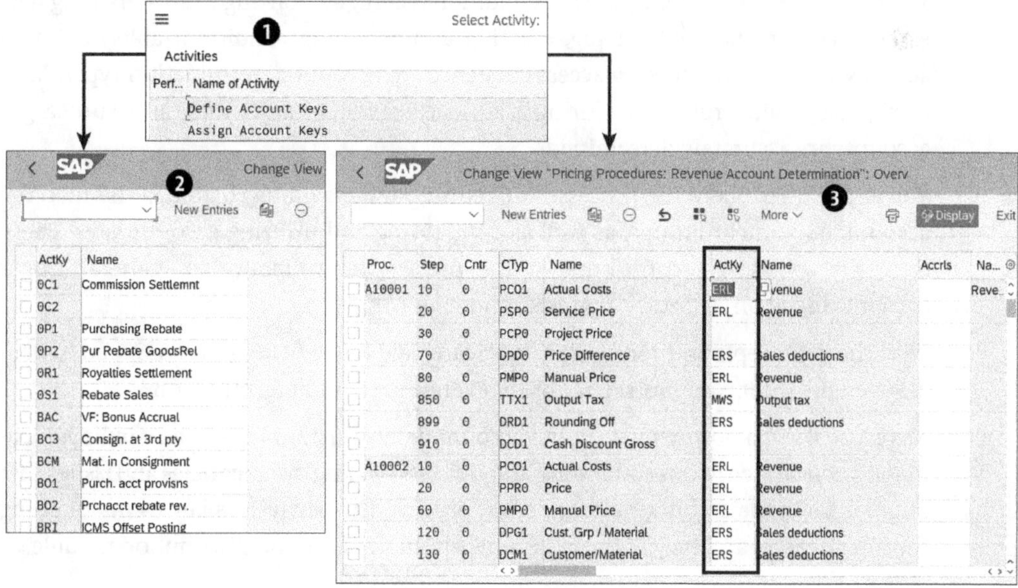

Figure 2.23 Define and Assign Account Keys

When you start this configuration transaction, you must first choose what you want to do ❶. Start by defining the account keys as shown in ❷. An account key is just a 3-digit code with a description. Then, assign this key to sales pricing conditions in the different sales pricing procedures via the accounting key (**ActKy** column), as shown in Figure 2.23 ❸. Notice in our example that, for pricing procedure **0018**, the sales price conditions are all assigned to the key **ERL** while the sales discount conditions are assigned to the key **ERS**. As a result, you can assign different general ledger accounts for sales revenue on account key ERL and for sales discounts on account key ERS.

Note

Insert an account key in the **Accruals** column, as shown in Figure 2.23 ❸, only for conditions that should post accruals, for example, rebate accruals. This selection works the same way as the other account keys used in normal posting.

After assigning an account key to conditions in pricing procedures, you can assign general ledger accounts to the account keys, as described in the Assign General Ledger Accounts section.

With this comprehensive exploration of the condition technique for account determination in SAP S/4HANA, we've covered a critical aspect of the sales and distribution module. We started by understanding the flexibility this technique offers, allowing for the assignment of general ledger accounts based on various fields in the sales order. This level of detail ensures that your business can meet diverse account determination requirements in SAP S/4HANA and tailor its financial recordings to specific scenarios.

You now know how to configure the condition technique, starting from activating necessary fields in the field catalog, creating and assigning condition tables, defining access sequences, and linking access sequences to account determination types. Each step plays a vital role in ensuring that sales revenue, discounts, and surcharges accounts are accurately determined.

You also learned about the processes of defining and assigning customer and material account assignment groups, as well as configuring and utilizing account keys. These elements collectively contribute to the precise determination of general ledger accounts based on customer, material, and sales conditions.

With these concepts and configurations clearly laid out, you are now equipped to handle even the most complex scenarios in revenue account determination.

Next, you'll learn how to analyze the account determination in the sales billing document, a crucial step to understanding and troubleshooting discrepancies and issues that might arise in the account determination process. This analysis is key to maintaining the integrity and accuracy of financial data within the sales and distribution module.

Account Determination Analysis: Sales Revenue Accounts

Now, let's analyze how general ledger accounts were determined in a sales billing document. First, display the billing document using any of the billing document display transactions, for instance, Transaction VF03, as shown in Figure 2.24.

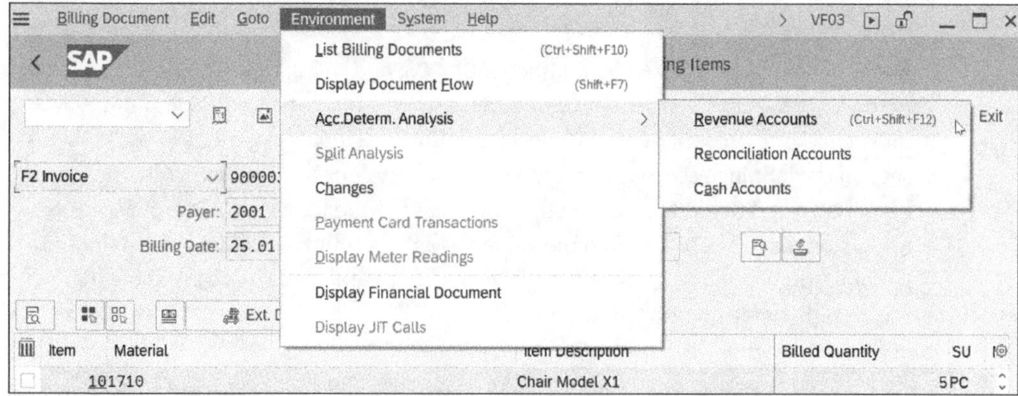

Figure 2.24 Menu for Account Determination Analysis in Sales Billing Documents

Display the account determination analysis by following the menu path shown in Figure 2.24. This step will open the account determination analysis screen shown in Figure 2.25.

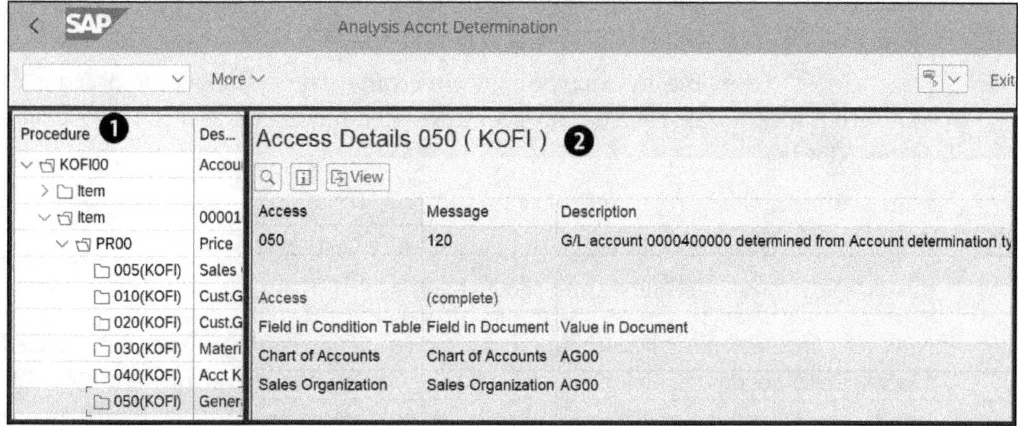

Figure 2.25 Account Determination Analysis in Sales Billing Documents

As shown in Figure 2.25 ❶, this account determination procedure used **KOFI00**, which is automatically determined based on the billing document type. Notice that some pricing conditions have an accounting impact, such as **PR00**. Heres, we see the different condition tables the system accessed, in the sequence configured. For the price condition **PR00**, the system checked the condition tables in the sequence **005**, **010**, **020**, and so on. A general ledger account couldn't be determined until access **050**. No general ledger was determined in the higher priority. By clicking on every table, you can see its details ❷. Notice the different inputs determined from the billing document, such as the chart of accounts and the account assignment groups. You can also see the final general ledger account determined, namely, **0041000000**.

> **Note**
>
> As shown in Figure 2.25 ❶, the numbers 001, 005, and 010 are *not* the condition table numbers but the access sequence numbers. To find the condition table number, you must check the account determination configuration step, as described in the previous section on defining access sequences and account determination types. Also, be aware not all the condition tables assigned to the access sequence are shown—only the ones the system had to access to find the general ledger account. Once the system finds the general ledger account, it doesn't check the rest of the tables and doesn't show them in the account determination analysis.

Now, you should understand the accounts determined in any sales billing documents and troubleshoot the issue when the wrong account is determined.

In some cases, the billing document is posted in sales without an accounting entry. For example, this problem can arise if the account determination for one of the accounts is completely missing, if the general ledger account is blocked for posting, and for many other reasons. This bad situation can go unnoticed, as the user posting the billing document won't see any errors. The billing document is posted correctly from a sales point of view and can be printed out to the customer, but the accounting entry was not posted. Most of the time, the accounting team is not aware of the sales transaction at all, which is why this problem usually goes unnoticed and can cause many issues, such as the following:

- Inaccurate financial reporting for sales
- Delayed revenue recognition, causing negative impact on cash flow
- Wrong customer receivables and aging analysis

Now, let's analyze billing documents that are missing accounting entries, often called *blocked sales documents*, learn how to find the causes of these errors. Running this report frequently to fix any errors is important, and this report must be included in our period-end closing routing.

Release Blocked Billing Documents for Accounting

You can generate a report of the billing documents missing accounting entries, analyze their block reasons, and release them to accounting. For this task, use Transaction VFX3 or follow the menu path **Logistics • Sales and Distribution • Billing • Billing document • Blocked Billing Docs.** Figure 2.26 shows the selection screen.

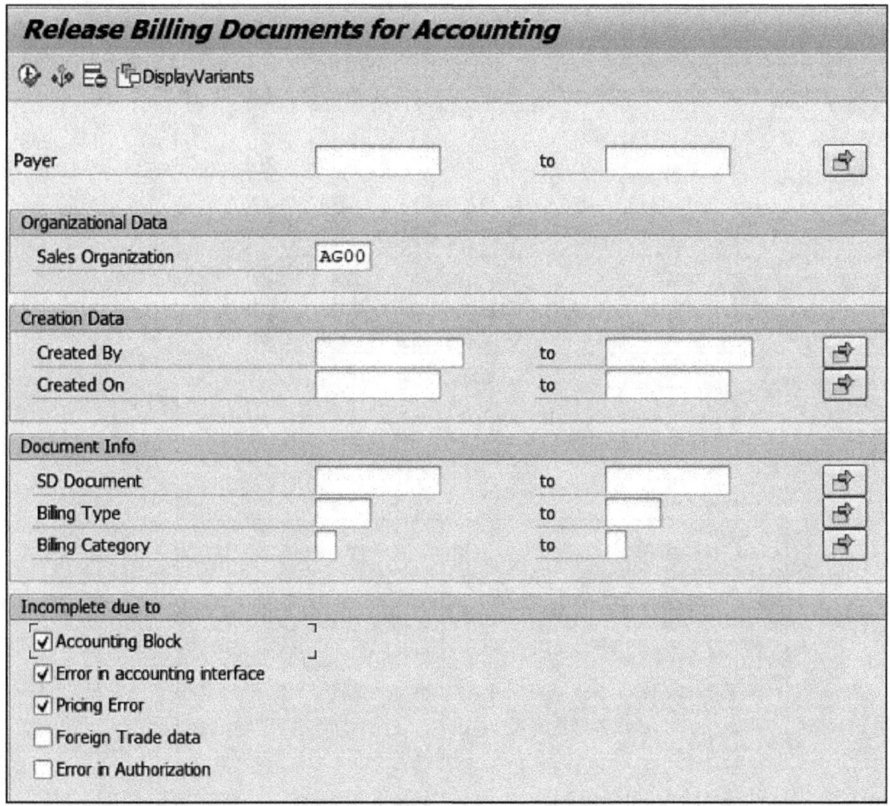

Figure 2.26 Release Blocked Billing Documents: Selection Screen

In the **Incomplete due to** section, choose from among the five available reasons for the billing document block:

- **Accounting Block**: If we have checked the option posting block in the billing type configuration in the SAP configuration via Transaction VOFA.
- **Error in accounting interface**: If a general ledger account is not determined due to missing accounting determination configuration.
- **Pricing Error**: If error arises in pricing conditions, for example, a mandatory condition is missing or a price has the value zero.
- **Foreign Trade data**: If we have activated the incompletion checks for certain fields in foreign trade, but these fields are incomplete.
- **Error in Authorization**: If the user who created the billing documents is missing an authorization object that is needed to post the document to accounting.

After maintaining the selection screen as needed, execute the transaction to see a list of blocked documents, as shown in Figure 2.27.

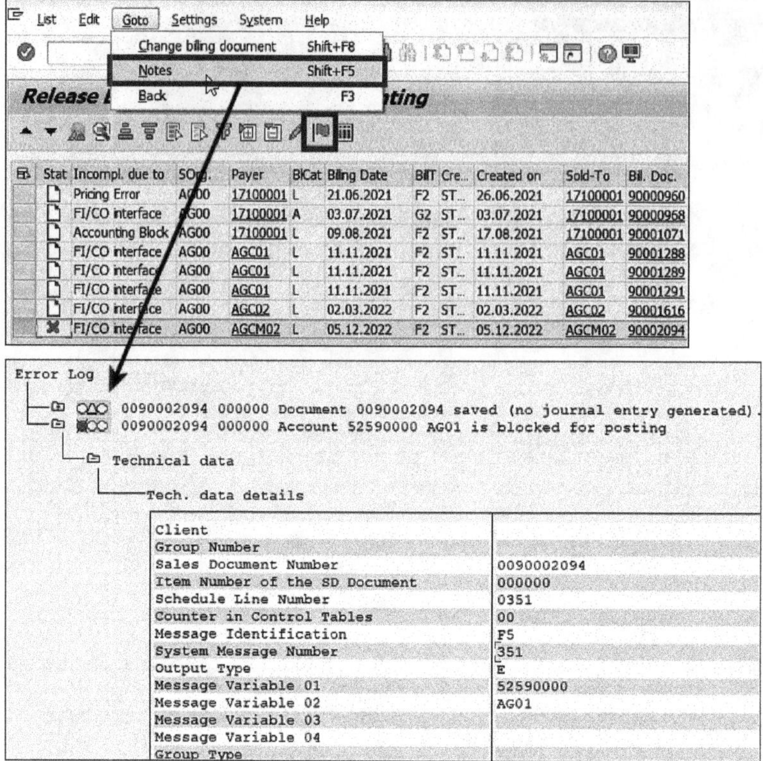

Figure 2.27 Release Blocked Billing Documents: Analysis Screen

The line with the status icon have not yet been processed in this session. You can select any line and click on the release icon to release a document to accounting. If the document is released to accounting without errors, the status changes to successful ; otherwise, the status changes to failed . In the case of failure, go to the menu **Got to • Notes**, as shown in Figure 2.27, to view the error's details. In our example, as shown in Figure 2.27, the accounting document posting failed because one of the general ledger accounts determined is blocked for posting.

After fixing the error, repeat the same release process again to release the accounting document.

Now, you're equipped to handle any issues in account determinations for sales billing documents. You can check the account determination analysis to understand how any account is determined and correct the configuration if needed. You can also check if for sales billing documents that were not posted to accounting, analyze the reason for their blocks, fix these issues, and release the accounting postings.

With this step, you now know how to configure and troubleshoot the accounting determination configuration for sales revenues, discounts, and surcharges in sales billing documents. Let's move to the next account in shown earlier in Table 2.3, namely, the AR account.

2.1.5 Accounts Receivable

The AR account is determined by default as the reconciliation account maintained in the customer business partner master data. This is explained in detail in Chapter 4.

The sales and distribution module also has additional optional configuration steps to determine alternative AR reconciliation accounts based on many of the fields available in the sales process. This feature is useful, for example, if you have a business requirement to determine different reconciliation accounts for the same customer based on sales offices, sales organizations, distribution channels, or material account assignment groups. Many other fields are available as well.

All the configuration steps for alternative reconciliation account determination can be found in the SAP configuration menu, as shown in Figure 2.28.

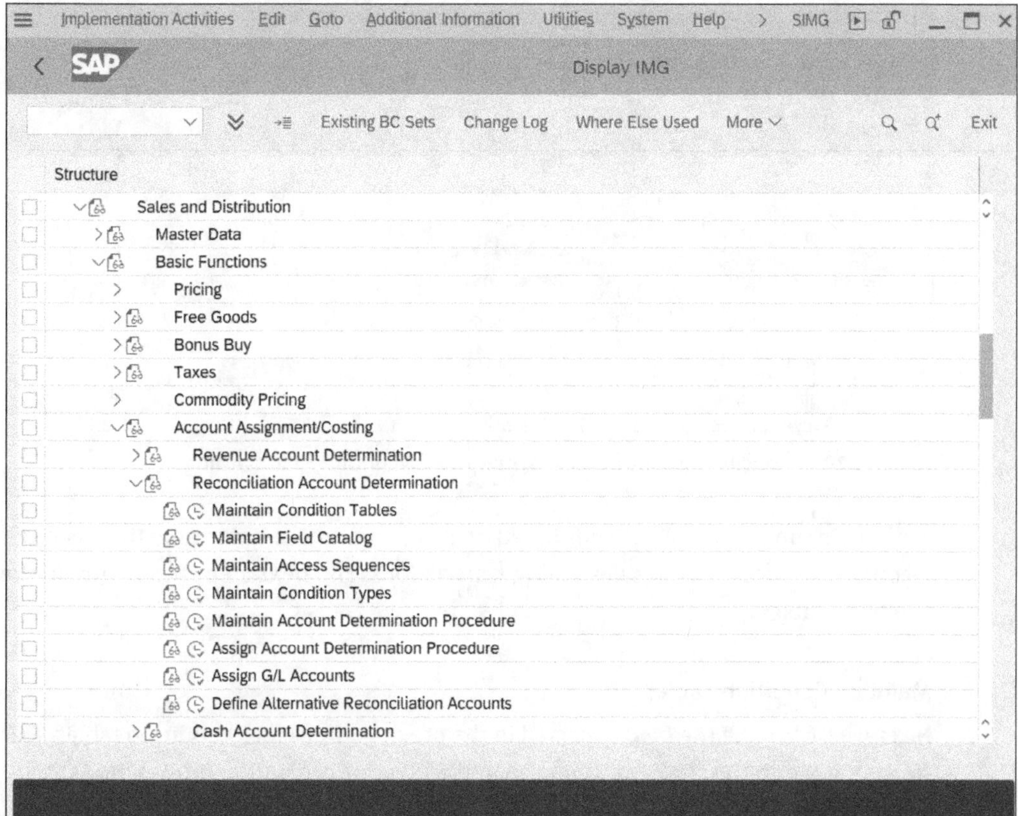

Figure 2.28 Alternative Reconciliation Accounts Determination Configuration Menu

As shown in Figure 2.28, the configuration steps are quite similar to the steps described in Section 2.1.4. Note that the condition tables, access sequences, account determination types, and account determination procedures that are defined for alternative reconciliation accounts are different from those defined for revenue accounts but follow the same concept. In the following sections, we'll walk through each configuration step.

Maintain Field Catalog

First, let's activate the fields that will be used in these condition tables, which is similar to the configuration step in the revenue account determination described earlier in Section 2.1.4, when we changed the field catalog for condition tables. Follow the menu path **Sales and Distribution • Basic Functions • Account Assignment/Costing • Reconciliation Account Determination • Maintain Field Catalog**. The configuration screen is shown in Figure 2.29.

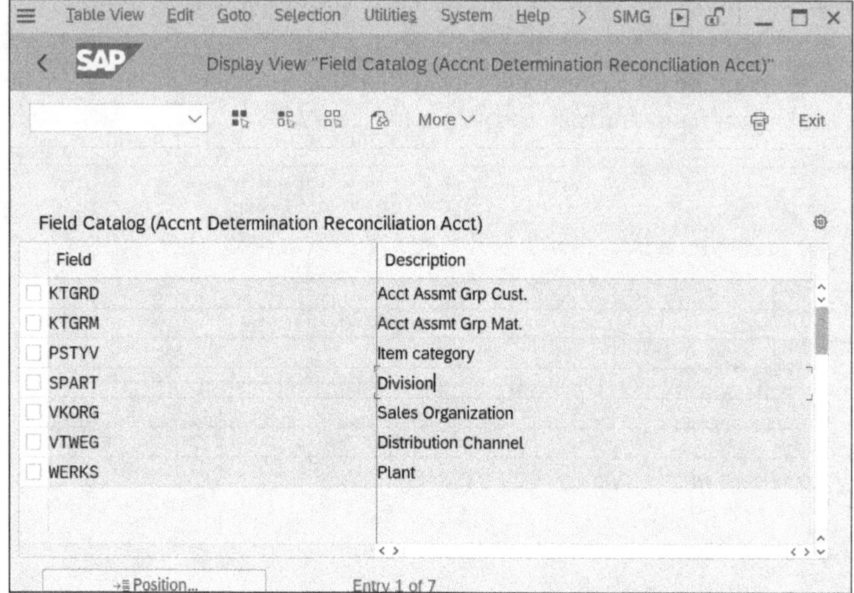

Figure 2.29 Reconciliation Account Determination: Maintain Field Catalog

Now, maintain the different fields based on how you want to determine the customer reconciliation accounts in sales billing transactions. The next step is to maintain the condition tables.

Maintain Condition Tables

Next, we'll combine the fields selected in the previous steps into condition tables. This step is similar to the configuration step in the revenue account determination discussion earlier in Section 2.1.4.

Follow the menu path **Sales and Distribution • Basic Functions • Account Assignment/Costing • Reconciliation Account Determination • Maintain Condition Tables.** The configuration screen is shown in Figure 2.30.

Notice the definition of condition table 008, which includes the fields **Sales Organization** and **Distribution Channel**.

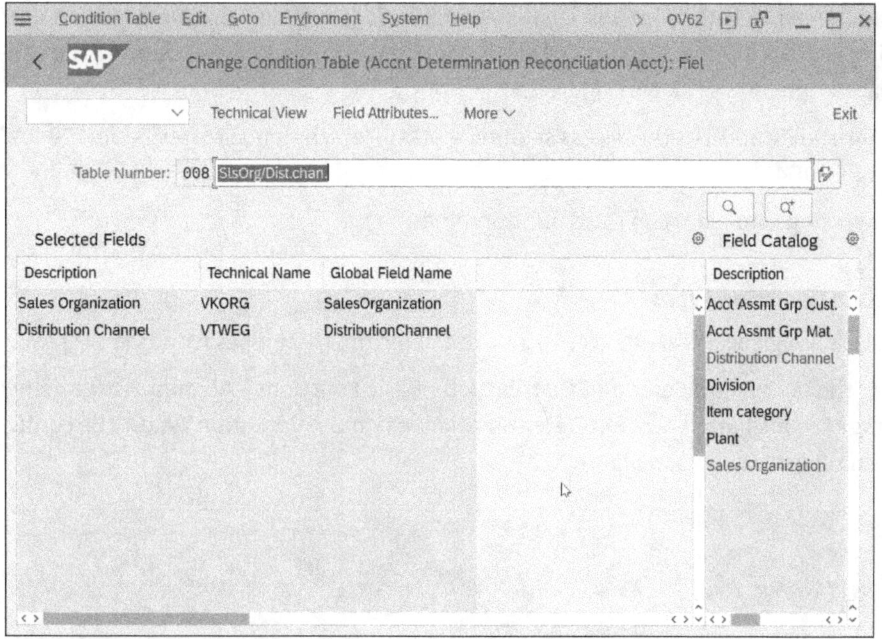

Figure 2.30 Reconciliation Account Determination: Maintain Condition Tables

Next, let's maintain the access sequences.

Maintain Access Sequences

Maintain the sequence in which SAP will access the condition tables to determine the general ledger accounts, which similar to the configuration step in the revenue account determination described earlier in Section 2.1.4: Define Access Sequences and Account Determination Types.

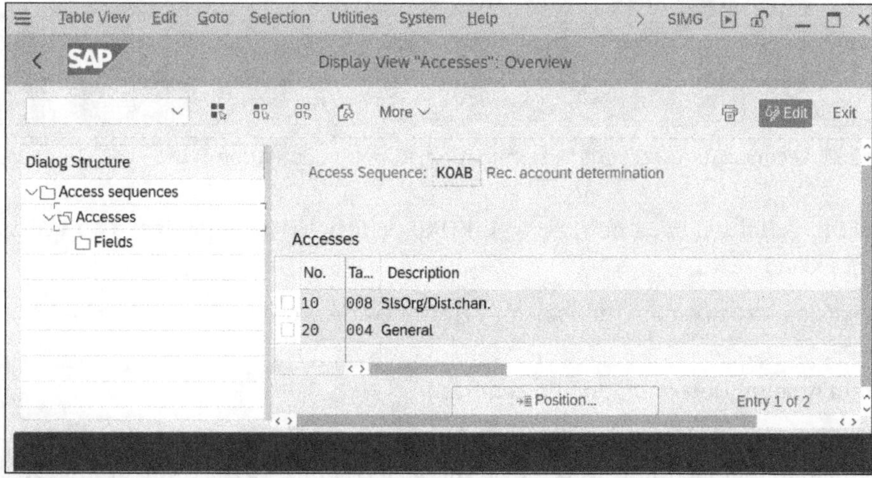

Figure 2.31 Reconciliation Account Determination: Maintain Access Sequence

Follow the menu path **Sales and Distribution • Basic Functions • Account Assignment/ Costing • Reconciliation Account Determination • Maintain Access Sequences.** The configuration screen is shown in Figure 2.31.

Notice the definition of the access sequence **KOAB**, which first accesses table **008** and then table **004**.

The next step is to maintain the condition types.

Maintain Condition Types

Now, let's assign access sequences to account determination types (condition types).

Follow the menu path **Sales and Distribution • Basic Functions • Account Assignment/ Costing • Reconciliation Account Determination • Maintain Condition Types.** The configuration screen is shown in Figure 2.32.

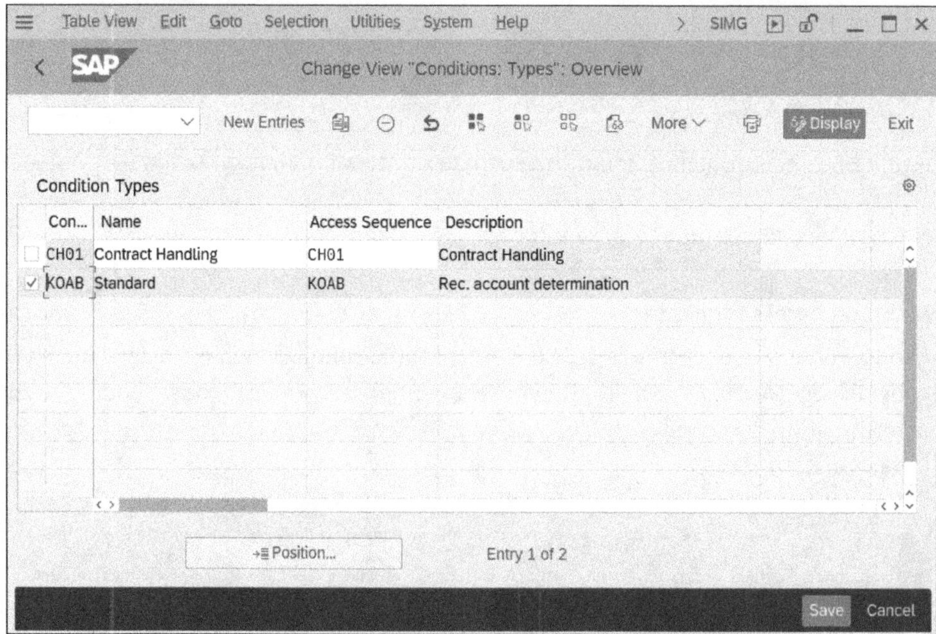

Figure 2.32 Reconciliation Account Determination: Maintain Condition Types

Notice the definition of condition type **KOAB**, which has been assigned to access sequence **KOAB**.

Our next step is to maintain the account determination procedures.

Maintain Account Determination Procedures

In this step, we'll maintain account determination procedures and assign account determination types to these procedures. This step is similar to the configuration step in revenue account determination:

- **Maintain condition types**

  First, we'll assign the access sequences to account determination types (condition types).

  Follow the menu path **Sales and Distribution • Basic Functions • Account Assignment/ Costing • Reconciliation Account Determination • Maintain Condition Types**. The configuration screen is shown in Figure 2.32.

  Notice the definition of condition type KOAB, which has been assigned to access sequence KOAB.

  The next step is to maintain the account determination procedures.

- **Maintain account determination procedures**

  Follow the menu path **Sales and Distribution • Basic Functions • Account Assignment/ Costing • Reconciliation Account Determination • Maintain Account Determination Procedure.** The configuration screen is shown in Figure 2.33.

  Notice how procedure **KOFIAB** has been assigned to account determination type **KOAB.**

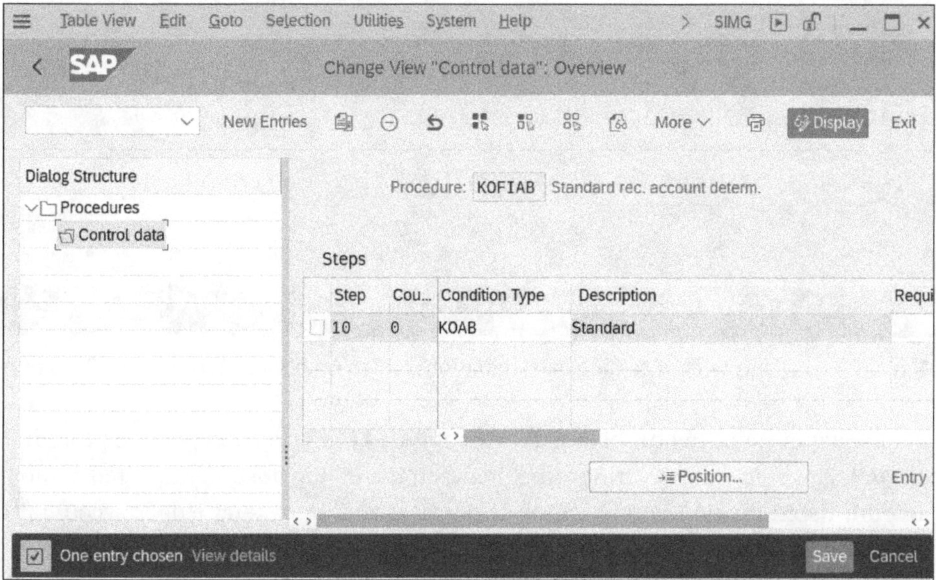

Figure 2.33 Reconciliation Account Determination: Maintain Account Determination Procedure

Next, we can assign procedures to billing document types.

Assign Account Determination Procedures

In this step, we'll assign the account determination procedures to billing document types.

Follow the menu path **Sales and Distribution • Basic Functions • Account Assignment/Costing • Reconciliation Account Determination • Assign Account Determination Procedure.** The configuration screen is shown in Figure 2.34.

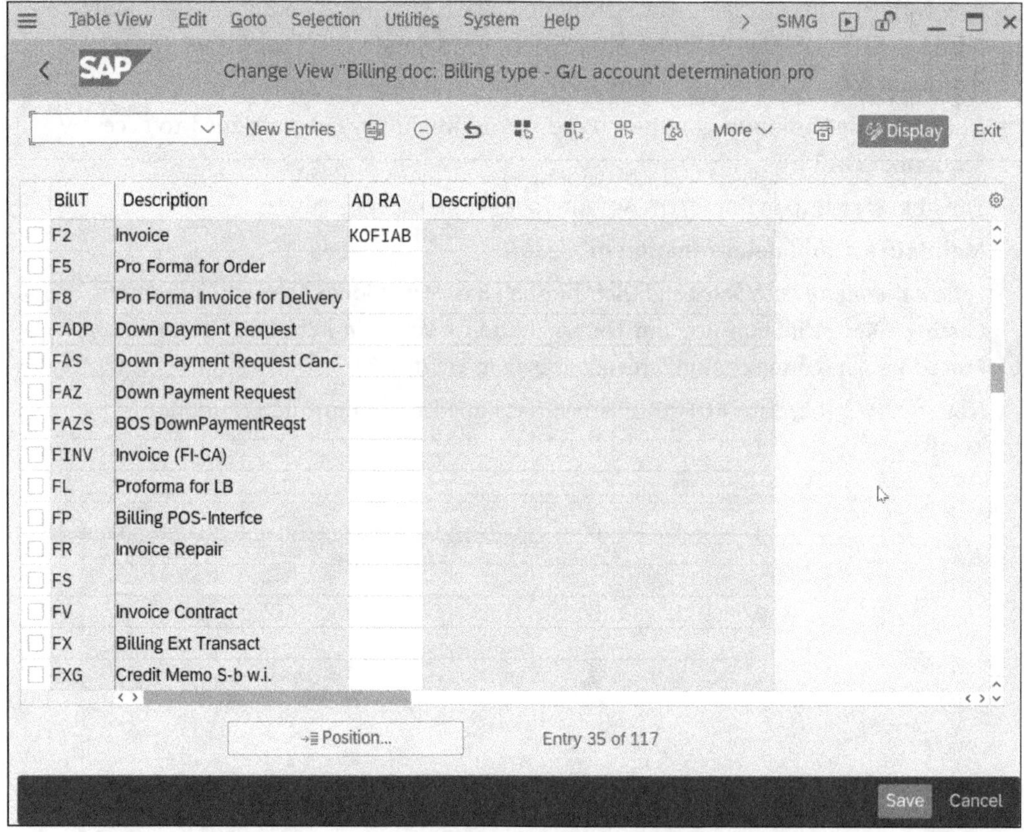

Figure 2.34 Reconciliation Account Determination: Assign Account Determination Procedure

In our example, notice how billing document type **F2** has been assigned to procedure **KOFIAB**. This step is for creating a link between the billing document posted and our account determination configuration through assigning general ledger accounts to condition tables.

Assign General Ledger Accounts

Now, you can assign alternative reconciliation accounts to condition tables. For this step, use Transaction OV64 or follow the menu path **Sales and Distribution • Basic Functions • Account Assignment/Costing • Reconciliation Account Determination • Assign G/L Accounts.** The configuration screen shown in Figure 2.35.

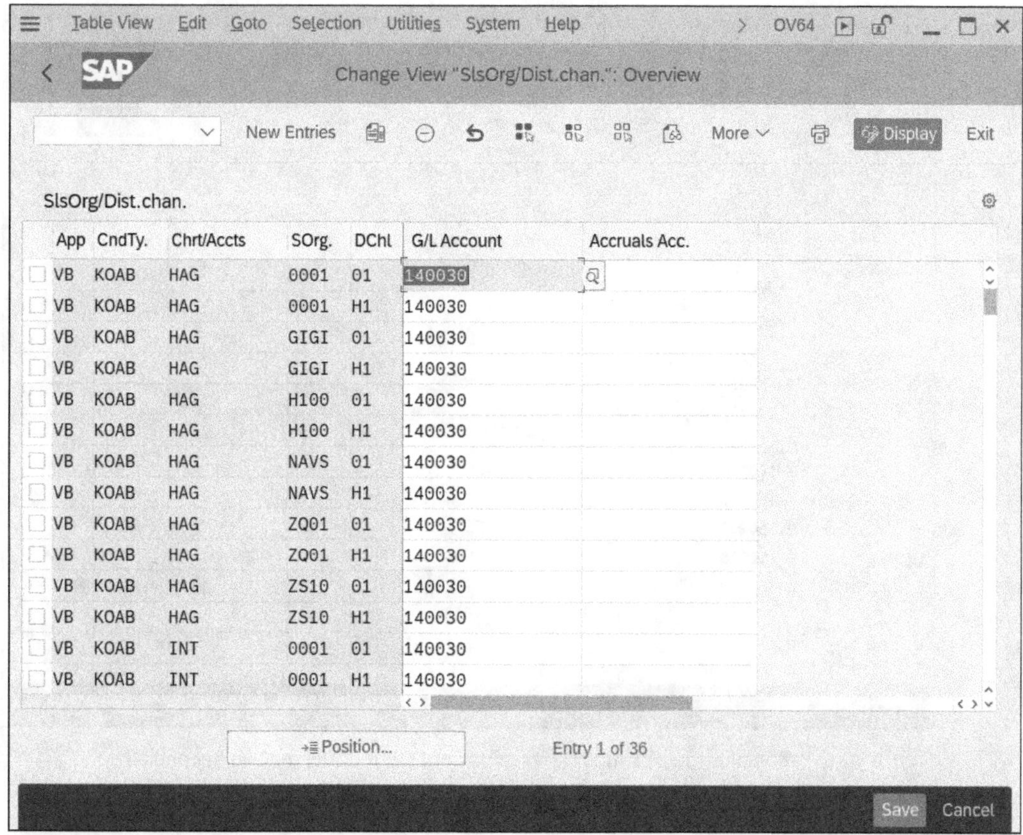

App	CndTy.	Chrt/Accts	SOrg.	DChl	G/L Account	Accruals Acc.
VB	KOAB	HAG	0001	01	140030	
VB	KOAB	HAG	0001	H1	140030	
VB	KOAB	HAG	GIGI	01	140030	
VB	KOAB	HAG	GIGI	H1	140030	
VB	KOAB	HAG	H100	01	140030	
VB	KOAB	HAG	H100	H1	140030	
VB	KOAB	HAG	NAVS	01	140030	
VB	KOAB	HAG	NAVS	H1	140030	
VB	KOAB	HAG	ZQ01	01	140030	
VB	KOAB	HAG	ZQ01	H1	140030	
VB	KOAB	HAG	ZS10	01	140030	
VB	KOAB	HAG	ZS10	H1	140030	
VB	KOAB	INT	0001	01	140030	
VB	KOAB	INT	0001	H1	140030	

Figure 2.35 Reconciliation Account Determination: Assign General Ledger Accounts

Now, only one mandatory step remains, which is to define the accounts assigned in this step as alternative reconciliation accounts for the standard accounts maintained in the customer business partner master data, as described next.

Define Alternative Reconciliation Accounts

This configuration step can be accessed by following the menu path **Sales and Distribution • Basic Functions • Account Assignment/Costing • Reconciliation Account Determination • Define Alternative Reconciliation Accounts.** The configuration screen is shown in Figure 2.36.

On this screen, insert the standard reconciliation account that's determined by default from the customer business partner master data. Then, insert the allowed alternative reconciliation accounts that can replace this standard account.

In our example shown in Figure 2.36, the default reconciliation account in the **G/L Acc** column is **140000**, which we allow to be replaced by the accounts in the **Alt. G/L** column, namely, **147001**, **147100**, or **147300**.

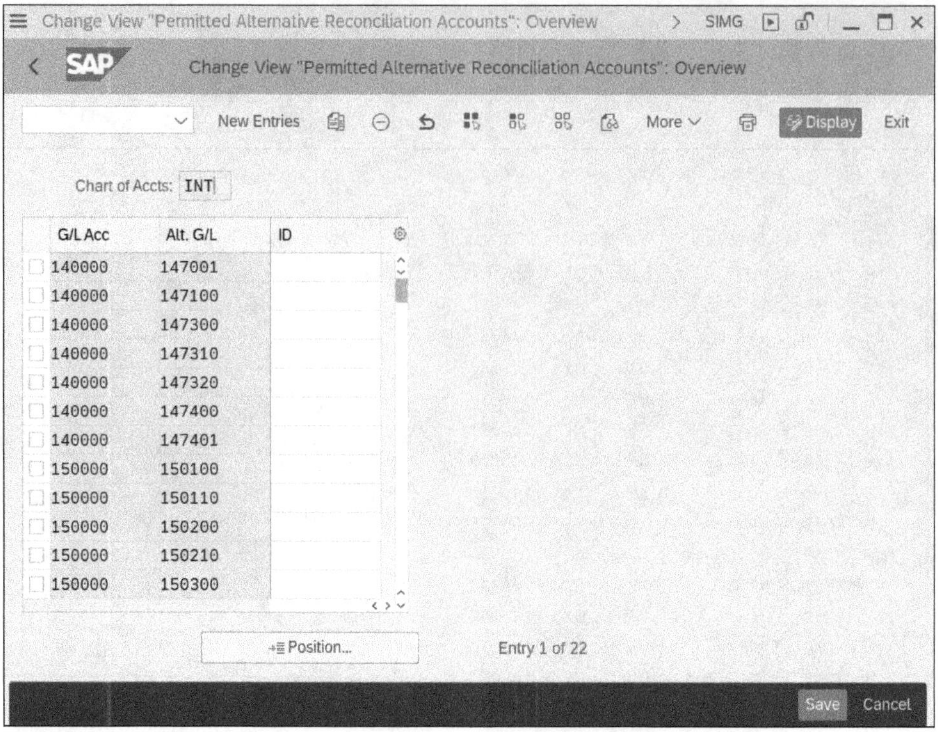

Change View "Permitted Alternative Reconciliation Accounts": Overview

Chart of Accts: INT

G/L Acc	Alt. G/L	ID
140000	147001	
140000	147100	
140000	147300	
140000	147310	
140000	147320	
140000	147400	
140000	147401	
150000	150100	
150000	150110	
150000	150200	
150000	150210	
150000	150300	

Entry 1 of 22

Figure 2.36 Reconciliation Account Determination: Define Alternative Reconciliation Accounts

Now, you understand how to configure the account determination for alternative reconciliation accounts in the sales process. You can determine different reconciliation accounts for the same customer based on other sales fields such as the distribution channel. This maximum flexibility in customer reconciliation account determination guarantees matches to different business requirements.

If the system fails to determine an alternative reconciliation account due to any missing configuration steps, then it will continue with the standard reconciliation account determined from the customer business partner master data, and it won't issue any error messages. You must always compare the postings in the alternative reconciliation accounts against the sales billing figures. If any discrepancies arise, then the account determination configuration may have some errors. In this case, you must analyze the account determination in the billing document to uncover the error. Let's look into how to do this task next.

Accounts Determination Analysis: Alternative Reconciliation Accounts

To analyze the account determination of alternative reconciliation accounts in any billing document, the steps are similar to the steps explained in Section 2.1.4, on account determination analysis.

Display the billing document via Transaction VF03 and then start the account determination analysis, as shown in Figure 2.37.

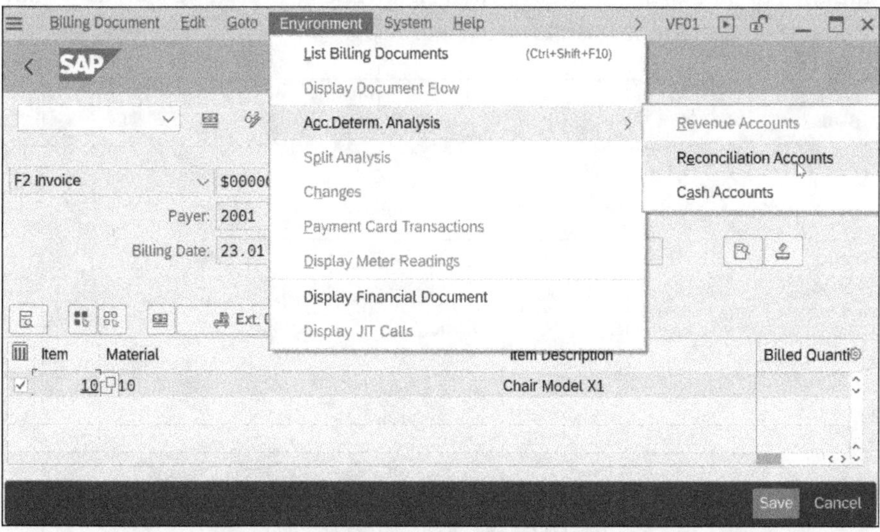

Figure 2.37 Reconciliation Account Determination: Start the Account Determination Analysis

On the next screen, you'll see the reconciliation account determination analysis, as shown in Figure 2.38.

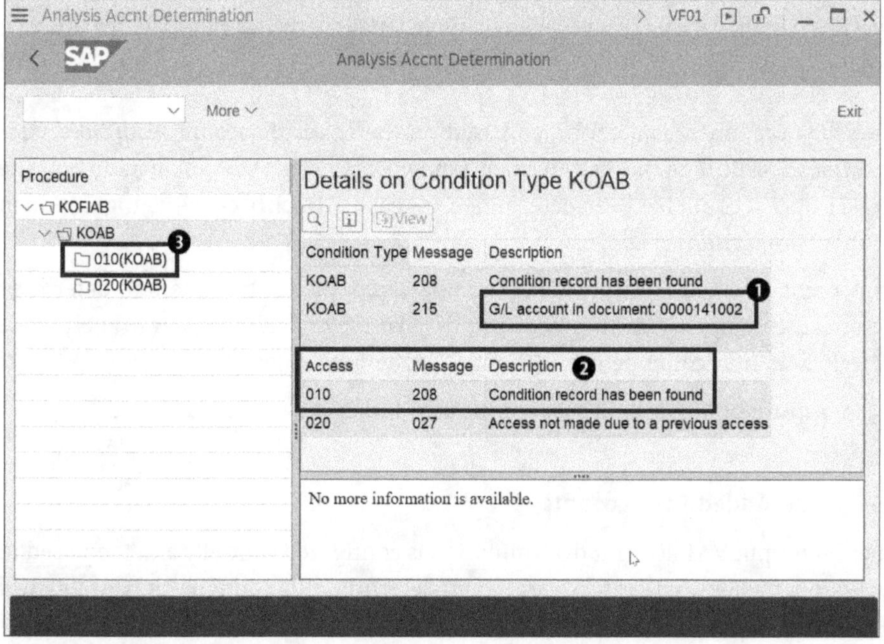

Figure 2.38 Reconciliation Account Determination: Result of the Account Determination Analysis

In our example, the general ledger account determined **141002**, as shown in ❶. As shown in ❷, note how this account has been determined through the access step **010**. You can then double-click the access step in Figure 2.38 ❸ to access details about how this account was determined, as shown in Figure 2.39.

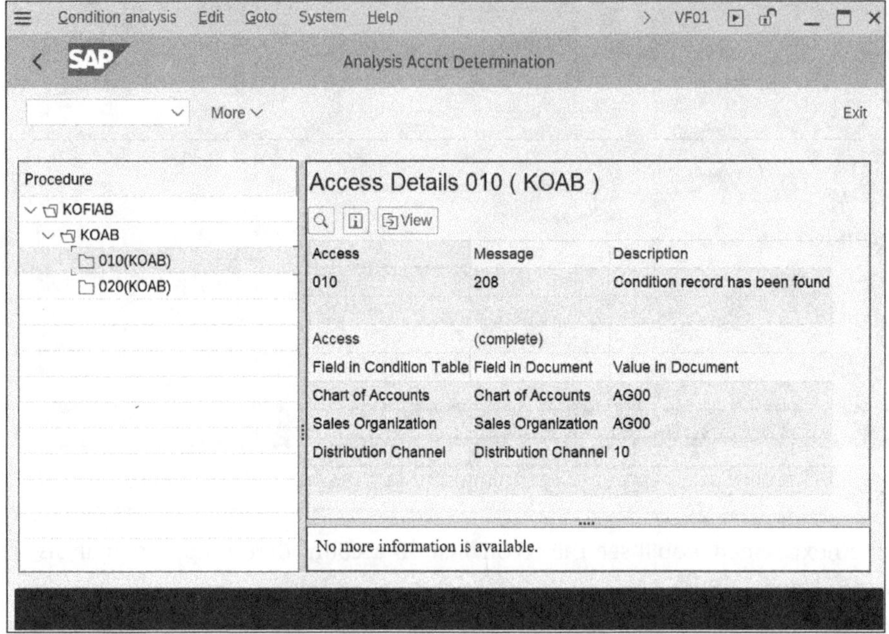

Figure 2.39 Reconciliation Account Determination: Details of the Account Determination Analysis

Notice how account was determined based on the chart of account **AG00**, the sales organization **AG00**, and the distribution channel **10**. Now that you understand how the general ledger account has been determined, it's easy to track back the configuration in case any issues arise.

At this point, you completely understand how to configure the system to determine different customer reconciliation accounts based on sales fields and how to analyze the billing document account determination to resolve any issues.

Now, let's move on to the next account listed in Table 2.3—the VAT account.

2.1.6 Value-Added Tax Accounts

Input and output VAT account determination is configured centrally and then used in the different business processes. The account determination configuration for VAT will be explained in Chapter 4, Section 4.3.

2.1.7 Withholding Tax Accounts

WHT are configured centrally and then used in the different business processes. The account determination configuration for WHT will be explained in Chapter 4, Section 4.4.

With this detailed exploration of the standard sales from stock process in SAP S/4HANA, we've comprehensively covered each phase of this fundamental sales procedure. From the initial customer inquiry to the final billing, we've explored each step to understand the associated accounting entries and the general ledger accounts involved. This included a deep dive into the details of inventory management, COGS, sales revenue and discounts, and customer reconciliation accounts. We've seen how SAP S/4HANA efficiently manages the different process steps to ensure accurate and timely financial reporting.

In this process, we've also dived into the configuration aspects within SAP for account determination related to this process. This understanding is crucial to ensure that financial postings accurately mirror the actual sales transactions, thereby maintaining financial integrity and compliance with accounting standards.

As we wrap up this section, we've laid a solid foundation for understanding the complete sales cycle in SAP S/4HANA, particularly for standard sales from stock. Our next step is to explore the sales from stock process with COGS posting at billing. This variant of the standard sales process introduces an interesting twist, where the COGS is posted not at the time of goods issue but at the billing stage. This adjustment in the accounting process offers unique insights and challenges, which we'll explore in detail in the following section.

2.2 Sales from Stock with COGS Posting at Billing

This process follows the same flow as the sales from stock business process described in Section 2.1, but with some changes in the accounting entries to post the COGS value only at the time of posting the sales billing document. Let's look into more details.

2.2.1 Business Process Overview

In the standard sales from stock process, the COGS account is posted in the goods issue accounting entry as explained in Table 2.1, and the sales revenue account is posted in the sales billing accounting entry, as explained in Table 2.3.

According to the accounting matching principle, companies must report expenses at the same time as the revenues they are related to, which means that the COGS and sales revenue values related to a sales process should be reported in the same financial period.

In some business cases, a long time passes between the goods issue and the billing transactions. This difference may mean that the COGS value is posted in a different period from the sales revenue value. In these cases, we can configure the account determination in a way that the COGS value is posted at the sales billing accounting entry along with the sales revenue value. This non-standard configuration has some limitations and is not recommended, but let's discuss it anyway to show how account determination can change to match different business cases.

The business process flow is the same as the sales from stock business process described in Section 2.1, but now, our accounting entries for goods issue and sales billing are different. Let's look into these accounting entries.

Goods Issue

The goods issue accounting entry will no longer be posted to the COGS account, but instead, it will post to an intermediary account that we can call *goods in transit*. The accounting entry is shown in Table 2.5.

Debit	Credit	Debit Amount ($)	Credit Amount ($)
Goods in transit		1000	
	Inventory		1000

Table 2.5 Accounting Entry of Post Goods Issue: Goods in Transit

Goods in transit is a balance sheet account used as an intermediary account to replace the COGS account during post goods issue. This account will be cleared against COGS at the customer billing accounting entry. The value posted is the cost of the items issued.

Sales Billing

The sales billing will also be posted to the COGS account and the goods in transit account. The accounting entry will be as shown in Table 2.6.

Debit	Credit	Debit Amount ($)	Credit Amount ($)
AR		1350	
Sales discount		250	
WHT		50	
	Output VAT		150
	Sales revenue		1500
COGS		1000	

Table 2.6 Accounting Entry of Sales Billing with COGS

Debit	Credit	Debit Amount ($)	Credit Amount ($)
	Goods in transit		1000

Table 2.6 Accounting Entry of Sales Billing with COGS (Cont.)

All the accounts used in this accounting entry are the same as explained in the sales from stock business process described in Section 2.1. The difference is the COGS and goods in transit accounts:

- **COGS**
  - This account is the same account used in the sales from stock business process described in Section 2.1.
  - The value posted is calculated from the internal cost sales condition, which is condition **VPRS** in the standard configuration. The value of this condition is calculated based on the cost of the items in the billing documents.
- **Goods in transit**
  - The same account used in goods issue.
  - The value posted is equal to the COGS value, which should be equal to the value posted to goods in transit at goods issue.

Note

Many configuration issues can cause sales condition VPRS to have different value from the item cost, especially in complicated cases involving split valuations. This discrepancy may cause incorrect values to be posted to COGS and will leave a balance in the goods in transit account because the credit posted at billing is not equal to the debit posted at goods issue. Therefore, you should be careful when configuring your system to use COGS posting at billing.

Next, let's show you how to change the account determination configuration to post to the COGS account at the sales billing document and to post to the goods in transit account.

2.2.2 Goods in Transit Account at Goods Issue

To post to the good in transit account at goods issue, follow the same steps as described in Section 2.1.2. In this case, however, maintain the goods in transit account number instead of the COGS account number via Transaction OMWB. Assign the goods in transit account to the transaction key **GBB** and the account modification **VAY**, as shown in Figure 2.40.

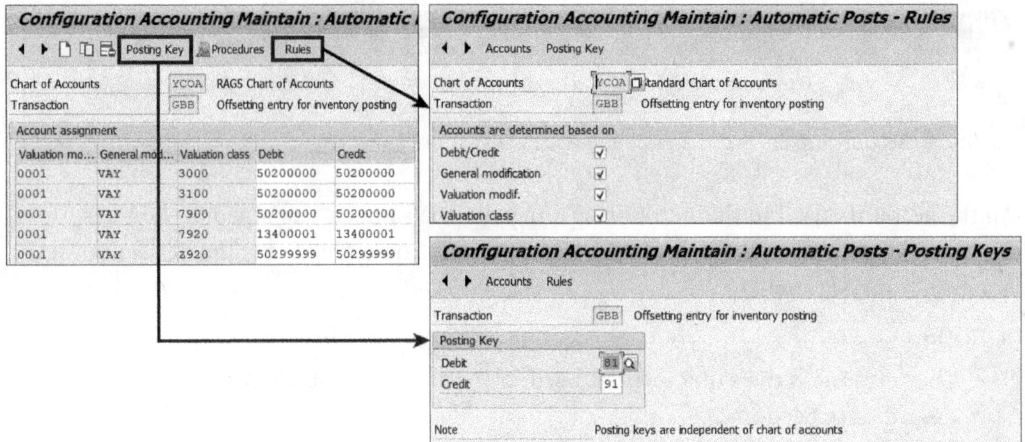

Figure 2.40 Assign Accounts, Rules, and Posting Keys: Goods in Transit at Goods Issue

The next step is to configure the account determination of the goods in transit account and the COGS account at sales billing.

2.2.3 COGS and Goods in Transit Accounts at Sales Billing

The account determination for these two accounts at sales billing can be configured by using the same technique described in Section 2.1.4. Simply create and assign an account key to the sales pricing condition **VPRS** then assign the general ledger accounts to a condition table that includes the account key.

The condition VPRS is the standard sales pricing condition that shows the item internal cost. The same configuration explained in this section can also be used with any customized condition copied from condition VPRS. Let's look into the detailed configuration steps.

Define and Assign Account Keys

This configuration step follows the same steps outlined in Section 2.1.4 where we defined and assigned account keys. You can either define a new account key or use an existing one. Then, assign this key to the internal cost condition **VPRS** in the **Accrual Acc.** column, as shown in Figure 2.41.

Figure 2.41 shows the account key definition ❶ and the assignment of the account key ❷ to the internal cost condition **VPRS** in the **G/L Account** (GIT) and **Accruals Account** (COGS) columns ❸.

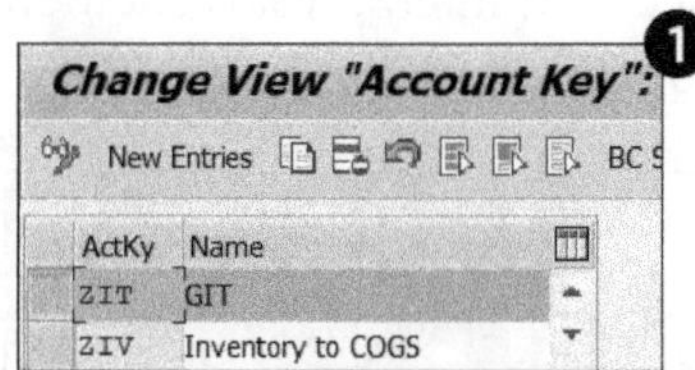
Change View "Account Key": ❶

ActKy	Name
ZIT	GIT
ZIV	Inventory to COGS

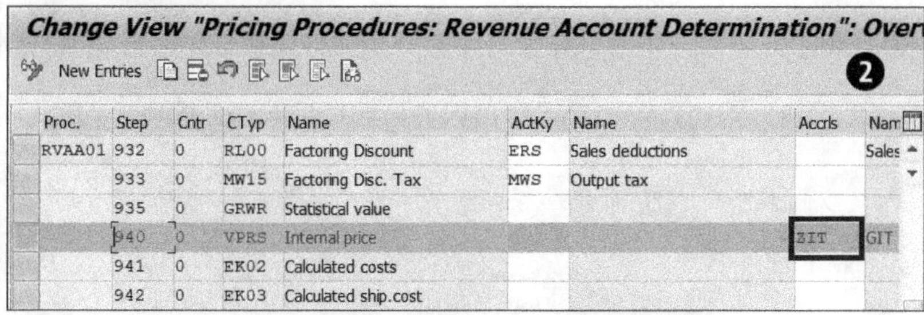
Change View "Pricing Procedures: Revenue Account Determination": Over ❷

Proc.	Step	Cntr	CTyp	Name	ActKy	Name	Accrls	Nam
RVAA01	932	0	RL00	Factoring Discount	ERS	Sales deductions		Sales
	933	0	MW15	Factoring Disc. Tax	MWS	Output tax		
	935	0	GRWR	Statistical value				
	940	0	VPRS	Internal price			ZIT	GIT
	941	0	EK02	Calculated costs				
	942	0	EK03	Calculated ship.cost				

Change View "Cust.Grp/MaterialGrp/AcctKey": Overview ❸

Cust.Grp/MaterialGrp/AcctKey

App	CndTy.	Chrt/Accts	SOrg.	AAGC	AA...	ActKy	G/L Account	Accruals Acc.
V	KOFI	YCOA	AG00	01	02	ZIT	13400001	51500000

Figure 2.41 Define and Assign Account Keys and General Ledger Accounts: Goods in Transit and COGS at Billing

Next, we can assign the general ledger accounts to this account key.

Assign General Ledger Accounts

This configuration step follows the same process described in the Section subsection in Section 2.1.4. You can assign the general ledger accounts as needed to condition tables that include the **Account Key** field, as shown in Figure 2.41 ❸. Assign the goods in transit account in the **G/L Account** column; this account will be posted in the credit side of the accounting entry. Next, assign the COGS account in the **Accrual Account** column; this account will be posted in the debit side of the accounting entry.

The last step is to allow the internal cost condition (i.e., condition VPRS) to be posted to an accrual account; otherwise, the COGS account that we added in the **Accrual Account** column in this step won't be considered in the posting.

Allow Accruals in Pricing Condition

To allow any sales pricing condition to be posted to an accrual account, open Transaction V/06 (Define Condition Types). You can also follow the menu path **Sales and Distribution • Basic Functions • Pricing • Pricing Control • Define Condition Types**. Then, choose the **Set Condition Types for Pricing** option. To allow the internal cost condition to post to an accrual account, select the **Accruals** checkbox, as shown in Figure 2.42.

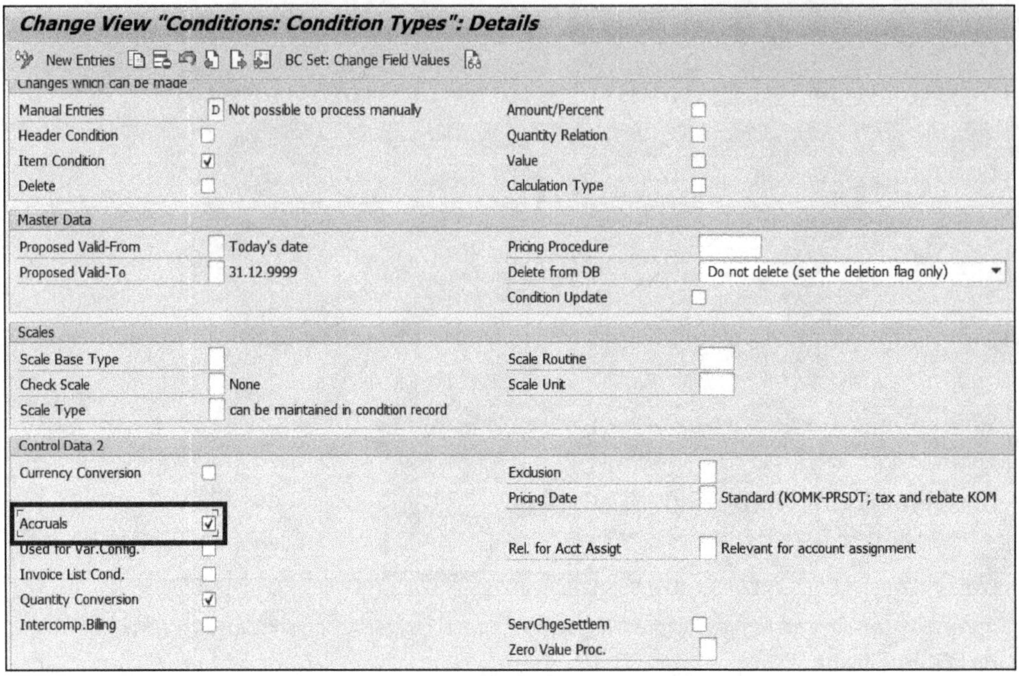

Figure 2.42 Allow VPRS Condition Type to Post to an Accrual Account

This is the last configuration step needed to activate the account determination of the goods in transit account at goods issue instead of a COGS account determination as well as to activate the account determination of goods in transit and COGS at sales billing.

As we conclude the section on sales from stock with COGS posting at billing, we've dove into a unique variant of the standard sales process. This approach, where the COGS is posted only at the time of sales billing, aligns the COGS with the corresponding revenue, adhering to the *accounting matching principle*. This method, though less conventional, is crucial for certain business scenarios where the time gap between goods issue and the billing is significant.

We have explored the key changes in accounting entries for goods issue and sales billing, including the use of a goods in transit account as an intermediary at goods issue, and the subsequent posting to both the goods in transit and COGS accounts during sales billing. The process demonstrated SAP S/4HANA's flexibility in accommodating different accounting requirements and ensuring accurate financial reporting.

With the detailed configuration steps outlined, from adjusting the account determination for goods in transit at goods issue to setting up COGS and goods in transit accounts at sales billing, we've equipped ourselves with the knowledge to implement this process effectively in SAP. This knowledge is essential for businesses that require COGS to be recognized at the time of revenue realization.

Now, as we move forward, we transition to the next process in SAP S/4HANA—namely, sales from stock with valuated stock-in-transit (VSIT). This process introduces another layer of complexity, involving valuated stock management during the sales process. In the following section, we'll explore how this process differs from the standard sales from stock, particularly focusing on the accounting implications of managing valuated stock in transit and how SAP S/4HANA facilitates this sophisticated sales process.

2.3 Sales from Stock with Valuated Stock in Transit

Expanding upon the foundational sales from stock process, this section introduces the concept of VSIT, a crucial element for certain business scenarios within SAP S/4HANA. VSIT becomes particularly significant in situations dictated by specific incoterms, where our responsibility for the goods extends beyond the conventional point of goods issue from the warehouse. This scenario typically arises with agreements where delivery obligations extend to the customer's warehouse or to a designated port in the receiving country.

The inclusion of VSIT in the sales process addresses a key business need—the need to maintain visibility and control over inventory even after it has left the company's premises. This visibility is especially important for financial and inventory accuracy, as it allows for continued valuation and accounting of the stock while in transit. This process ensures that inventory is recorded in our books, reflecting the actual physical movement and ownership of goods only when they reach their final agreed-upon destination.

The proof of delivery step, exclusive to this process, is pivotal and is not just a logistical confirmation but a significant accounting event. The proof of delivery process is executed upon the physical delivery of the products to the designated location, serving as the trigger to adjust inventory accounts. At this point, the inventory is now debited from our books, accurately aligning with the transfer of ownership to the customer. This step ensures compliance with the agreed-upon terms and provides an accurate reflection of inventory levels and financial obligations.

In the broader context of this chapter, understanding VSIT is essential for businesses that operate under complex delivery terms and require accurate tracking of goods until the final handover. This topic exemplifies SAP S/4HANA's capability to adapt to diverse logistical and financial requirements, thus providing a comprehensive solution for managing complex sales processes. This process not only enhances inventory and financial accuracy but also aligns with international trade practices, making it a vital component in our global business landscape.

Next, we'll look at this process flow in detail and explore the differences in accounting entries from the sales from stock process described in Section 2.1; then we'll show you how to configure the account determination for these accounting entries.

2.3.1 Business Process Overview

This process is similar to sales from stock process described in Section 2.1 but with the additional step of proof of delivery. This business process flow is shown in Figure 2.43.

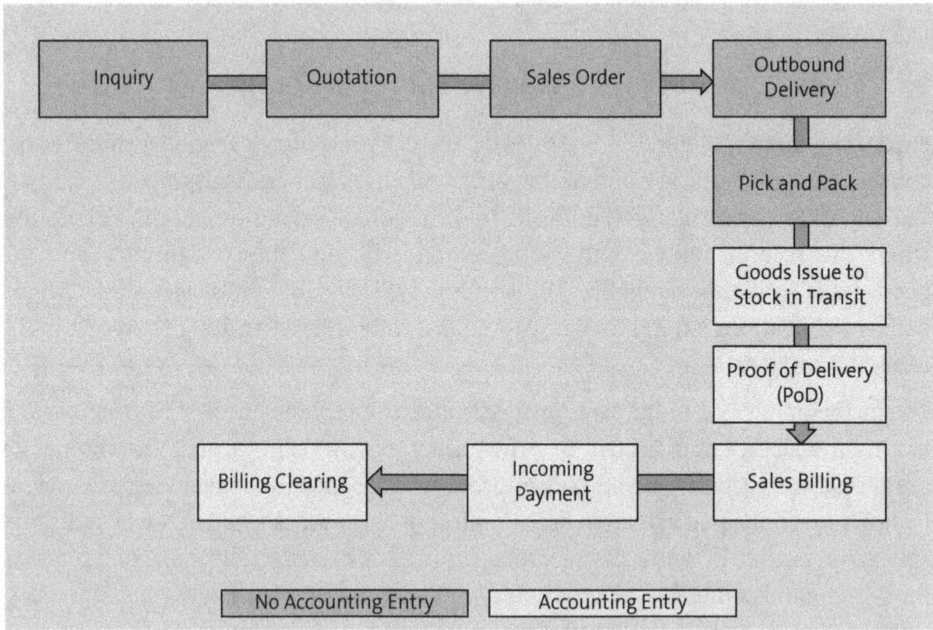

Figure 2.43 Standard Sales with Valuated Stock in Transit Process Flow

The steps from inquiry to pick and pack are the same as the sales from stock process. Let's have a quick look the step that differ from the sales from stock process and the corresponding differences among accounting entries. The first difference is in the goods issue step.

Good Issue to Stock in Transit

At this point, we've issued the sold items from our warehouse, but we're still the owners of the items, and we're responsible for any damage that may happen until we deliver them to the agreed-upon location. The accounting entry is shown in Table 2.7.

Debit	Credit	Debit Amount ($)	Credit Amount ($)
Stock in Transit		1000	
	Inventory		1000

Table 2.7 Accounting Entry of Goods Issue to Valuated Stock in Transit

In this entry, we move the sold items from the general unrestricted stock to the sales order stock in transit. We can always track the quantity and value of our stock in transit in the inventory reports. In our example, the inventory account is the same as described in the sales from process. The stock in transit account is a balance sheet account with the same definition as the inventory account. The value posted at this point is the COGS.

Proof of Delivery

Once the items are delivered to the agreed-upon location, we're ready to transfer the ownership to the customer. At this point, we can process the proof of delivery and remove the stock from our books. The accounting entry is shown in Table 2.8.

Debit	Credit	Debit Amount ($)	Credit Amount ($)
COGS		1000	
	Stock in transit		1000

Table 2.8 Accounting Entry of Proof of Delivery

In this entry, we remove the stock from the stock in transit and post the value to the COGS, which is the same account as in the standard sales process. The stock in transit account should have the same entries as shown earlier in Table 2.7.

The remaining process steps—sales billing, incoming payment, and payment clearing—are the same as the sales from stock process. After the proof of delivery step in the process, as shown in Figure 2.43, the next steps are all the same as the standard sales from stock process outlined in Section 2.1.

With this task, you now know the difference in the sales process that's introduced by adding the VSIT feature. We explored the different accounting entries included in the process. All the accounts used are the same as the normal sales from stock process, except the VSIT account.

Now, let's configure the account determination for the valuated stock in transit account.

2.3.2 Valuated Stock in Transit Account

This account is determined in the same way as described in Chapter 1, Section 1.1.2, when we defined valuation classes. Simply assign a valuation class to every material in the **Accounting 1** view of the material master data, which you can access via Transaction MM03. On the same screen, you also have the option of assigning a second valuation class in the **VC: Sale Ord. Stk** field. This valuation class is used to determine the general ledger accounts for sales order stock, which is the case when you post the goods issue to valuated goods in transit. (The stock is reserved in transit for a specific sales order.) The **VC: Sale Ord. Stk** field is shown in Figure 2.44 and has the value **3101**.

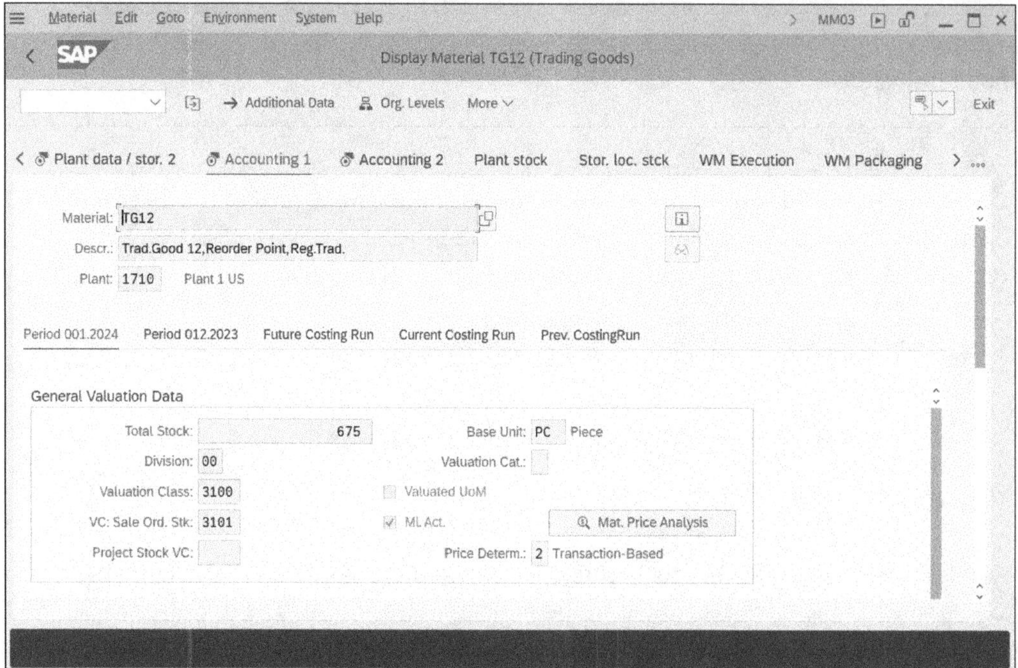

Figure 2.44 Valuation Class for Valuated Stock in Transit

> **Note**
>
> If the **VC: Sale Ord. Stk** field is empty, the normal valuation class is used to determine the general ledger account for the accounting entry described in Section 2.3.1 (goods issue to stock in transit), which means both the debit and credit will go to the same account: the inventory account.

As described in Chapter 1, Section 1.1.2, in the Configure Automatic Posting section, we must assign general ledger accounts to this valuation class for the key **BSX**, as shown in Figure 2.45.

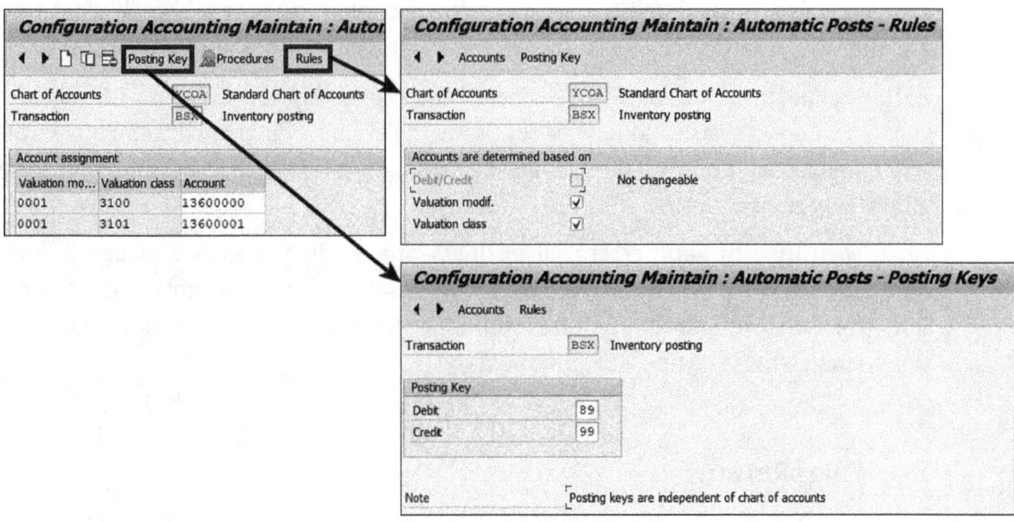

Figure 2.45 Assign Accounts, Rules, and Posting Keys for VSIT Inventory

Figure 2.45 shows the valuation class **3101** has been assigned to the inventory account **13600001** (stock in transit), which is different from the normal stock account **13600000**, which is assigned to the valuation class **3100**. You also must assign general ledger accounts for the account key GBB with the account modification **VAY** as shown in Figure 2.46.

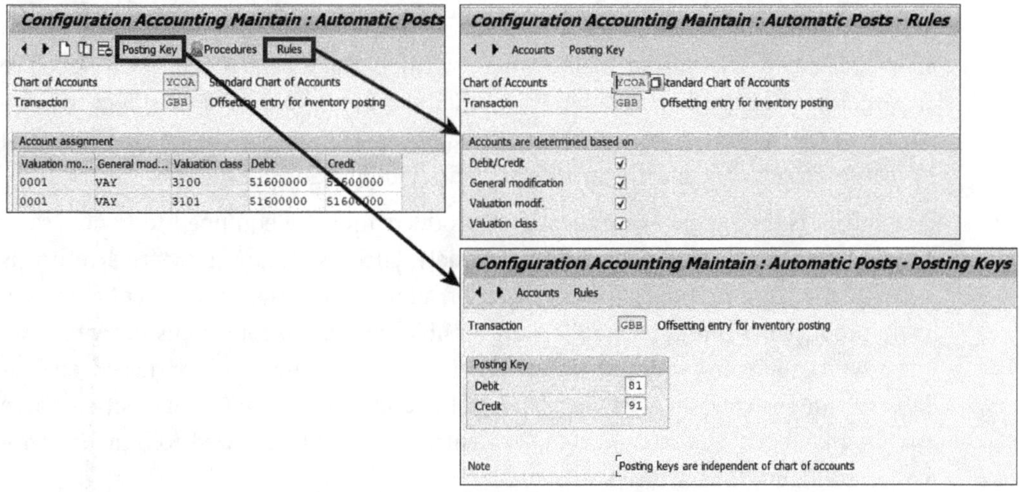

Figure 2.46 Assign Accounts, Rules, and Posting Keys for VSIT COGS

Figure 2.46 shows the assignment of the VSIT valuation class **3101** to the COGS account to be used in the accounting entries shown earlier Table 2.8. Normally, the business requirement is for this COGS account to be same as the normal COGS account used in standard sales. Thus, as shown in Figure 2.46, both the standard valuation class **3100** and the VSIT valuation class **3101** are assigned to the same COGS account **51600000**.

At this point, you understand the business process overview for the sales with valuated stock in transit process and how to configure the account determination. You're now well equipped to configure the account determination for this process and to analyze and fix any errors.

Next, we'll turn to another crucial business process in the sales and distribution domain: the sales returns process. This process uses the same concepts as the normal sales from stock process but requires some specific account determination configuration to match these business requirements.

2.4 Sales Returns

Any company that sells products must also have processes for sales returns handling. Sales returns can arise in multiple cases. For example, when you accept the sales return, you can ask the customer to send the products back and then issue a refund to the customer, or you can send a product replacement. In some cases, you may ask the customer to directly scrap the products, instead of returning them, and then issue a refund or a replacement. These different cases require different accounting approaches and, therefore, different account determination configurations. This process can be managed in SAP S/4HANA through the *advanced returns management process*.

In the dynamic and customer-centric world of sales, the handling of sales returns is an inevitable and crucial aspect. In SAP S/4HANA, the sales returns process is designed to manage the various scenarios that arise when a product sold does not meet the customer's expectations or requirements. This process not only addresses the logistical aspects of returns but also the intricate financial implications that follow.

Any business that engages in the sale of products must be equipped to handle sales returns efficiently. Scenarios can range widely, such as instances where a return is accompanied by a refund to the customer or, alternatively, the provision of a replacement product. In some situations, a more feasible approach is for the customer to scrap the product, following which a refund or replacement is issued. Each of these scenarios presents unique challenges in terms of logistics and accounting. The diversity in these cases demands distinct accounting treatments, necessitating tailored account determination configurations in SAP S/4HANA.

In SAP S/4HANA, the advanced returns management process is adeptly designed to handle these varied scenarios. This robust process ensures that returns are managed effectively, aligning logistical operations with precise accounting treatments. By utilizing this process, businesses can ensure that sales returns are not just a logistical activity but are integrated seamlessly into the financial workflow. This integration is vital for maintaining accurate financial records and ensuring customer satisfaction, making the sales returns process a key component in the realm of sales and distribution within SAP S/4HANA.

In this section, we'll explore the various facets of the sales returns process, understanding how SAP S/4HANA facilitates different return scenarios, and the corresponding account determination configurations. Our focus will be on comprehending how SAP S/4HANA manages the complexities associated with sales returns, ensuring that each return is accounted for accurately and efficiently, both logistically and financially.

Let's start with an overview of the business process and the accounting entries.

2.4.1 Business Process Overview

The sales returns process starts with the creation of a sales returns order that's usually created with a reference to a sales order. Then, based on the return sales order, you'll create an inbound delivery, a goods receipt (GR), and a customer credit memo. Figure 2.47 shows an overview of the sales returns process.

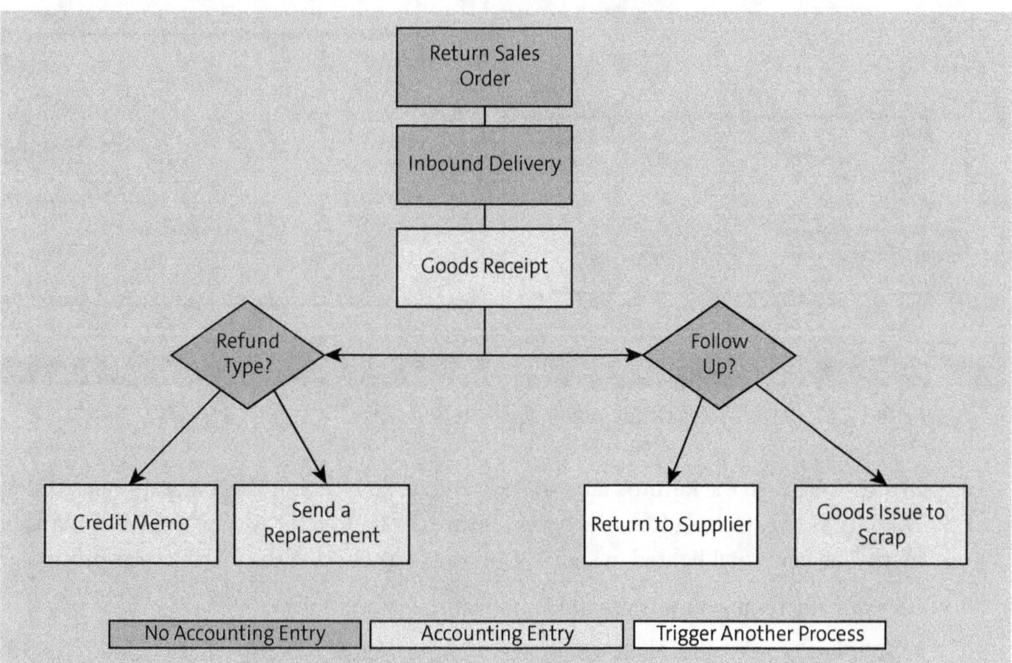

Figure 2.47 Sales Returns Process Overview

Let's learn how account determination works in every step of this process next.

Returns Sales Order

This return order is usually created with reference to a sales order and copies its details, including the materials being returned, the quantities, and the sales prices that will be used in the credit memo. You can also use other fields and then include them in account determination condition tables to determine specific general ledger accounts based on the field values, as described earlier in Section 2.1.4.

The returns sales order also determines the follow-up activities in the returns process. For example, whether you send a replacement material instead of a credit note and whether returned items should be directly scrapped after receipt. The values you select control how the business process will continue and, therefore, what accounting entries will be posted.

Figure 2.48 shows the advanced returns management tab in a sales returns order.

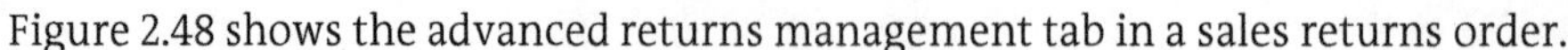

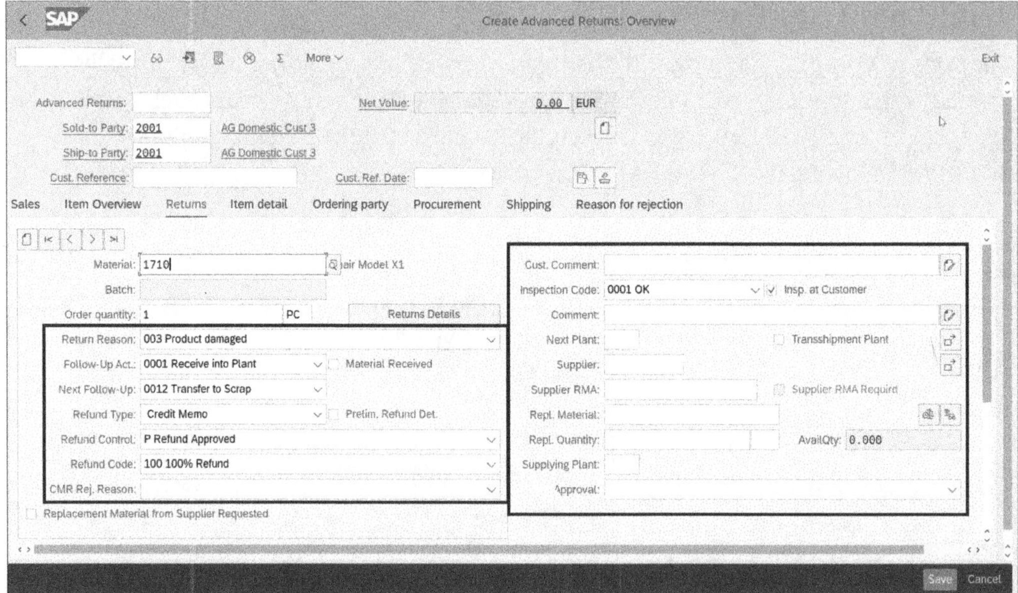

Figure 2.48 Return Sales Order: Advanced Sales Returns

In SAP S/4HANA, the **Returns** tab can be found in Transaction VA01 (Create Sales Order). This tab is only available for order types that are configured for advanced returns management. The fields highlighted in Figure 2.48 control the rest of the sales returns process.

Let's quickly review how these fields impact the process flow:

- **Follow-Up Act.** and **Next Follow-Up**: These fields control what happens to the returned materials. For example, we've selected the follow-up activity **Receive into Plant**, and the next follow-up is **Transfer to Scrap**. Thus, once you save the order, SAP will process these two activities. Many other follow-up activities are available. For example, you can choose a follow-up activity to send the returned items back to your supplier or to send the items to another plant.
- **Refund Type, Refund Control**, and **Refund Code**: These fields control how the customer will be compensated. For example, through a credit memo or a replacement

material. If you issue a credit memo, then you must also select the refund code, which controls the refund percentage. The refund code controls when the credit memo will be issued and whether approval is required.

- **CMR Rej. Reason, Cust. Comment,** and **Inspection Code**: These fields control the quality inspection result. These fields can be left empty if the quality inspection result is not known at the time of return order creation, which means you'll insert quality inspection results later.
- **Next Plant**: If the follow-up activity is to send the returned items to another plant, then in this field, maintain the receiving plant code. SAP will automatically create a stock transfer order to this plant.
- **Supplier**: If the follow-up activity is to send the material back to the supplier, maintain the supplier in this field, and SAP will automatically create a return purchase order.
- **Repl. Material, Repl. Quantity,** and **Supplying Plant**: If the refund type is to send a replacement material, then in this field, maintain the replacement material code, quantity, and the plant from which the material will be issued.

This return sales order creation step on its own generates no accounting entries but does trigger the next process step.

The next step is to create an inbound delivery with reference to the returns order or check the inbound delivery created automatically by the advanced returns management.

Inbound Delivery

The inbound delivery includes the details of the returned items, the quantities, storage locations where the items will be received, and other details needed to post the material GR. This step has no accounting entries.

The next step is to post the GR with reference to the inbound delivery.

Goods Receipt

In this step, you'll post the receipt of the returned materials in inventory management and accounting. This step can have different impacts based on the process triggered in the return sales order, as this impacts the movement types used in the GR posting.

The movement type determines whether the items are to be received in quality inspection blocked stock. If the items are received in quality blocked stock, we can see the items in our inventory reports separately from the other accepted items, and no accounting entry is posted when we post the GR. When these returned items are released after passing the quality inspection, the accounting entry is posted. If the items fail the quality inspection, then the sales return is rejected, and no accounting entry will be posted.

The accounting entry posted in the GR to unrestricted stock is shown in Table 2.9.

Debit	Credit	Debit Amount ($)	Credit Amount ($)
Inventory		1000	
	COGS		1000

Table 2.9 Accounting Entry of GR of Sales Returns

We're already familiar with these accounts because this step is a reversal of the same accounting entry posted at the time of issuing the goods for the original sales order (shown earlier in Table 2.1).

The value posted to these accounts is equal to the current stock value when the GR is processed and not the value that was available at the time of post goods issue, which is the standard system behavior.

After the GR, the rest of the process is determined based on what you've selected in the **Refund Type** field and the **Follow-Up Act.** field.

The **Refund Type** field specifies how to compensate the customer for the returned materials. You can either issue a credit memo to refund the customer or send a replacement for the materials returned.

The **Follow-Up Act.** field determines what happens to the returned materials. This choice can cause other material movements and accounting entries.

Let's look at how the process continues based on the **Refund Type** field and some of the available **Follow-Up Act.** field options that are triggered automatically after GR, based on the return order.

Refund Type: Issue Credit Memo

In the return order, if you've selected the **Refund Type: Credit Memo** option, then when you save the return order, the system will either create a credit memo request that can be later converted to a credit memo or will create a credit memo directly and post the accounting entry. This last part depends on the option you select in the **Refund Control** field.

When the credit memo is posted, the financial entry shown in Table 2.10 will be used.

Debit	Credit	Debit Amount ($)	Credit Amount ($)
Sales returns		1000	
	AR		1000

Table 2.10 Accounting Entry of Credit Memo

In our example, the sales returns account can either be the sales revenue account (used earlier in Section 2.2.1 in the section) or a separate account with a similar definition. We'll cover how to configure a separate account for sales returns shortly. The value posted here is the value we want to refund to the customer, which is copied from the original sales order. The AR account is the same customer account used in Section 2.2.1.

Now, let's turn to the other **Refund Type** field option available: **Replacement Material**.

Refund Type: Replacement Material

In the return order, if you've selected the **Refund Type: Replacement Material** option, then after saving the return order, the process of creating an outbound delivery is created for the replacement material. Then, you'll post the goods issue with reference to this outbound delivery. The goods issue will trigger the same accounting entry as described earlier in Section 2.2.1. In this process, you're sending a replacement material to the customer instead of a refund through a credit memo.

Now, you understand the different options available to compensate a customer. Next, let's consider how returned materials are handled through the **Follow-Up Act.** field.

Follow Up: Goods Issue to Scrap

To scrap the materials returned from a customer, select this option. Now, the GR transaction will also trigger another material movement to issue the materials received to scrap, and another accounting entry will be created, as shown in Table 2.11.

Debit	Credit	Debit Amount ($)	Credit Amount ($)
Scrap expense		1000	
	Inventory		1000

Table 2.11 Accounting Entry of Goods Issue to Scrap

Now, let's look more closely at the accounts involved in this accounting entry:

- **Inventory**
  - The inventory balance sheet account.
  - The value posted is equal to the inventory item cost (standard or moving average) multiplied by the quantity.
  - This account is determined as explained in Chapter 1, Section 1.1.2.
- **Scrap expense**
  - A consumption expense account can have any description you want, such as "Material scrap."
  - The value posted is equal to the inventory item cost (standard or moving average) multiplied by the quantity.

- This account can either be created as an "operating expense or revenue account" with the cost element category 1 (expense) or as a 'non-operating expense or revenue account" with no cost element.
- If the account is created as an operating expense account, then you must configure a default cost center for this cost element.
- We'll show you how to configure this account shortly.

Another option for the follow-up activity is to send the returned materials back to your supplier instead of scrapping the materials. In this case, the follow-up action: return to supplier.

Follow Up: Return to Supplier

If you select this follow-up activity, then saving the return sales order will trigger a full purchase returns process.

At this point, we've covered the sales returns business process in SAP S/4HANA. We've covered key steps from initiating a return order to handling final follow-up actions. This process, detailed in its approach, showcases SAP's capability to adapt to various return scenarios and integrate them seamlessly into the financial workflow.

Key aspects in this process include creating return orders, managing inbound deliveries, posting GRs, and deciding on customer compensation (i.e., through refunds or replacements). Each step, governed by the initial return order setup, directly influences the subsequent accounting entries, thus ensuring accuracy in financial reporting.

As we conclude this overview, we hope it's clear that SAP S/4HANA's sales returns process is a comprehensive solution, bridging the operational and financial aspects of handling returns. SAP S/4HANA stands out in its ability to cater to different business needs while maintaining financial integrity.

Next, let's explore configuring the account determination of the general ledger accounts used in the accounting entries in this business process.

2.4.2 Sales Returns Accounts

The first account in this process is the sales returns account shown in Table 2.10.

The account determination for the sales returns account follows the same steps as outlined in Section 2.1.4. As shown in Table 2.10, in the sales returns process, you can either debit the normal sales revenue account or debit a sales returns account, depending on your business requirements.

To determine a different account for sales returns, find a field that's different in the return sales order from the normal sales order and include this field in a condition table. The most common field in this context is the sales order item category. Figure 2.49 shows an example of the condition table that can be used.

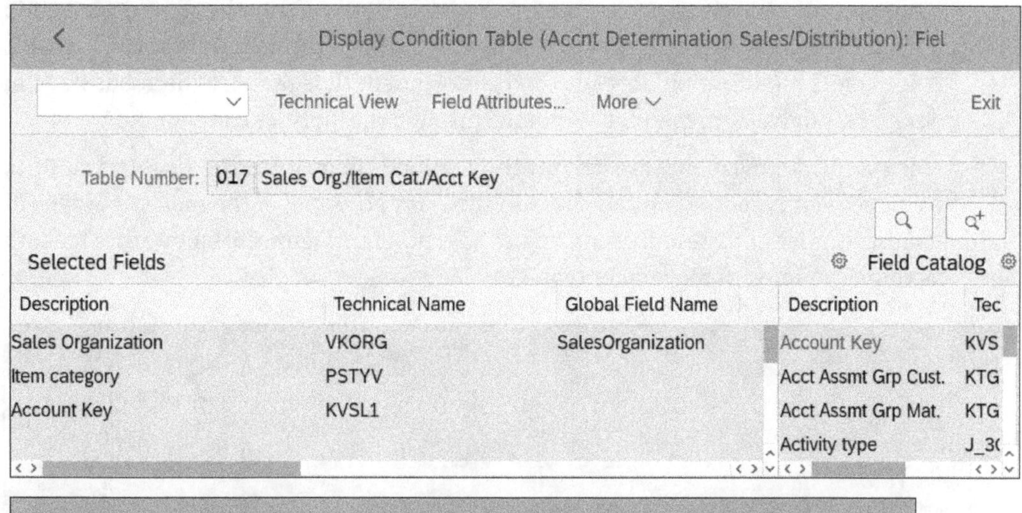

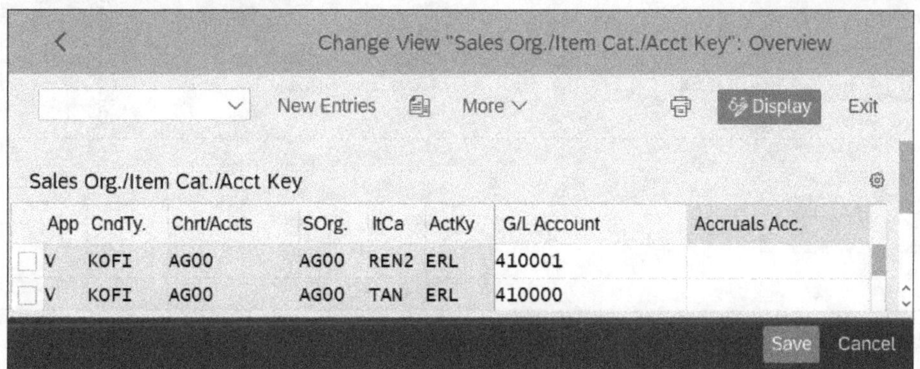

Figure 2.49 Account Determination of the Sales Returns Account

This condition table allows you to configure different accounts for the sales revenue based on the sales item category. Figure 2.49 shows that the item category **TAN** is assigned to sales revenue account **410000** while item category **REN2** is assigned to sales returns account **410001**.

The sales item category is determined automatically in the sales order based on the configuration of the sales and distribution module. We won't get into these details since this configuration object is not related to account determination. Item category TAN is the standard item category determined in a normal sales order (order type OR), while the item category REN2 is the standard item category determined in an advanced sales returns order (order type RE2).

Now that you understand how to configure a sales returns account different from the sales revenue account, let's move to the next account in the process.

2.4.3 Scrap Expense Account

The next account in this process is the scrap expense account, shown in Table 2.11.

Configuring the scrap expense account determination follows the steps described in Chapter 1, Section 1.1.2. The scrap account is determined through transaction key GBB. The general modification depends on the movement type used for the goods issue to scrap. This movement type is determined through the logistics configuration.

The standard movement type determined for goods issue to scrap from blocked stock is movement type 555. You can check the account grouping of the movement type in Transaction OMJJ, as described in Chapter 1, Section 1.1.2. Figure 2.50 shows the standard account grouping of movement type 555.

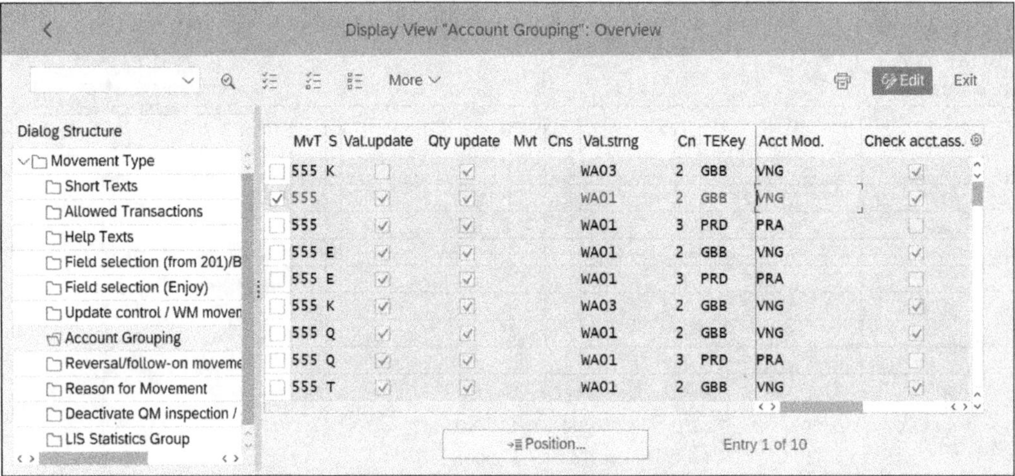

Figure 2.50 Movement Type 555: Account Grouping

As shown in Figure 2.50, the account determined for this movement type in our case will be based on the transaction key GBB (the **TEKey** field) and account modification VNG (the **Acct Mod.** field). The next step is to assign the general ledger account as explained in Chapter 1, Section 1.1.2.

With this step, we conclude our exploration of the advanced sales returns process in SAP S/4HANA, focusing on the crucial aspect of account determination. We hope you've gained a comprehensive understanding of how SAP handles the financial impact of different return scenarios. This deep dive into account determination configurations highlights the precise ways in which SAP S/4HANA ensures that every return scenario is accurately reflected in financial postings, from credit memos to replacement goods and scrapping returned items.

The detailed configurations we've examined so far underscore the system's ability to adapt to various business needs while maintaining financial accuracy and compliance.

This adaptability is key to managing the complexities inherent in the sales returns process and is instrumental in ensuring that your financial records are precise and reliable.

With this thorough understanding of account determination in the context of sales returns, you are now equipped to transition smoothly into another essential sales process: third-party sales. This process introduces a different set of challenges and considerations, such as the direct shipment of products from a supplier to a customer. As we embark on this new topic, we'll explore how SAP S/4HANA manages and integrates the operational and financial aspects of third-party sales transactions, highlighting the system's flexibility and comprehensiveness in handling diverse sales scenarios.

2.5 Third-Party Sales

Third-party sales, a distinct process in SAP S/4HANA, addresses situations where businesses sell products that they do not physically stock. In this process, a product sold to a customer is directly supplied by a third-party vendor, bypassing the seller's inventory. This unique scenario, where the stock never enters the seller's warehouse, requires a specific approach to accounting and logistics within the SAP system.

This process begins when a sale is made to a customer for an item that the business does not currently hold. Subsequently, the business initiates a purchase order to a supplier, requesting the direct delivery of the item to the customer. This approach is particularly relevant for businesses that operate with lean inventory models or that deal in products that are bulky, costly to store, or have rapid obsolescence rates.

The key challenge in third-party sales lies in the need to accurately track and record the financial transactions associated with these sales. Even though the product does not physically pass through the seller's inventory, the seller is still responsible for the customer invoice and the supplier invoice. Additionally, the COGS must be accurately posted to reflect the economic realities of the transaction.

In the upcoming sections, we'll explore the business process overview of third-party sales, examining the specific steps from the initiation of the sales order to the final billing and payment. Then, we'll provide a detailed exploration of the unique account determination configurations required in SAP S/4HANA to manage the financial aspects of these transactions effectively.

2.5.1 Business Process Overview

An overview of the third-party sales process flow is shown in Figure 2.51.

The third-party sales process is a combination of the sales from stock process described in Section 2.1 and the normal purchasing process described in Chapter 1, Section 1.1. Thus, you should already understand most of the steps shown in Figure 2.51.

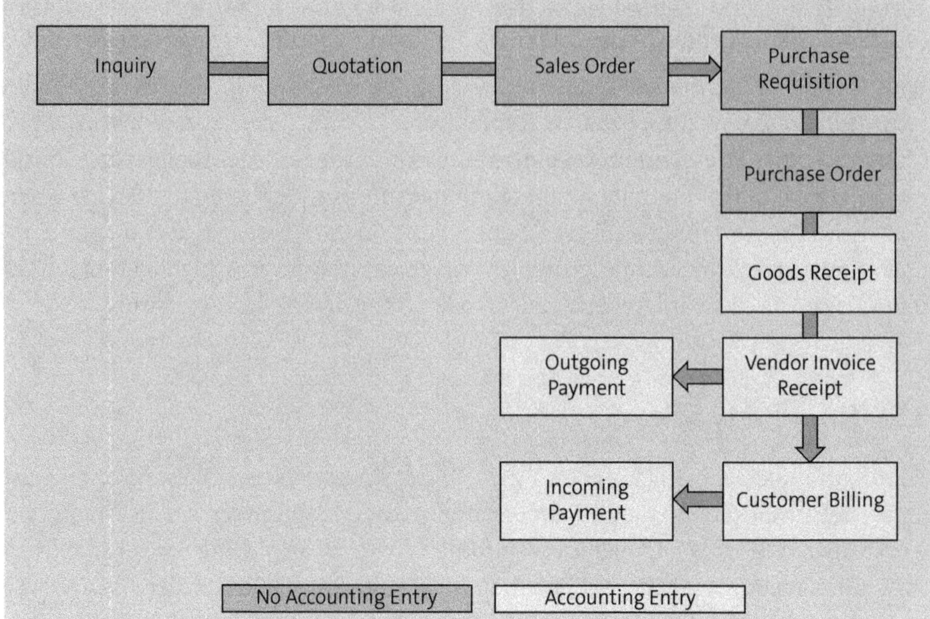

Figure 2.51 Third-Party Sales Process Flow

Still, let's examine the various steps of this process. For the steps that are the same as processes we've explained, we'll refer to the previous sections for more details:

1. **Inquiry**
   This process is the same as the sales from stock process outlined in Section 2.1.
2. **Quotation**
   This process is the same as the sales from stock process outlined in Section 2.1.
3. **Sales order**
   You'll create a normal sales order like sales from stock process outlined in Section 2.1. The difference is that you'll use materials that are marked as third-party sales items, which triggers the purchase requisitions to acquire these items from our suppliers.
4. **Purchase requisition**
   This process is the same as the normal purchasing process outlined in Chapter 1, Section 1.1. This purchase requisition can be converted to a purchase order manually or automatically.
5. **Purchase order**
   This process is the same as the normal purchasing process outlined in Chapter 1, Section 1.1. This purchase requisition can be converted to a purchase order manually or automatically. The only difference is that the item category is S (third-party).
6. **Goods receipt**
   The items are directly shipped from supplier to customer, so you physically won't

have any goods received into your warehouse. Once the supplier confirms sending the items to the customer, you should post this GR to update your accounting records. The accounting entry is shown in Table 2.12.

Debit	Credit	Debit Amount ($)	Credit Amount ($)
Third-party sales consumption (COGS)		1000	
	Goods receipt/invoice receipt (GR/IR)		1000

Table 2.12 Accounting Entry: Third-Party Sales GR

If you consider separate sales and purchasing processes, you would post a GR from supplier first where you debit the inventory account against the GR/IR account. Then, you would post a goods issue to customer where you debit the COGS account and credit the inventory account. In the case of third-party sales, you post only one entry with the net accounting impact of both these entries, as shown in Table 2.12.

This entry is specific to the third-party sales process, so let's look into these accounts next:

- **Consumption (COGS)**
  - Similar to the COGS account used in the sales from stock process outlined in Section 2.1.
  - This account must be created as a cost element category 1 since the cost will be assigned to the sales order as a cost object.
  - The value posted is equal to the stock valuation value (standard cost or moving average).
  - The account determination for this account will be explained in the next section.
- **GR/IR**
  - The same account used in the normal purchasing process outlined in Chapter 1, Section 1.1.

7. **Vendor IR**
   This process is the same as the normal purchasing process outlined in Chapter 1, Section 1.1.
8. **Customer billing**
   This process is the same as the sales billing process outlined in the Sales Billing section of Section 2.2.1.
9. **Incoming and outgoing payments**
   The process for incoming payments is the same as the sales from process outlined in Section 2.1, and the process for outgoing payments is the same as the normal purchasing process explain in Chapter 1, Section 1.1.

Now, you should have a clear understanding of the third-party sales process in SAP S/4HANA. You now know how products are efficiently moved from suppliers directly to customers, bypassing traditional stock handling. This process optimizes logistics while ensuring the financial transactions are tracked and recorded accurately.

As we conclude this overview, the next vital step is to dive into the account determination for these transactions. This configuration ensures that, despite the absence of physical stock handling, the financial entries for customer billing and supplier invoicing accurately reflect the transaction's economic activities.

2.5.2 Third-Party Sales Consumption Account

The first account involved in this process is the third-party sales consumption account, shown in Table 2.12.

This account is determined following the same concept of Section 2.1.2 in the sales from stock process. The difference is that account modification GBB is used for the consumption account of third-party sales as determined based on other configuration steps. After configuring the determination of the account modification, you can assign the account to this account modification in Transaction OBYC, as explained in Section 2.1.2. Figure 2.52 shows the determination steps of the general ledger account for the third-party sales consumption account.

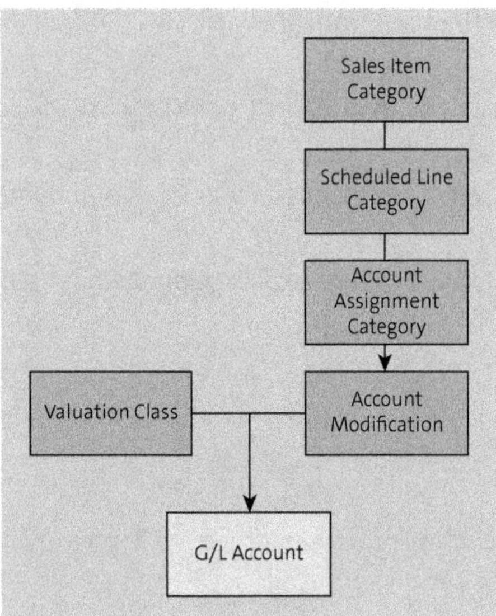

Figure 2.52 Third-Party Sales Consumption Account Determination Logic

Now, let's consider the steps shown in Figure 2.52 in detail: The sales item category and scheduled line category are determined in the sales order through the sales and distribution configuration. These values are found in the sales order item details, as shown in Figure 2.53.

Change Standard Order 730: Item Data
More
Exit
Sales Document Item: 10
Item category: TAS Third Party Item
Material: 252
Third Party Material
Shipping
Billing Document
Tax Exemption License
Excise duty
Conditions
Account Assignment
Schedule lines
Partner
Fixed Date and Qty:
Order Quantity: 100 PC
Delivery Time:
Delivered qty: 0
Sales
Shipping
Procurement
Purch. requisition
Quantities/Dates
P. Delivery Date
Order Quantity
Rounded qty
Confirmed Qty
Sa...
Delivery Block
Delivered qty
Sch...
Purchase Re
D 07.11.2023
100
100
100PC
CS
Save
Cancel

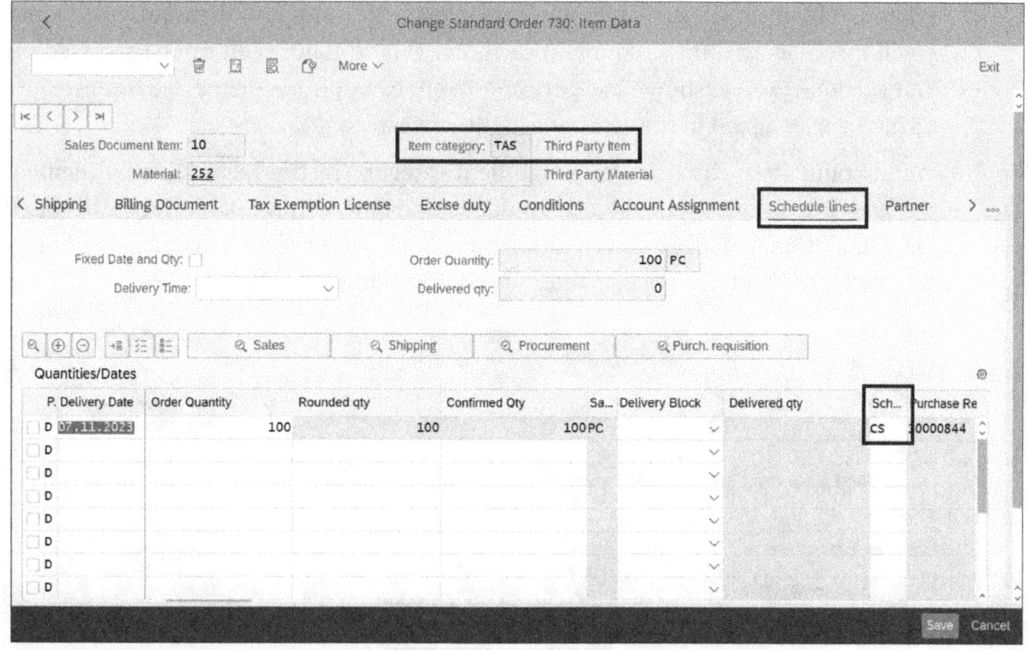

Figure 2.53 Third-Party Sales Item Category and Schedule Line Category

Figure 2.53 shows the details of **Sales Document Item 10** under the **Schedule lines** tab. The sale item category is **TAS,** and the schedule line category is **CS**, which are the standard values for third-party items.

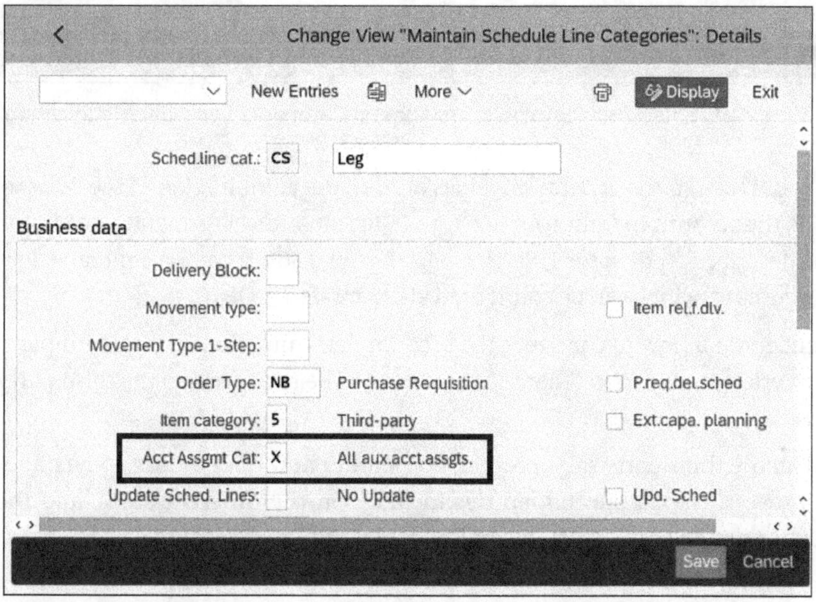

Change View "Maintain Schedule Line Categories": Details
New Entries
More
Display
Exit
Sched.line cat.: CS
Leg
Business data
Delivery Block:
Movement type:
Item rel.f.dlv.
Movement Type 1-Step:
Order Type: NB
Purchase Requisition
P.req.del.sched
Item category: 5
Third-party
Ext.capa. planning
Acct Assgmt Cat: X
All aux.acct.assgts.
Update Sched. Lines:
No Update
Upd. Sched
Save
Cancel

Figure 2.54 Third-Party Sales: Schedule Line Category Assignment to Account Assignment Category

When configuring the schedule line category in the sales and distribution module, you'll insert an account assignment category. This configuration can be accessed via Transaction VOV6, as shown in Figure 2.54. In our example, the scheduled line category **CS** has been assigned to the account assignment category **X**.

When configuring the account assignment category in the materials management module, you can assign the account modification. This configuration can be accessed via Transaction OME9, as shown in Figure 2.55. In our example, account assignment category **X** has been assigned to account modification **VAX**.

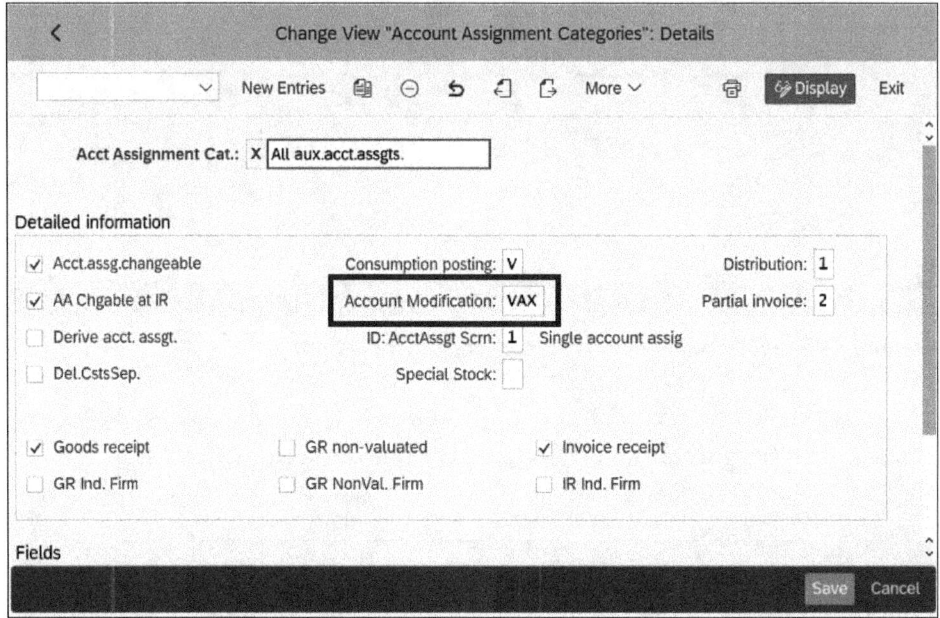

Figure 2.55 Third Party Sales: Assign Account Assignment Category to Account Modification

The last step is to assign the general ledger account to the combination of the transaction key GBB, the account modification VAX, and the valuation class maintained in the material, as described in Chapter 1, Section 1.1.2. Figure 2.56 shows the general ledger account assignment, which can be configured via Transaction OBYC.

Now, you understand how to configure the account determination for the third-party sales consumption account. No other accounts need to be to address in the third-party sales process.

As we wrap up the third-party sales process, you should now have a clear understanding of the process, the accounting entries included in the different steps, and the account determination configuration required for the different accounts.

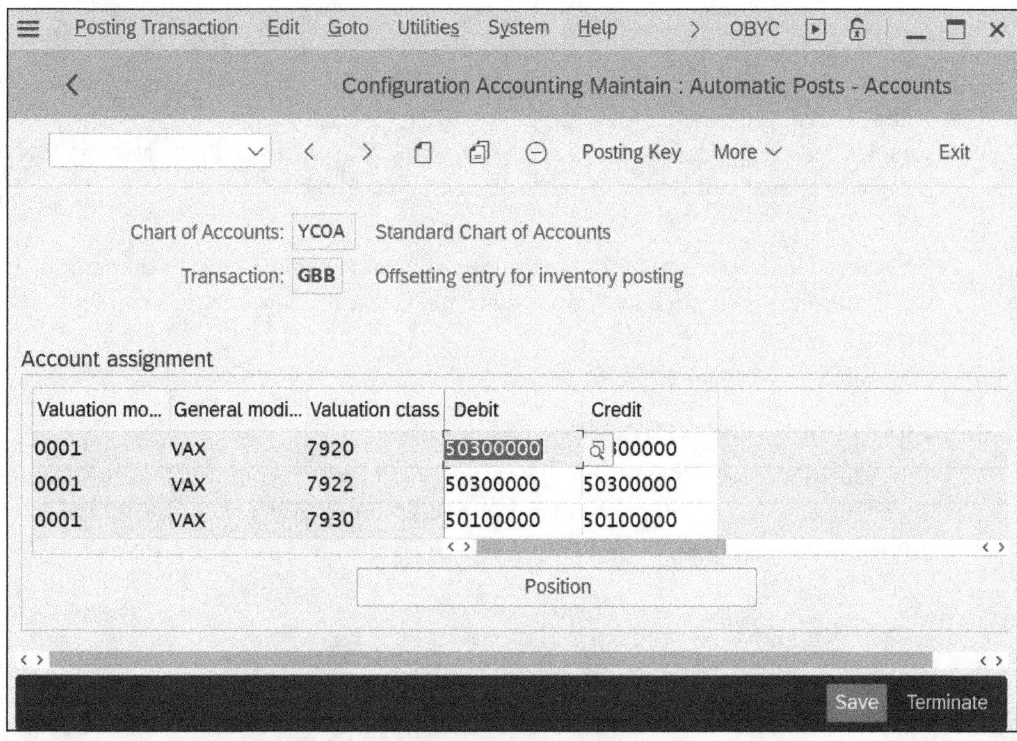

Figure 2.56 Third-Party Sales: Assign General Ledger Account

Let's now shift our focus to another critical process in the sales and distribution area: cash sales. This process represents a unique transaction type where immediate payment is an integral part of the sales cycle, contrasting with the deferred payment structure of standard sales. In the upcoming sections, we'll dive into cash sales in detail, by examining its operational flow and the specific accounting entries it generates, as well as the configuration steps necessary for accurate account determination in this immediate payment scenario.

2.6 Cash Sales

The cash sales process is used when a customer directly purchases and retrieves goods, settling the payment on the spot. Once the order is logged, the delivery process is initiated immediately. This transaction allows for an immediate printing of the cash invoice from the order itself, and billing is directly associated with the order entry. Unlike other order types, such as rush orders or standard orders, cash sales do not result in receivables because the payment is promptly recorded against a cash account instead of billing the customer.

This process can also be used for point of sale (POS) scenarios.

Note

Cash sales are different from rush sales. In cash sales, the payment is received immediately, and no AR posting is created at all, whereas rush sales involve invoicing the customer and posting to AR, with payment typically received later. Rush sales has the same steps as standard sales, but the steps occur more quickly.

Let's consider an overview of the cash sales process and explore how it's different from standard sales; then, we'll look into additional account determination configuration steps.

2.6.1 Business Process Overview

The cash sales process is similar to the standard sales process but usually starts immediately by creating a cash sales order. Figure 2.57 shows an overview of the cash sales process flow.

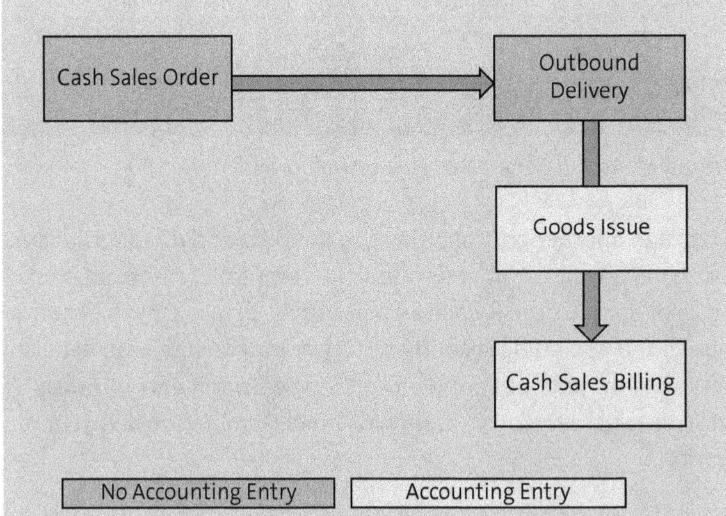

Figure 2.57 Cash Sales Process Flow

Once the cash sales order is saved, the outbound delivery is created automatically. Then, you can post the goods issue with reference to the outbound delivery either manually or through a schedule background job. In this process, you'll also post the billing document with reference to the sales order and not the outbound delivery, which means you can issue the sales billing document immediately after saving the sales order.

Note

The standard sales order type for cash sales is order type BV.

As shown in Figure 2.57, the initial steps (sales order, outbound delivery, and goods issue) are the same as described in the sales from stock process in Section 2.1. The accounting entry for goods issue is also the same. The only difference is the accounting entry of the billing document. The billing document for cash sales is created with reference to the cash sales order and posts the account entry shown in Table 2.13.

Debit	Credit	Debit Amount ($)	Credit Amount ($)
Cash		1650	
	VAT		150
	Sales revenue		1500

Table 2.13 Accounting Entry of Cash Sales Billing

This accounting entry is similar to the entries shown earlier in Table 2.3 for the standard sales process. The difference that, in the cash sales process, you post to cash accounts instead of AR accounts. You can also apply VAT and discounts as in the sales from stock process.

Now, you should understand the cash sales process and the different account entries involved. You should clearly understand how this process is different from standard sales from stock from the accounting point of view.

Next, let's explore configuring the account determination of the cash accounts used in the cash sales process.

2.6.2 Cash Accounts

The cash accounts follow the same concepts as described earlier in Section 2.1.4, but they have their own catalog fields and condition tables.

Figure 2.58 shows an overview of the account determination logic for cash accounts using the condition technique. The determination starts with the billing document type used in the sales billing documents and the chart of accounts assigned to the billing company code as inputs and ends with the output of the determined general ledger accounts. We already covered these objects and how they work in Section 2.1.4, but the configuration steps are different. By the end of this section, you'll understand how to configure each of these objects to determine the correct cash accounts.

All the configuration steps for the sales cash accounts can be performed via the menu path **Sales and Distribution • Basic Functions • Account Assignment/Costing • Cash Account Determination.** The various configuration steps are shown in Figure 2.59.

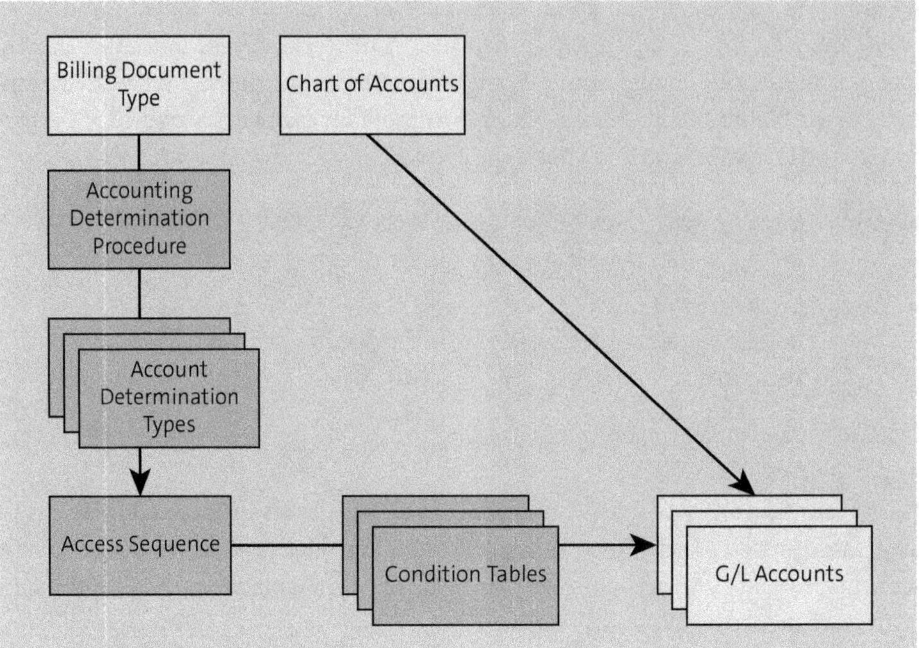

Figure 2.58 Sales Cash Account Determination

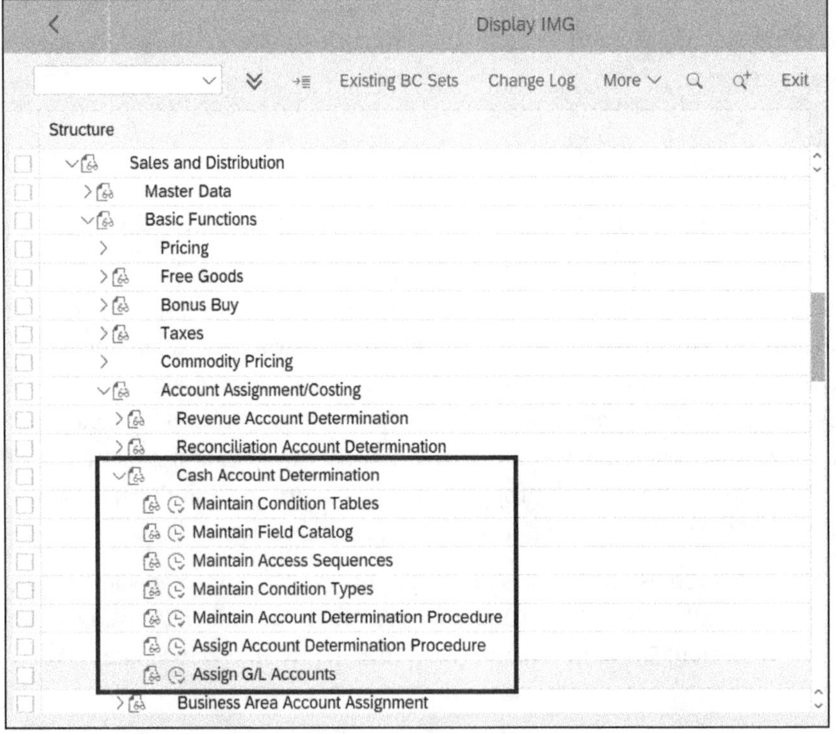

Figure 2.59 Sales Cash Account Determination Configuration Steps

Since these configuration steps work the same way as explained earlier in Section 2.1.4, we'll refer to that section when needed in the cash sales configuration steps. Let's now describe the configuration steps in order.

Maintain Field Catalog for Cash Accounts

This step was described earlier when we changed the field catalog for condition tables in Section 2.1.4. The field catalog for the sales cash account determination is different from the catalog of the sales revenue accounts.

SAP Table

The fields activated for the sales cash determination account can be viewed in table T681F. Figure 2.60 shows the value "VC" in the **Application** column for cash settlements ❷, while the fields for the sales revenue account are the value "V" in the **Application** column for sales and distribution ❶.

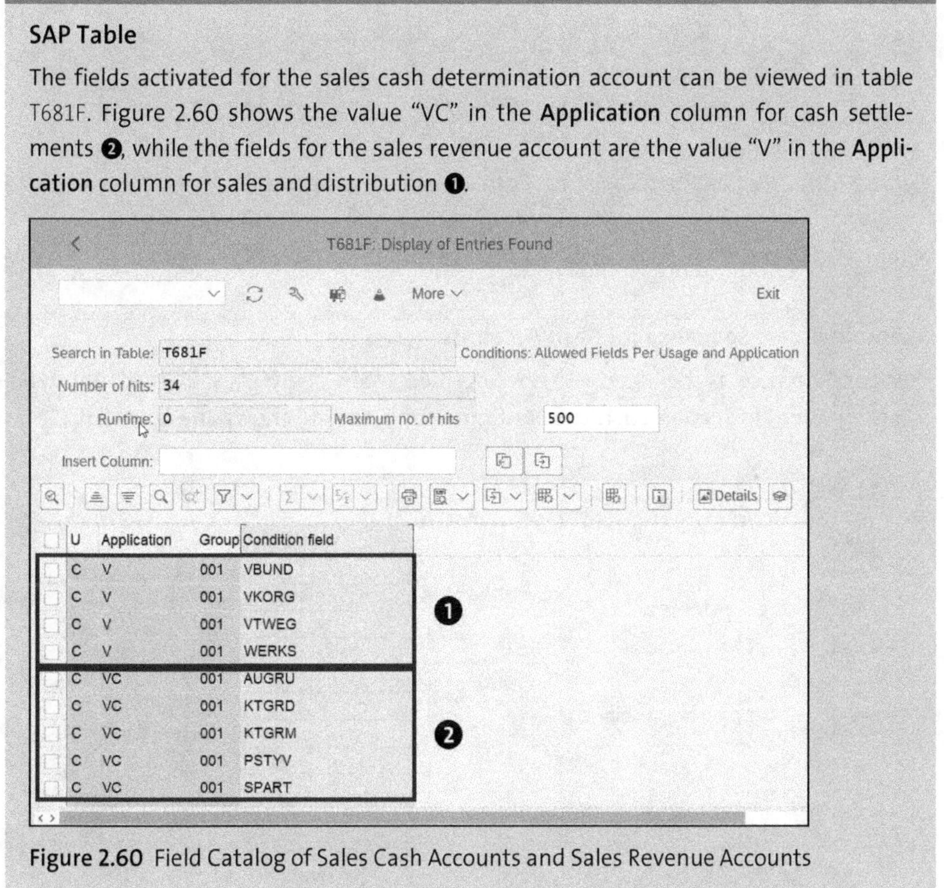

Figure 2.60 Field Catalog of Sales Cash Accounts and Sales Revenue Accounts

After activating the fields needed, now it's time to maintain the condition tables.

Create Condition Tables for Cash Accounts

This configuration is the same as when we created condition tables for revenue account determination in Section 2.1.4. The condition tables for the cash accounts are different from those for the revenue accounts. An example condition table is shown in Figure 2.61.

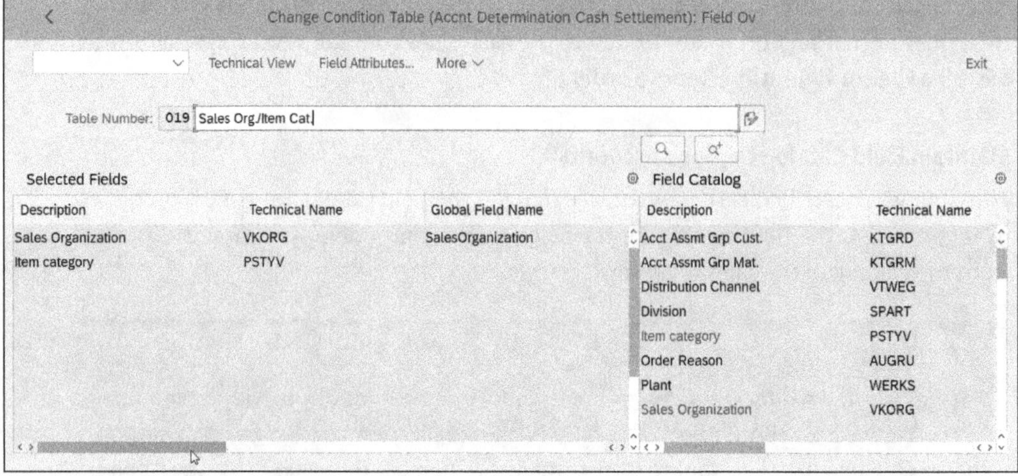

Figure 2.61 Create Condition Tables for Cash Account Determination

After creating the condition tables, you can maintain the access sequence.

Maintain Access Sequence for Cash Accounts

This configuration is the same when we defined access sequences and account determination types in Section 2.1.4. The configuration steps are shown in Figure 2.62.

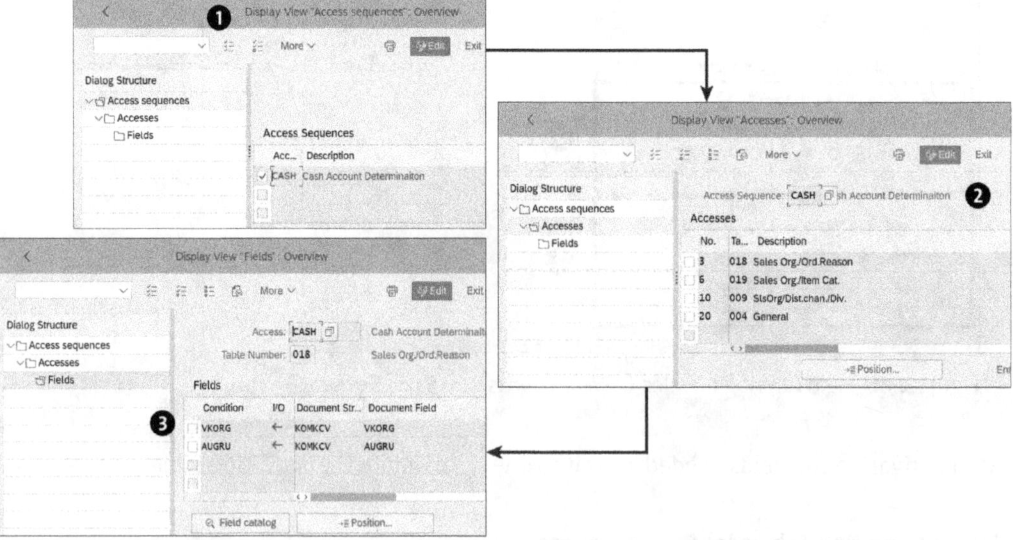

Figure 2.62 Define Access Sequence for Cash Account Determination

In this case, first create the access sequence in ❶, then switch to the **Accesses** screen, and maintain the condition tables access sequence in ❷. You can also see the fields included in each condition table in ❸ by switching to the **Fields** screen.

For cash account determination, you only need one access sequence. In the next step, you'll assign this access sequence to a condition type.

Maintain Condition Types for Cash Accounts

This configuration is the same when we defined access sequences and account determination types in Section 2.1.4. In the **Maintain Condition Types** configuration step, assign the access sequence to a condition type, as shown in Figure 2.63.

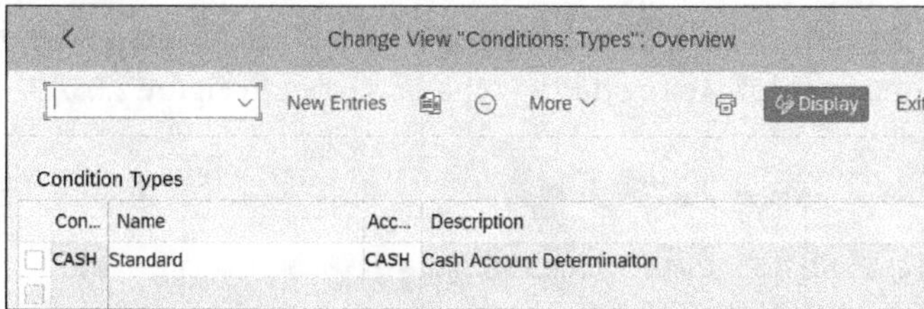

Figure 2.63 Sales Cash Account: Maintain Condition Types

The next step is to assign this condition type to an account determination procedure for cash accounts.

Maintain Account Determination Procedure for Cash Accounts

This configuration is the same when we defined account determination procedures in Section 2.1.4. You only need one account determination procedure for the cash accounts. The configuration is shown in Figure 2.64.

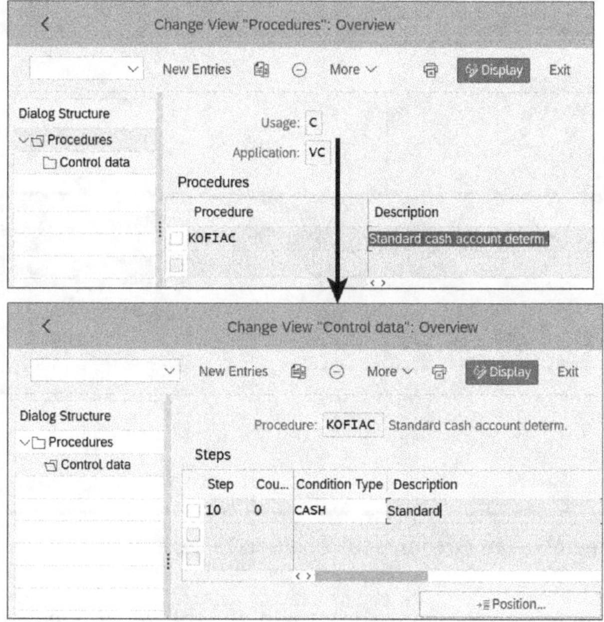

Figure 2.64 Sales Cash Accounts: Define Account Determination Procedure

In this case, maintain an accounting determination procedure and then assign the condition type to it.

The next step is to assign the account determination procedure to billing types.

Assign Account Determination Procedure for Cash Accounts

This configuration is the same when we assigned account determination procedures for AR in Section 2.1.5.

The standard billing type for cash sales is billing type BV. On this screen, assign your cash account determination procedure to the billing type, as shown in Figure 2.65.

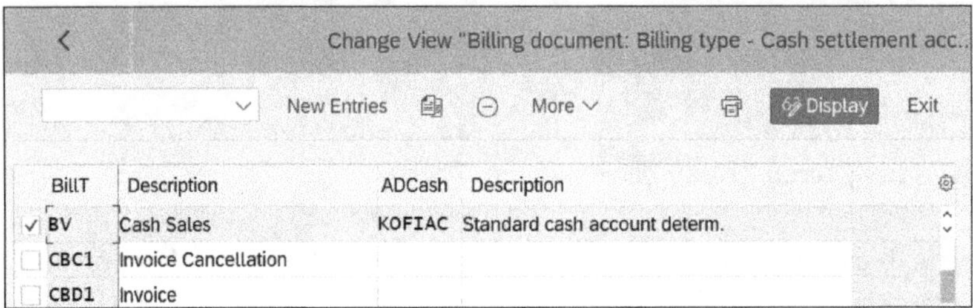

Figure 2.65 Assign Cash Account Determination Procedure to Billing Type

The last step is to assign the cash accounts to the condition tables.

Assign General Ledger Accounts for Cash Accounts

Now, you can assign the cash accounts to various condition tables, as shown in Figure 2.66.

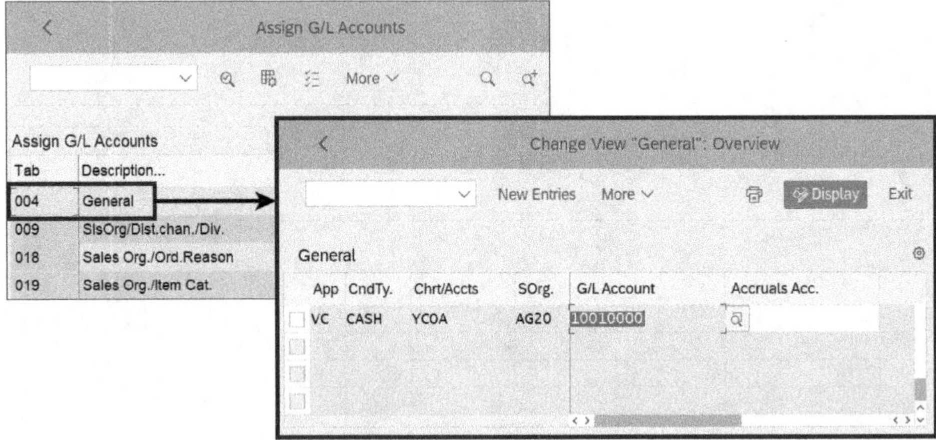

Figure 2.66 Cash Sales Assign General Ledger Accounts to Condition Tables

In this case, first select the condition table and then maintain the general ledger accounts inside the table.

Now, the configuration is complete. Whenever you post a billing document with the billing type BV, the system will automatically determine the cash accounts. If for any reason the system cannot determine a cash account, it will post to the AR reconciliation account like in the standard sales process. You can see this kind of transaction during a billing document's account determination analysis.

You're now fully equipped to configure the account determination process for cash accounts in the cash sales process. If you find that the cash accounts are not determined properly in the sales billing documents, then you can analyze the account determination to find the issue. Next, let's see how you can analyze the sales cash account determination in any billing document.

Account Determination Analysis: Cash Accounts

To display the steps the system performed to determine the cash accounts in any cash billing document, display the billing document via Transaction VF03. Start the account determination analysis for the cash accounts, as shown in Figure 2.67.

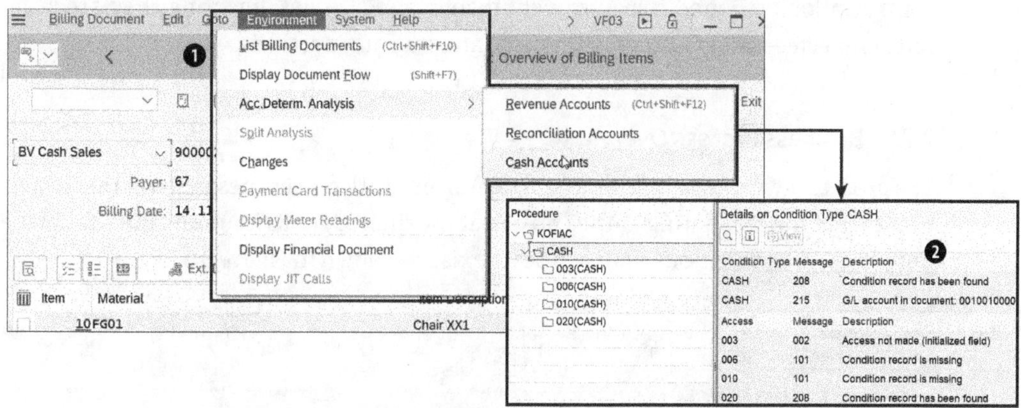

Figure 2.67 Cash Account Determination Analysis in Sales Billing Documents

First, as shown in ❶, in the **Environment** menu, start the account determination analysis by selecting the **Acc.Determ. Analysis** and **Cash** accounts. The account determination analysis for the cash accounts is shown in ❷.

Notice that the system determined the general ledger account **0010010000** from access sequence step **020**, which is assigned to the condition table 004, as per our configuration.

Now, you know how to determine all the accounts needed for the cash sales process and how to analyze the account determination for any errors.

With this step, we conclude our journey into the cash sales process. You should now understand the process and its accounting entries and how to configure the account determination for these accounts.

Let's now turn our attention to another important sales process: the free-of-charge (FoC) sales process.

2.7 Free-of-Charge Sales

The FoC sales process is essential when goods are provided to customers without any direct payment, often for promotional purposes, to increase customer satisfaction, or to fulfill warranties. In the FoC sales process, even though no revenue is generated, the movement of goods and their associated cost implications must be accurately recorded in your company's financial system. This scenario requires specific account determination settings within SAP S/4HANA to handle these transactions appropriately. The same process is also involved when you need to charge a customer only for the sales output tax, but you're otherwise giving the product away for free, a rather common business requirement.

Next, let's explore the business process and the specialized account determination configurations required for effectively managing FoC sales, ensuring they are accurately reflected in the business's financial and inventory records.

2.7.1 Business Process Overview

The process flow is exactly same as the sales from stock process described earlier in Section 2.1, but the accounting entries posted at the time of goods issue and invoice can be different. Three variations of this process exist, as shown in Figure 2.68.

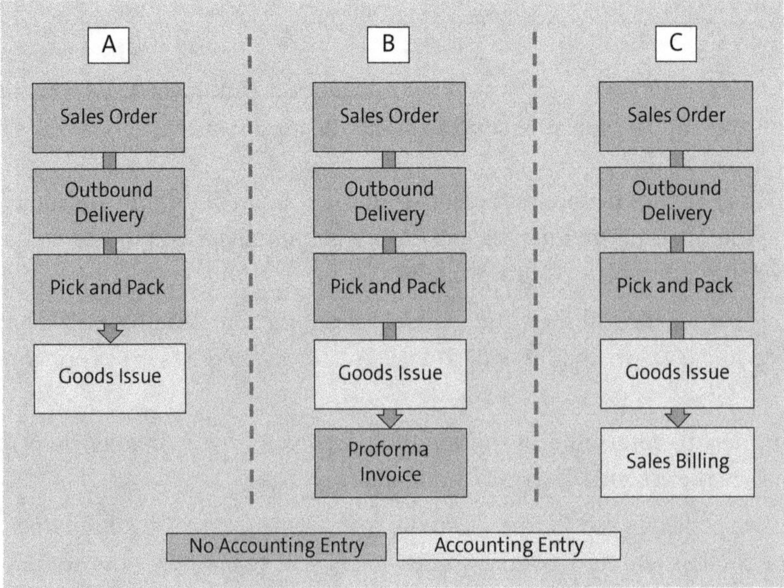

Figure 2.68 FoC Sales Process Flow

As shown in Figure 2.68, the first four steps of all the three process flow variations (A, B, and C) are the same. These steps are similar to the standard sales from stock process. Let's quickly look at these steps and add some comments specific to the FoC process.

Sales Order

A sales order can contain both standard items and FoC items together. For example, if you are selling 100 units of product A to a customer for the normal price but also adding 15 units of the same product for free, then you can create two line items in the sales order: one for the paid items and one for the FoC items

> **Note**
>
> You can decide to sell any product FoC; you don't need to create a separate material code for FoC sales. Whether the item is FoC or paid for depends on the inputs added in the sales order line item.

After creating the sales order, the next step is creating the outbound delivery and then picking and packing. These two steps are the same as in the sales from stock process described in Section 2.1 without any differences. Let's move on to the following step: post goods issue.

Goods Issue

This step is the first step where an accounting entry is generated in the process. The accounting entry can be the same as the goods issue process described earlier in Section 2.1.1 for the standard sales from stock process.

The accounting entry can also be different if you have a business requirement to post the COGS of the FoC items to a separate account from the standard COGS account of the paid items. This account would be an expense (P&L) account. For the sake of our example, we'll call this account FoC COGS. The accounting entry for the post goods issue of the FoC items is as shown in Table 2.14.

Debit	Credit	Debit Amount ($)	Credit Amount ($)
COGS		1000	
FoC COGS		150	
	Inventory		1150

Table 2.14 Accounting Entry of Post Goods Issue of FoC Items

Now, let's quickly look at the meaning of these accounts and the values posted to each of them. The account determination configuration is structured in the following ways:

- **COGS**
  - The standard COGS account for paid items.
  - Same as the COGS process described in Section 2.1.1.
  - The value posted is the cost of the products sold to the customer for a price.
- **FoC COGS**
  - The COGS account determined for the FoC items.
  - Same definition as the standard COGS account.
  - The value posted is the cost of the products sold to the customer FoC.
- **Inventory**
  - The same account as the inventory process described in Section 2.1.1.

As shown in Figure 2.68, after the post goods issue step, three options are available for the next step. Let's look at each of these three options and when they can be used.

Free-of-Charge Sales Billing

The three options available in the process include the following:

- **End the process after the step: post goods issue**
  Select this option if you don't want to post any further accounting entries in the process, and you don't want to print a billing document for the customer. Using this option also means no sales output tax will be posted.
- **Create proforma invoice without accounting entry**
  Select this option if you don't want to post any further accounting entries in the process, but you want to print a billing document for the customer. A proforma invoice involves a special sales billing type that has no accounting impact. Using this option also means no sales output tax will be posted.
- **Create sales billing with an accounting entry**
  With this option, you have an accounting entry and a billing document you can print, and you can post sales output tax.

Let's say you decide on the third option; the accounting entry is shown in Table 2.15.

Debit	Credit	Debit Amount ($)	Credit Amount ($)
AR		100	
VAT			100
Sales discount		1000	
	Sales revenue		1000

Table 2.15 Accounting Entry of FoC Sales Billing

The first two options—end the process after the post goods issue or create proforma invoice without accounting entry—are selected if you don't want to post any further accounting entries in the process, which is usually the case if no sales output tax is included. The account determination configuration of the accounts used in the accounting entries shown in Table 2.14 and Table 2.15 were explained in other processes, except the FoC COGS account. Thus, let's focus on how to configure the account determination configuration of this account.

2.7.2 Determination of Free-of-Charge COGS Account

The FoC COGS account is a normal COGS account, but you only post the COGS of the FoC items to it, while the COGS of the other standard items go to the normal COGS account. Since this account is a COGS account, the account determination follows the same steps as the standard COGS accounts at goods issue in the sales from stock process described in Section 2.1.

The complication comes from the fact that the same material code can be sold one time as FoC and another as paid, which can even happen within the same sales order.

In one sales order, you can have one line item of material X that's paid and another line item of the same material that's FoC. In this way, you can tell the system that a line item is FoC by changing the item category in the sales order, as shown in Figure 2.69, which shows Transaction VA03 (Display Sales Order).

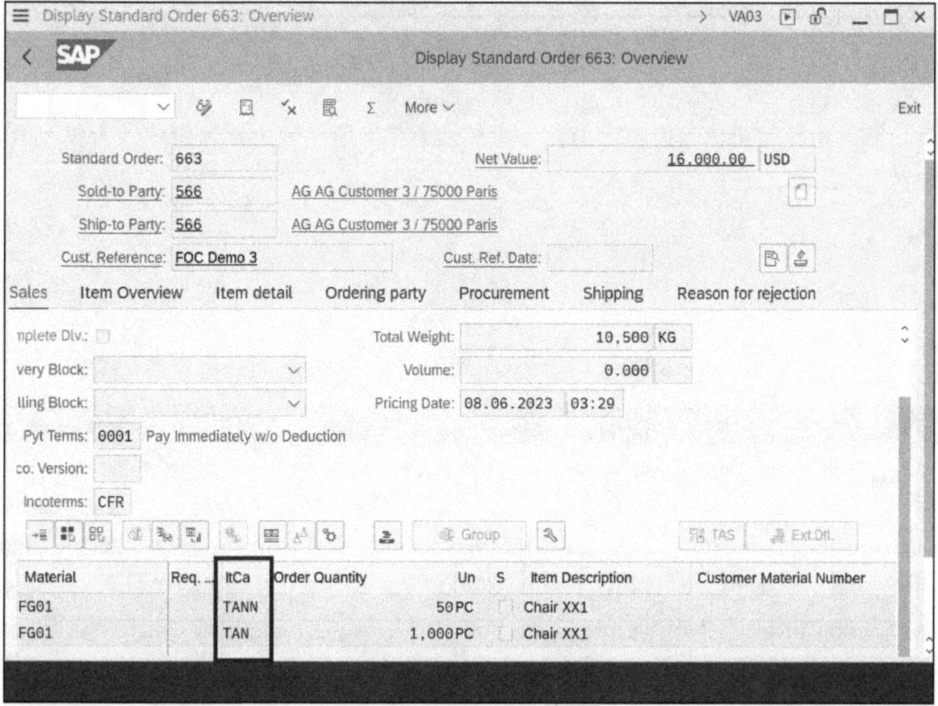

Figure 2.69 Sales Order: FoC and Normal Items

In our example, we have two line items in the sales order for the same material FG01. The first line item has the item category TANN (FoC) while the second line item has the item category TAN (standard item). When you post the goods issue of this sales order, you want the first line to post to COGS and the second line to post to FoC COGS.

What options are available to determine different COGS accounts?

- Use different valuation classes in the materials? we can't do this here as this is the same material code with the same valuation class sold as FoC and standard.
- Use different material movement types at goods issue? This is the valid option. If we can use different movement types, then we can determine different general ledger accounts as explained in in Chapter 1, Section 1.2, when we looked at inventory accounts and defined account groupings for movement types.

The movement type used at the time of post goods issue is determined through the schedule line-item category, which is maintained in the sales order line item, as shown in Figure 2.70.

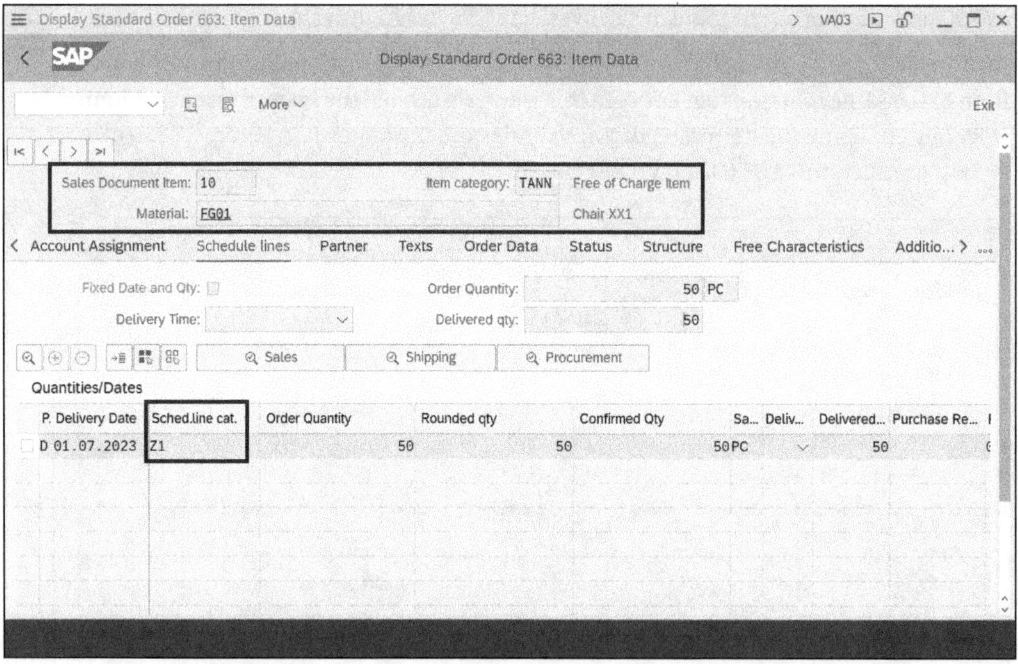

Figure 2.70 Sales Order Items: Schedule Line Category

Figure 2.70 shows Transaction VA03 (Display Sales Order) with the details of line item **10**, which has the item category **TANN**. Notice the **schedule line category Z1**.

Now, you can assign a movement type in the schedule line category configuration via Transaction VOV6, as shown in Figure 2.71.

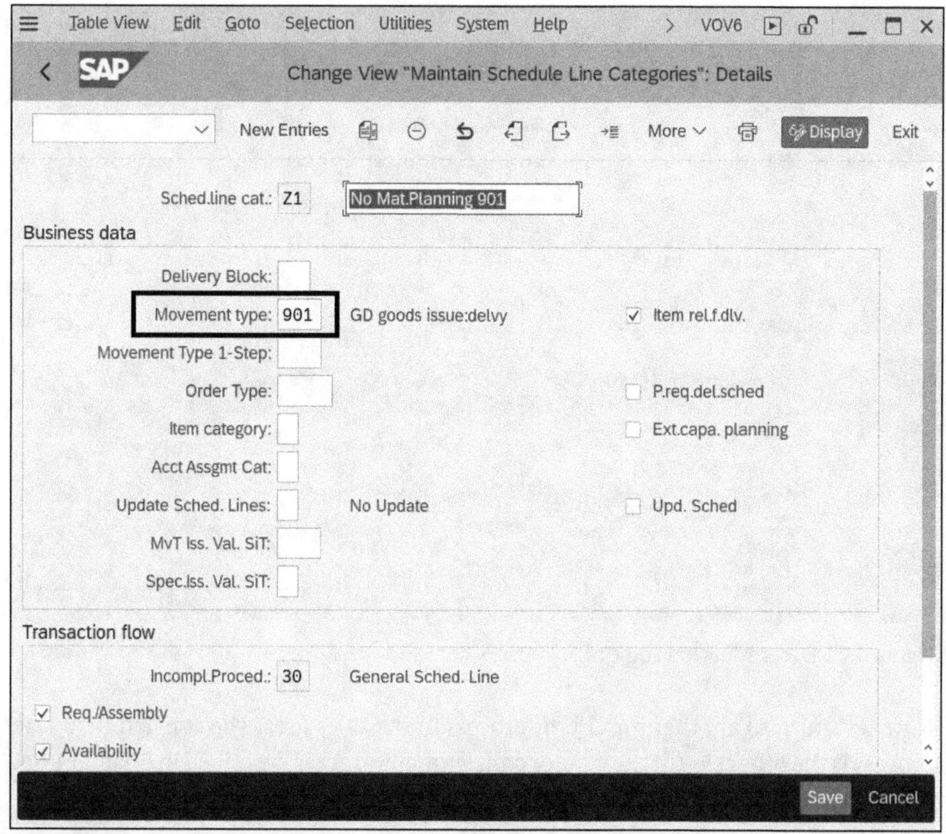

Figure 2.71 Assign Schedule Line Category to Movement Type

On this screen, you can see the definition of the schedule line category **Z1**, and the movement type 901 is assigned. As a result, when you process the post goods issue of this item (with this schedule line category), the movement type will be **901**.

The link between movement types and account determination was described earlier in Chapter 1, Section 1.2, when we looked at inventory accounts and defined account groupings for movement types.

Now, let's look at the link between the sales item category and the COGS account, as shown in Figure 2.72.

The FoC COGS account determination process is a great example of the flexibility of account determination using the transaction technique. You can use different items categories in a sales order to post to different COGS accounts.

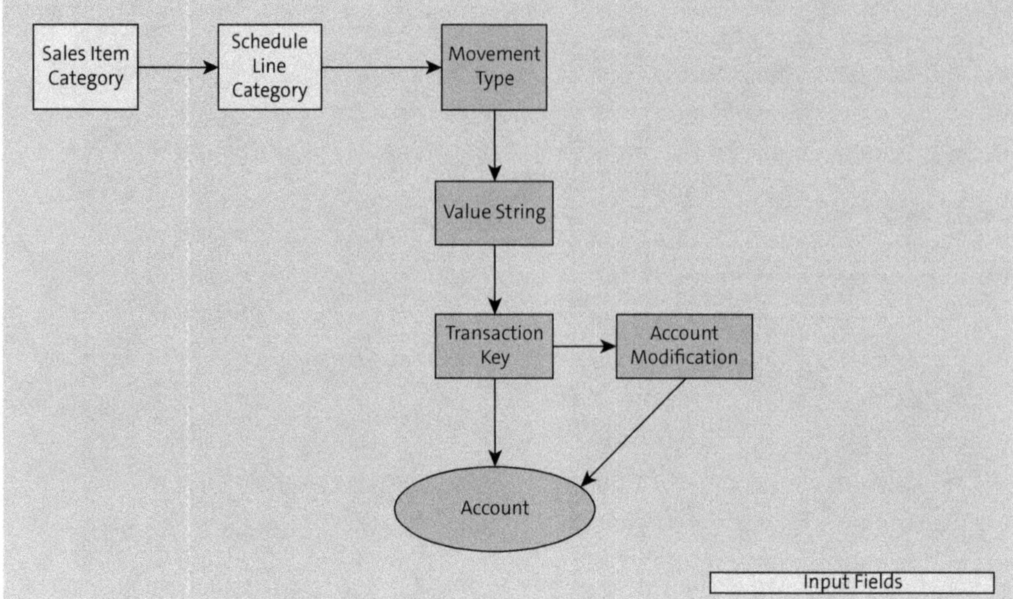

Figure 2.72 Determination Logic of FoC COGS Account

Now, let's turn to another important process in the sales area. The consignment sales process is the other side of the vendor consignment process described in Chapter 1, Section 1.5, when we discussed vendor consignment.

2.8 Consignment Sales (Customer Consignment)

In consignment sales, your company owns the products that are shipped to a customer until the products are consumed by the customer. The customer can either use the products for its own use (e.g., in production) or can sell the products to another customer. Thus, your customer doesn't owe your company anything until the products are consumed or sold to a third customer. Thus, the stock remains in the ownership and on the books of your company even though the products are physically located in your customer's warehouse, a rather common scenario.

An overview of the many subprocesses involved in the consignment sales business process is shown in Figure 2.73.

We'll explore each step further in the following sections. The first subprocess in the consignment process is performing the consignment fill-up, which is the process of filling up your customer's consignment stock.

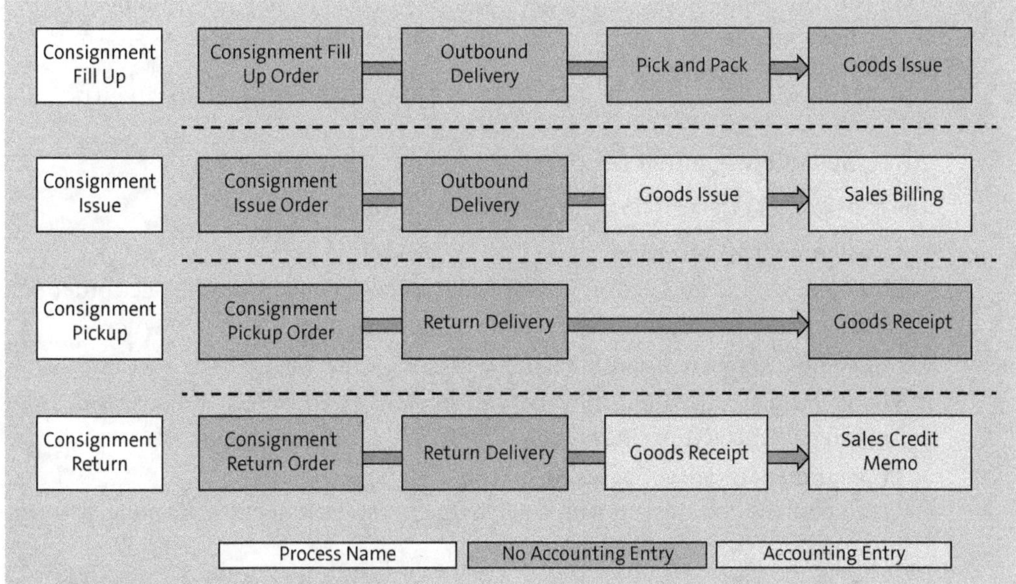

Figure 2.73 Sales Consignment Process Overview

2.8.1 Sales Consignment Fill-Up

The consignment fill-up process involves the following steps:

1. **Consignment fill-up: order**
   Similar to a standard sales order, this order will be created with the sales order type "consignment fill-up." This step has no accounting entry.
2. **Consignment fill-up: outbound delivery**
   You'll create an outbound delivery with reference to the fill-up order. This step has no accounting entry.
3. **Consignment fill-up: pick and pack**
   You'll prepare the products to be issued to the customer. This step has no accounting entry.
4. **Consignment fill-up: goods issue**
   You'll issue the products to your customer's location. This step has no accounting entry since the products are still in your ownership and are only issued to fill up the customer's consignment stock. This step will generate a material document changing the stock indicator from free available stock to the customer's consignment stock. You can thus display the consignment stock available for each customer.

The consignment fill-up process doesn't include the customer billing step since nothing has actually been sold yet. You're just changing the location of the items from your location to that of your customer.

After the consignment fill-up, you'll follow up periodically on your customer to check if they have sold any of your items. If so, the consignment issue process shown in Figure 2.73 will be triggered.

2.8.2 Sales Consignment Issue

The consignment issue process involves the following steps:

1. **Consignment issue: order**
   Similar to a standard sales order, this order will be created with the sales order type "consignment issue." This step had no accounting entry.
2. **Consignment issue: outbound delivery**
   You'll create an outbound delivery with reference to the consignment issue order. This step has no accounting entry.
3. **Consignment issue: goods issue**
   You'll post the goods issue, which will trigger a material document and an accounting entry:
   - The material document will remove the quantity sold from the customer's consignment stock to reflect the actual sale.
   - The accounting document will reduce the inventory value against the COGS account. This accounting entry is shown in Table 2.16.

Debit	Credit	Debit Amount ($)	Credit Amount ($)
COGS		1000	
	Inventory		1000

Table 2.16 Accounting Entry of Post Goods Issue for Sales Consignment Issue

   This entry is the same as a standard goods issue accounting entry in the standard sales from stock process described earlier in Section 2.1.1.

4. **Consignment issue: billing**
   Now, you'll create the customer billing document. This step will generate the accounting entry shown in Table 2.17.

Debit	Credit	Debit Amount ($)	Credit Amount ($)
AR		1350	
Sales discount		250	
WHT		50	
	VAT		150
	Sales revenue		1500

Table 2.17 Accounting Entry of Billing for Sales Consignment Issue

This entry is the same as the standard sales billing accounting entry in the standard sales from stock process described earlier in Section 2.1.1.

The next process is the consignment pick-up, shown in Figure 2.73. This process is used when the consignment customer wants to send some items back to you for any reason. For example, they could not sell the items.

2.8.3 Sales Consignment Pick-Up

The consignment pick-up process involves the following steps:

1. **Consignment pick-up: order**
   Similar to a standard sales order, this order will be created with the sales order type "consignment pick-up." This step has no accounting entry.
2. **Consignment pick-up: return delivery**
   You'll create a return outbound delivery with reference to the pick-up order. This step has no accounting entry.
3. **Consignment pick-up: goods receipt**
   You receive the products in your location. This step has no accounting entry but creates a material document to move the inventory from the customer's consignment stock to your own stock.

The last subprocess in the sales consignment business process is the consignment return, as shown in Figure 2.73.

2.8.4 Sales Consignment Return

This process is triggered when the consignment customer informs you that some items have been returned after being sold or used. This activity is a reversal of the sales consignment issue subprocess.

The sales consignment return process involves the following steps:

1. **Consignment return: order**
   Similar to a standard sales order, this order will be created with the sales order type "consignment return." This step has no accounting entry.
2. **Consignment return: delivery**
   You'll create a return outbound delivery with reference to the return order. This step has no accounting entry.
3. **Consignment return: goods receipt**
   You'll receive the items with reference to the return delivery. This step posts an accounting entry and a material document:
   - The material document will add the items returned to the customer's consignment stock to reverse the sales process.

- The accounting document will increase the inventory or consignment inventory value against the COGS account. The accounting entry will be as shown in Table 2.18.

Debit	Credit	Debit Amount ($)	Credit Amount ($)
Inventory		1000	
	COGS		1000

Table 2.18 Accounting Entry of GR for Sales Consignment Return

This step is a reversal of the accounting entry posted in the consignment issue (i.e., the goods issue step).

4. **Consignment return: credit memo**
   Now, you'll post the credit memo to the consignment customer. This step reverses the accounting entry posted in the consignment issue (i.e., the billing step). The resulting accounting entry is shown in Table 2.19.

Debit	Credit	Debit Amount ($)	Credit Amount ($)
Sales revenue/sales returns		1500	
VAT		150	
	AR		1350
	Sales discount		250
	WHT		50

Table 2.19 Accounting Entry of Credit Memo for Sales Consignment Return

The accounts in this accounting entry are the same as used for sales billing in the standard sales from stock process described earlier in in Section 2.1.1.

Sales Consignment Account Determination

The account determination configuration for all the accounts used in the sales consignment process was fully explained earlier in Section 2.1.1.

2.9 Sales Rebates

The sales rebate process is a commercial agreement where your customer is granted a reduction in the price of goods or services, typically as a reward for meeting specific sales targets. Essentially, a sales rebate is a form of discount but differs from immediate price reductions because this rebate is conditional and often retrospective. Rebates are commonly used as a marketing strategy or as an incentive mechanism in business-to-business (B2B) transactions.

In SAP S/4HANA, the management of sales rebates is seamlessly handled through settlement management and condition contracts. These rebates are recognized and accounted for only when the stipulated conditions are met. These conditions could involve reaching a predetermined purchase value of a particular product within a year. Notably, in scenarios where a rebate contract spans multiple periods, a common practice is to post accruals at the end of each period. These accruals are subsequently reversed during the final settlement of the contract.

Settlement management in SAP S/4HANA enables a streamlined approach to this process by utilizing condition contracts to define, manage, and settle rebate agreements effectively. As we dive more deeply into this process, we'll explore the specific accounting entries triggered at various stages and learn how to configure these entries to ensure accurate financial reporting and compliance. The upcoming sections will provide detailed insights into the process flow and the configuration of accounting entries for rebate processing in SAP S/4HANA.

2.9.1 Business Process Overview

An overview of the sales rebates process flow is shown in Figure 2.74. The steps that cause accounting entries are highlighted in green.

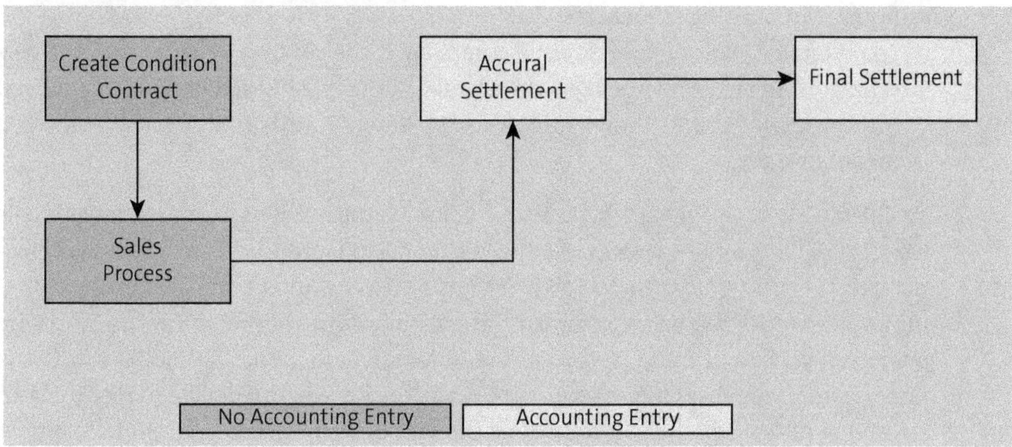

Figure 2.74 Sales Rebates Process Overview

Let's quickly look into each of these steps. We'll focus more on the steps that cause accounting entries.

Create Condition Contract

In SAP S/4HANA, the sales rebate process begins with the establishment of a condition contract in the settlement management module. This contract is a pivotal document that records all critical details of the rebate agreement with the customer.

Some key elements detailed in a condition contract include the following:

- **Material codes and product range**
  The contract specifies the particular products or material codes eligible for the rebate. This information is crucial for ensuring that only sales of these specified items count towards the rebate target.
- **Sales target and period**
  This information clearly defines the sales value threshold that the customer must reach within the validity period of the contract to qualify for the rebate. This target aligns with your company's strategic sales goals and incentivizes higher purchase volumes.
- **Validity and settlement dates**
  The contract delineates the timeframe within which the sales targets must be met. This information also specifies settlement dates, crucial for accounting and financial planning, marking when the rebate will be computed and disbursed.
- **Rebate percentage**
  The contract outlines the specific rebate percentages applicable upon reaching certain predetermined sales milestones. These percentages are strategically set to motivate customers to achieve higher sales volumes.
- **Accrual accounting for rebates**
  To maintain financial accuracy and compliance, the contract incorporates the percentage of the expected rebate to be accrued at the end of each period. This accrual is a forward-looking estimate, subject to adjustment based on actual sales at the end of the contract.

To illustrate how this rebate works with a quick example. Let's say your company sells electronic goods. To boost sales of a specific model of laptop, you offer a rebate agreement to our customers with the following terms: If a customer purchases laptops totaling a value of $50,000 between January 1st and June 30th, they will receive a 5% rebate on the total value of these laptops. If the total purchases reach $100,000 within the same period, the rebate increases to 10%. Additionally, you can decide to recognize an estimated rebate of 2% as an accrual at the end of each month during the contract period, reflecting the anticipated rebate obligation. This accrual will be reversed and adjusted according to the actual rebate amount when you conduct the final settlement, after June 30th.

Creating the condition contract itself doesn't generate any accounting entries.

After creating the condition contract, you can proceed with the relevant sales processes as usual.

Sales Processes

Any sales process that occurs within the validity period of the condition contract is tracked by settlement management. Nothing changes in the sales processes or in the accounting entries. You also won't see any changes in the pricing conditions of the sales order or the sales billing document.

One of the best features in the settlement management rebate solution in SAP S/4HANA is that it's completely independent from the sales process, which enables a lot of flexibility. For example, all the following scenarios are possible:

- You can start the year without any rebate program. But let's say that, in mid-January, you decide to start a rebate program. Now, we want all the sales that occurred since beginning of the year to be included in the target. You can create a rebate condition contract today with a validity period from beginning of the year, and at the end of the period, you would run the rebate settlement. The system will include all the sales within the validity period of the contract even if the transaction happened before the creation of the contract. No technical activities need to be performed to update the posted billing documents before contract creation because the settlement management solution is completely independent from sales transactions.
- You can, for example, create a condition contract with a 5% rebate, and after 2 months, you can change this rebate rate to 8% for any reason. When you run the first accrual or rebate settlement, the system will consider the changed percentage and adjust the rebate accruals accordingly. No changes are needed to the posted sales transactions.

These examples show how the settlement management rebate solution in SAP S/4HANA is much better than the classic solution, which required adding rebate conditions in every billing document. Previously, changing a rebate agreement once some sales transactions have been processed has been difficult. Also, in the classic solution, including sales transactions that happened before contract creation was extremely difficult.

Now, you understand that any sales process that occurs within the validity period of the contract will be included in the rebate calculation regardless of when the condition contract was created.

The next step is to process the rebate accrual calculation at the end of every financial period.

Accrual Settlement

In the rebate process, the actual rebate expense becomes your company's obligation only when certain targets are achieved. If the sales targets are not met, no rebate expense is incurred.

From an accounting perspective, and aligning with standard accounting principles, it's imperative you recognize expenses in the same period as the related revenue; this is called the *matching principle*. Therefore, even though the rebate discount is not immediately given to the customer, the potential rebate expense must be acknowledged in each period that corresponds to the revenue it relates to.

This accounting treatment is realized through the posting of a rebate accrual each month. This accrual is an estimate of the expected rebate expense, based on the current trajectory of sales. This value represents a proactive financial acknowledgment of the potential rebate liability that your company may face if the customer fulfills the rebate conditions.

The key concept behind "matching" in accounting is that expenses should be matched to the revenues they help generate. By accruing rebates monthly, you ensure that your financial records reflect this potential expense in matched with the sales that might trigger it.

This accrual is subject to adjustment and is ultimately reversed when you arrive at the final contract settlement phase when the actual rebate expense is determined based on the final sales figures. At this point, you would either give the rebate discount to the customer (or not) based on the sales target they achieved.

During the contract period at every period end (month end), you can post an accrual to account for the expected rebate expense in this period. This value posts the below financial entry, as shown in Table 2.20.

Debit	Credit	Debit Amount ($)	Credit Amount ($)
Rebate expense		3000	
	Accrued rebate		3000

Table 2.20 Accounting Entry: Sales Rebate Accrual Settlement

The value of the accrual is determined based on the accrual condition maintained in the condition contract (where we've maintained the accrual percentage) and the total value of rebate related purchases in the period. For example, if the total value purchased is $100,000 and the accrual percentage is 3%, then the accrual amount will be 100,000 × 3% = $3000.

Accrued rebate is a balance sheet account, while rebate expense is a P&L account.

Accrual settlement can be performed many times during every period end until the end of the contract when you would perform the final settlement.

The next step occurs at the end of the validity period of the condition contract, which is the final rebate settlement.

Final Settlement

The final settlement of a rebate condition contract is a crucial stage, marking the conclusion of the rebate period as defined in the contract. During this phase, the system performs a comprehensive evaluation to determine if the customer has successfully met the specified sales targets.

Two possibilities exist in this context:

- **The sales target is achieved.**

  If the customer fulfills the sales targets, the final settlement triggers the rebate expense process. This process involves reversing all previously accrued rebate amounts and posting the actual rebate amount to the rebate expense account. This posting reflects the total financial obligation your company owes to the customer as a rebate discount.

  The accounting entry in this case is split into two parts: The first part reverses all the previously accrued rebate, and the second part debits the actual rebate expense and credits the customer account. The accounting entry is shown in Table 2.21.

Debit	Credit	Debit Amount ($)	Credit Amount ($)
Accrued rebate		3,000	
	Rebate expense		3,000
Rebate expense		10,000	
	Customer AR		10,000

Table 2.21 Accounting Entry: Sales Rebate Final Settlement When Target Is Achieved

  The first two lines of this accounting entry, accrued rebate and rebate expense, are to reverse the accrued rebate posted earlier (shown earlier in Table 2.20) during all the previous periods of the contract. The value posted to these lines equals the total value of accrual rebate posted to this contract during all the periods.

  The third line of this accounting entry acknowledges the full amount of the rebate discount as an expense, and the fourth line deducts this amount from the customer account. The value posted in this fourth line equals the full value of the rebate discount that the customer receives according to the contract.

- **The sales target is not achieved.**

  In scenarios where the customer does not meet the sales targets, no rebate expense is incurred. Consequently, the system reverses any rebate accruals made during the contract period. This reversal ensures that your financial statements do not carry any rebate obligations that are no longer valid.

The accounting entry in this case debits the accrued rebate liability account and credits the rebate expense account. This entry reverses the previously recorded rebate expense, reflecting that the company no longer has an obligation to pay the rebate. The accounting entry is shown in Table 2.22.

Debit	Credit	Debit Amount ($)	Credit Amount ($)
Accrued rebate		3,000	
	Rebate expense		3,000

Table 2.22 Accounting Entry: Sales Rebate Final Settlement When Target Is Achieved

The value posted in this accounting entry equals the total value posted to the accrual rebate for this contract during all the previous periods.

Now, you understand the sales rebates process and how this process is handled by the settlement management module in SAP S/4HANA. You also understand all the accounting entries included in this process and the different accounts involved. Also, we showed you how the settlement management solution is completely independent from the sales processes and thus doesn't impact sales billing accounting entries or sales billing account determinations.

Now, let's show you how to configure the account determination for the sales rebate accrual account and the sales rebate expense account in settlement management.

2.9.2 Sales Rebate Accrual and Expense Accounts

In settlement management, account determination in rebate settlement follows the same technique as described earlier in Section 2.1.4. The rebate settlement document is similar to sales billing documents and has its own pricing procedure and pricing conditions. You can view these details in a rebate settlement document via Transaction WZR3 (Display Settlement Document), as shown in Figure 2.75.

On this screen, notice the pricing conditions of one of the line items in a rebate settlement document. This document is similar to a normal sales billing document, and thus, these account determination concepts are the same as described in Section 2.1.4. The rebate settlement account determination logic is shown in Figure 2.76.

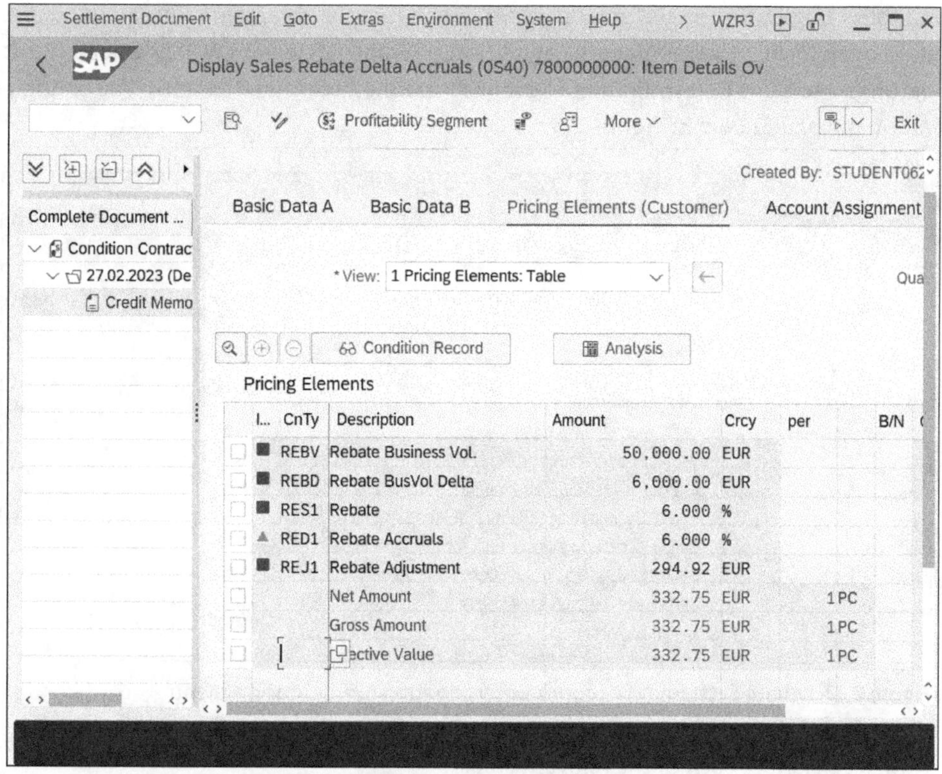

Figure 2.75 Sales Rebate Settlement Document Pricing Procedure

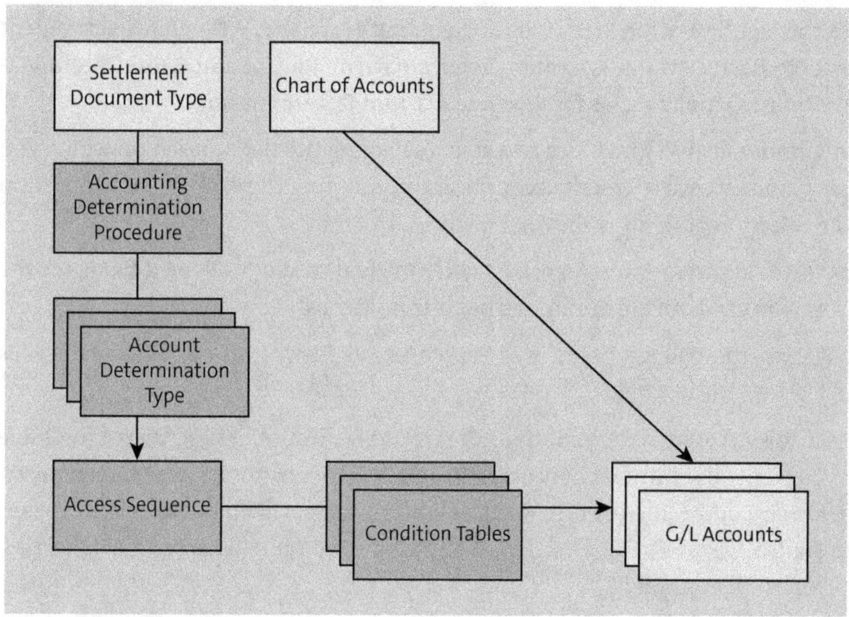

Figure 2.76 Rebate Settlement Account Determination Logic

All the configuration steps for rebate settlement account determination can be configured by following the menu path **Logistics • General • Settlement Management • Basic Settings • Account Determination • Revenue Account Determination (SD)**. The configuration menu is shown in Figure 2.77.

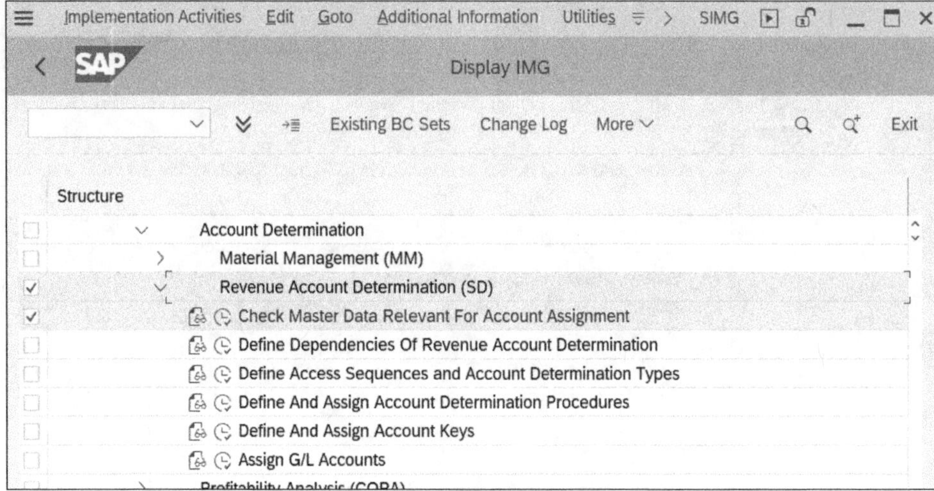

Figure 2.77 Rebate Settlement Account Determination Configuration Menu

Now, let's explore each of the relevant steps.

Define Dependencies for Revenue Account Determination

To access this configuration step, follow menu path **Logistics • General • Settlement Management • Basic Settings • Account Determination • Revenue Account Determination (SD) • Define Dependencies for Revenue Account Determination.**

This configuration activity includes two steps: selecting the fields based on which you want to determine the settlement accounts and combining these fields into condition tables. The selection of fields is shown in Figure 2.78.

To access the field activation screen, first click on **Field catalog: Allowed fields for the tables**. Now, you can activate or remove fields from the list.

After activating the needed fields, you can combine them into condition tables, as shown in Figure 2.79.

First, select the **Account determination: Create tables** option. Then, insert the table number you want (any number between 500 and 999). Then, on the next screen, select which fields should be included in your condition table. All the available fields are shown in Figure 2.79, at ❶. Double-click on any field to add it to **Selected Fields** section ❷ and include it in your table.

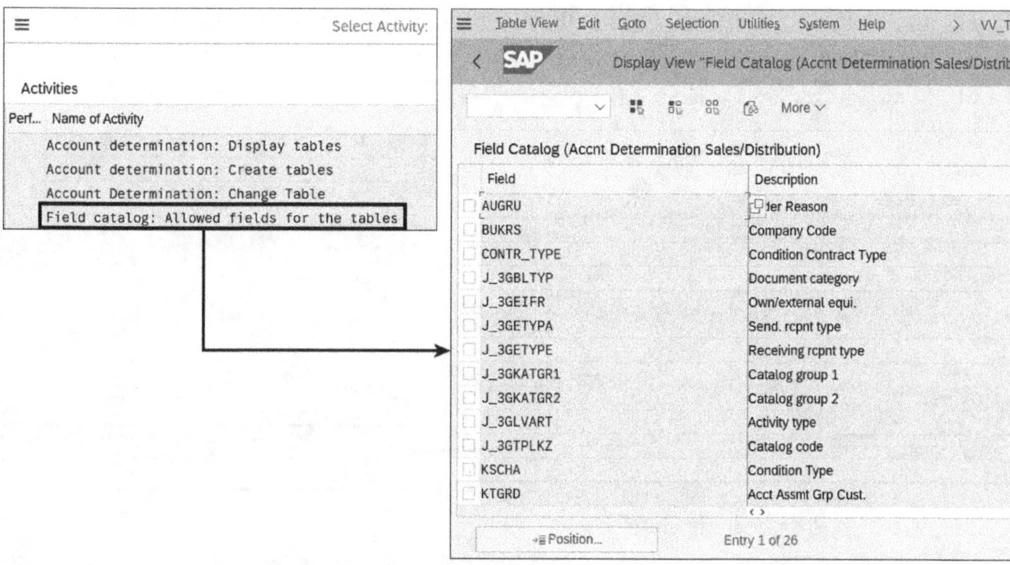

Figure 2.78 Rebate Settlement Account Determination Fields

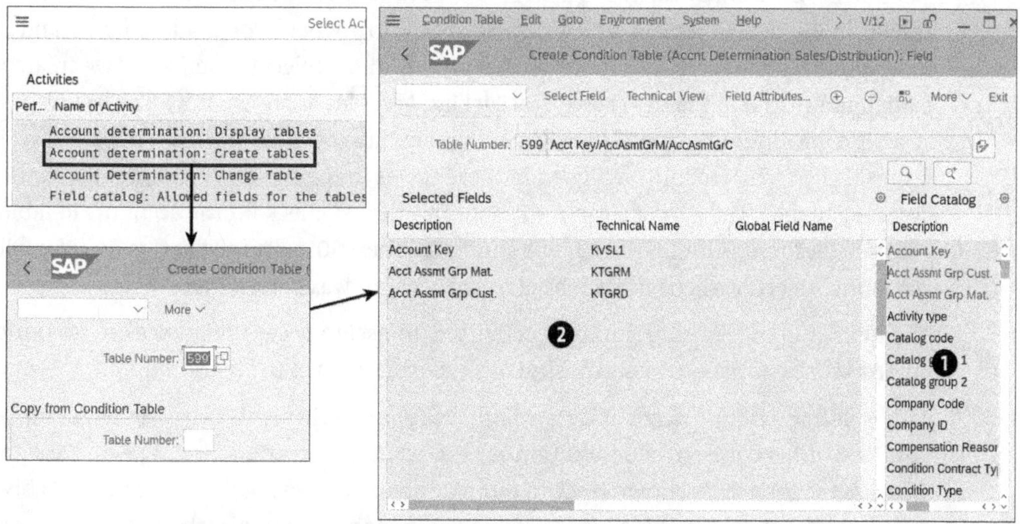

Figure 2.79 Rebate Settlement Account Determination Condition Tables

At this point, we've selected the fields needed for account determination and included them in some condition tables. The next step is to assign condition tables to access sequences.

Define Access Sequences and Account Determination Types

To access this configuration step, follow the menu path **Logistics • General • Settlement Management • Basic Settings • Account Determination • Revenue Account Determination (SD) • Define Access Sequences and Account Determination Types**.

In this activity, start by defining the access sequences and assigning condition tables to sequences, as shown in Figure 2.80.

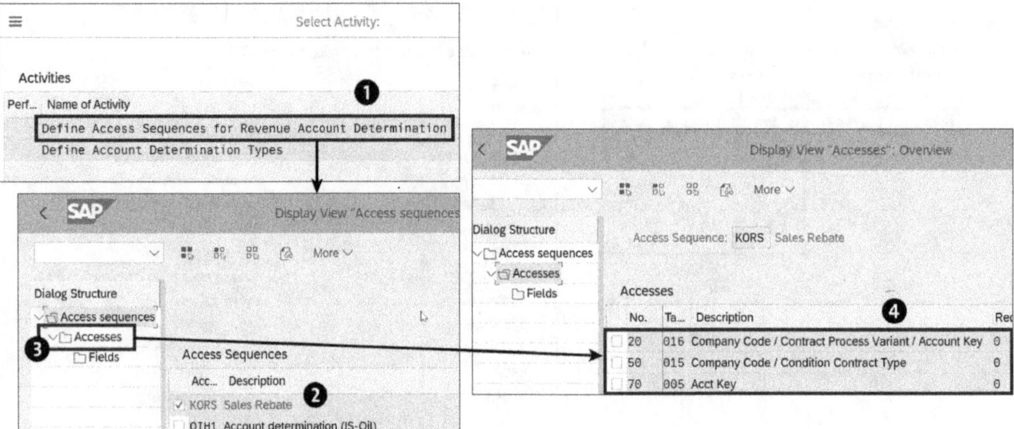

Figure 2.80 Rebate Settlement Account Determination Access Sequence

Next, select the **Define Access Sequences for Revenue Account Determination** option, as shown in ❶. This step takes you to the next window where you'll add a name and a description for the access sequence, as shown in ❷. Then, select the sequence we created earlier and switch to the next screen by clicking **Accesses** as shown in ❸. The last step is to assign the different condition tables in the sequence needed, as shown in ❹. The system will check the accounts assigned in these tables in the sequence maintained in the access sequence. Thus, for example, the system will first check for the accounts in table **016**; if no account is found, then it will check **015**; then **005**. If no account is found in all the tables, an error will be thrown because no account was determined.

The next step in this configuration activity is to assign access sequences to account determination types (also called condition types), as shown in Figure 2.81.

Now, select the **Define Account Determination Type** activity as shown in ❶. Then, add a new line with a code and a description for the account determination type and assign an access sequence, as shown in ❷. In our example, we're using the standard SAP condition type for sales rebates **KORS**, which is assigned to the standard SAP access sequence **KORS**.

At this point, you've created access sequences and assigned condition tables. Then, you created account determination types and assigned access sequences.

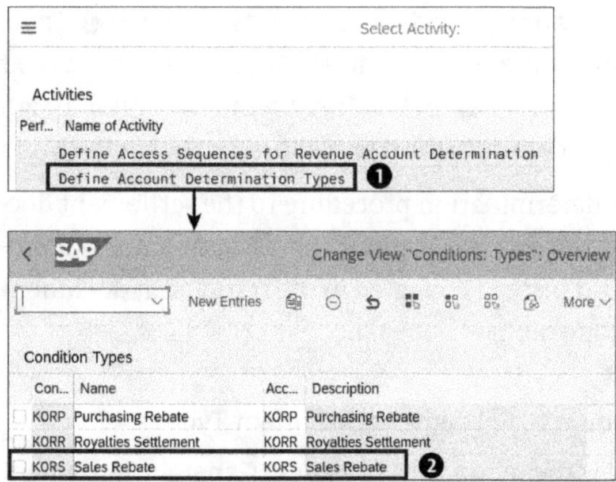

Figure 2.81 Rebate Settlement Account Determination Type

The next configuration step is to assign account determination types to account determination procedures.

Define and Assign Account Determination Procedures

To access this configuration step, follow the menu path **Logistics • General • Settlement Management • Basic Settings • Account Determination • Revenue Account Determination (SD) • Define and Assign Account Determination Procedures.** In this activity, you'll define the account determination procedures, as shown in Figure 2.82.

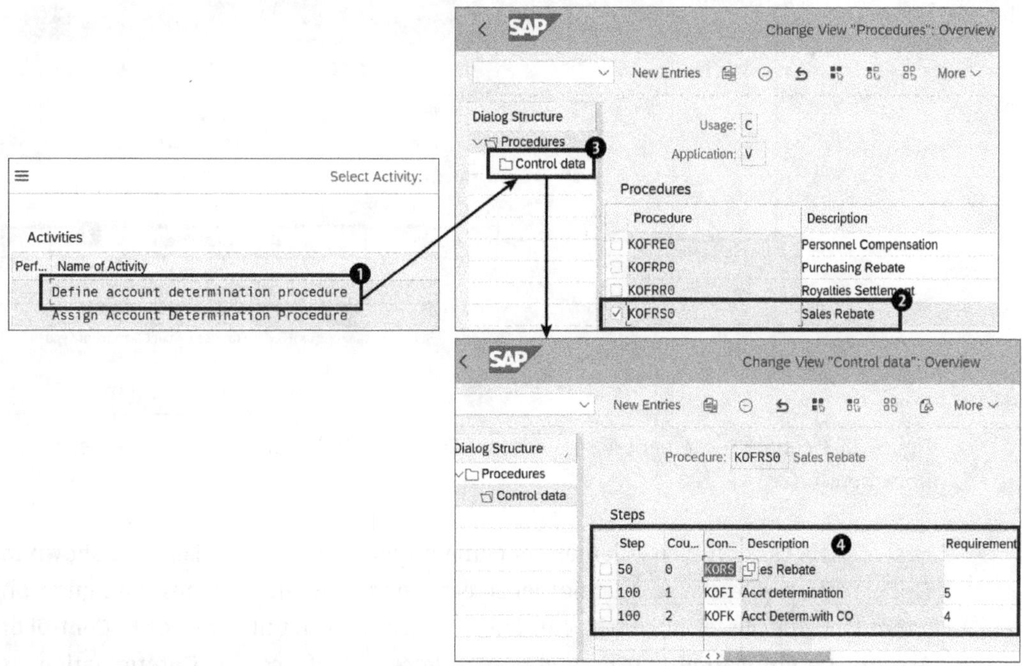

Figure 2.82 Rebate Settlement Account Determination Procedure

Forst, select the **Define account determination procedure** activity, as shown in ❶. Then, add a code and a description for the procedure, as shown in ❷. Then, switch to the next screen by clicking **Control data**, as shown in ❸. In this final window, you'll assign account determination types to account determination procedure, as shown in ❹.

The next step is to assign the account determination procedure to the settlement document type that will be used in the settlement transaction. This assignment is not performed in this configuration activity but instead is created in the settlement document type definition. Let's see how this is done next.

Assign Account Determination Procedure to Settlement Document Type

To access this configuration step, follow the menu path **Logistics • General • Settlement Management • Basic Settings • Settlement Documents • Settlement Document Types • Settlement Document.**

To assign an account determination procedure to a settlement document type, follow the steps shown in Figure 2.83.

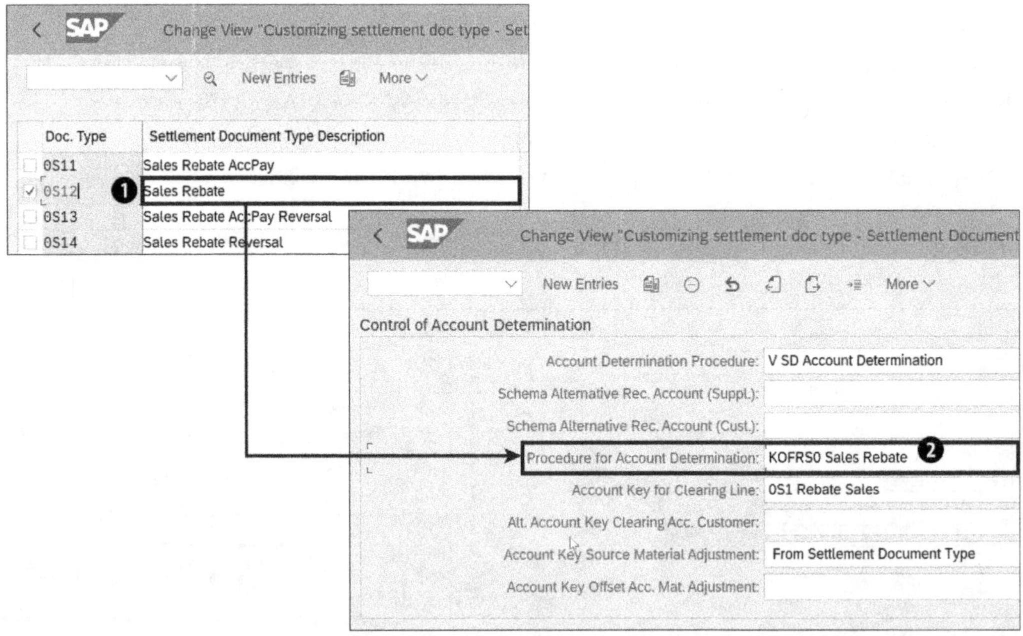

Figure 2.83 Rebate Settlement Assign Account Determination Procedure to Settlement Document Type

First, double-click on the settlement document type you want to change, as shown in ❶. This action will take you to another screen where you can see many configuration parameters for the settlement document type. Scroll down until you see to **Control of Account Determination** and then choose a **Procedure of Account Determination**, as shown in ❷.

With that, we've fully configured the account determination trail shown earlier in Figure 2.76. When we run the condition contract settlement, the settlement document type is determined automatically based on the configuration of the condition contract type. The settlement document type, in turn, leads to the account determination procedure, which leads to the account determination type, leading to the access sequences, leading to the condition tables where the accounts are assigned.

The last step is to assign general ledger accounts to condition tables.

Assign Rebate Settlement General Ledger Accounts

To access this configuration step, follow the menu path **Logistics • General • Settlement Management • Basic Settings • Account Determination • Revenue Account Determination (SD) • Assign G/L Accounts.**

In this step, start by selecting the condition table you want to maintain the general ledger in. Then, maintain the general ledger account assignment. The configuration step is shown in Figure 2.84.

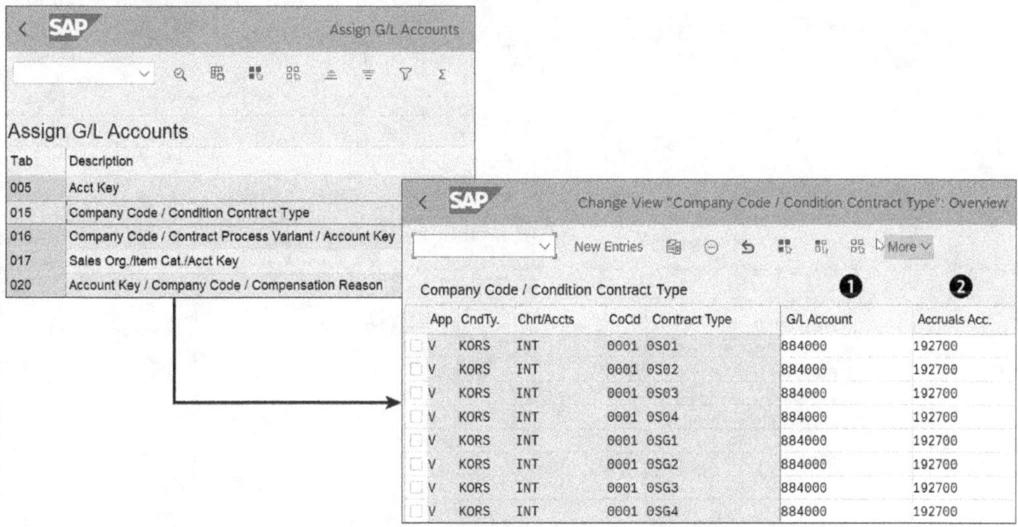

Figure 2.84 Rebate Settlement Account Determination Assign General Ledger Accounts

First, double-click on the condition table you want to maintain, for example, 015. This action takes you to the next screen where you'll see a column for each field assigned to the condition table. Notice the **G/L Account** column ❶, where you'll assign the rebate expense account, and the column **Accruals Acc.** ❷ where we maintain the rebate accrual account.

In our example shown in Figure 2.84, for the combination of application **V** (sales), **CndTy.** (account determination type) **KORS**, **Chrt/Accts INT**, and **Contract Type 0S01**, the rebate expense account is **884000**, and the rebate accruals account is **192700**.

Now we understand completely the link from the settlement document type until the general ledger accounts for rebate settlement shown earlier in Figure 2.76.

The next step is to learn how to analyze the account determination in any settlement document to be able to solve wrong account determination or any errors that happen due to wrong configuration. Let's look into the settlement document account determination analysis next.

Settlement Document Account Determination Analysis

To analyze the account determination of any settlement document, start by displaying the settlement document, which you can do via Transaction WZR3 if you know the settlement document number. You can also access all the settlement documents related to a condition contract in the document flow of the condition contract via Transaction WCOCO (Condition Contract Management Edit/Display). All the settlement documents related to the condition contract are shown in Figure 2.85, which is an example from Transaction WCOCO.

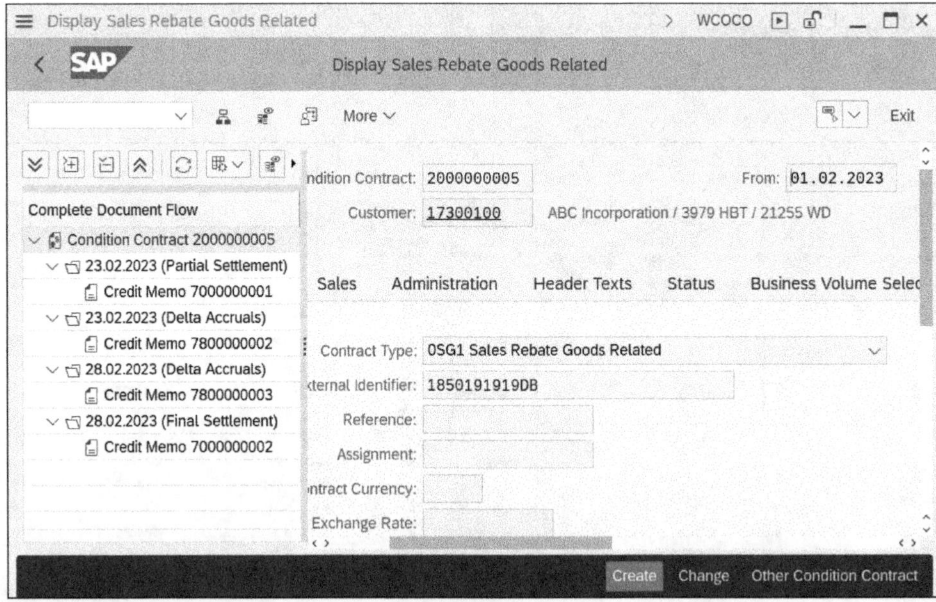

Figure 2.85 Condition Contract Display All Settlement Documents

In Transaction WCOCO, as shown in Figure 2.85, we're looking at the details of the condition contract **2000000005**. On the left, you can see the **Complete Document Flow** of the condition contract. You can see the type of each document. For example, the credit memo **70000000002** was posted for **Final Settlement**. To display this document, double-click a document number to display the settlement document, as shown in Figure 2.86.

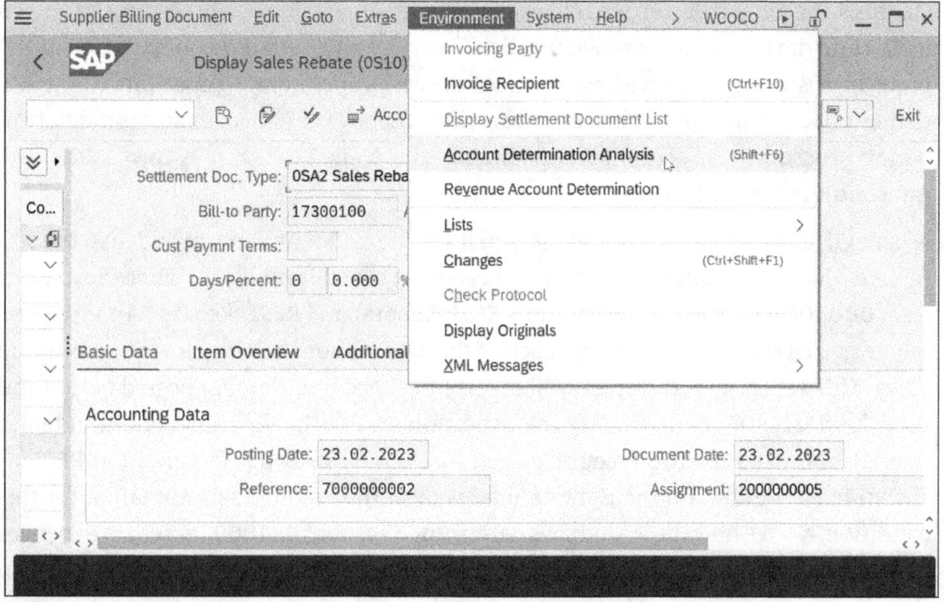

Figure 2.86 Rebate Settlement Document: Start Account Determination Analysis

In our example, we have the rebate **Settlement Document Type 0SA2**. To start the account determination analysis, in the top menu, select **Environment • Account Determination Analysis**. This action will open the account determination analysis screen, as shown in Figure 2.87.

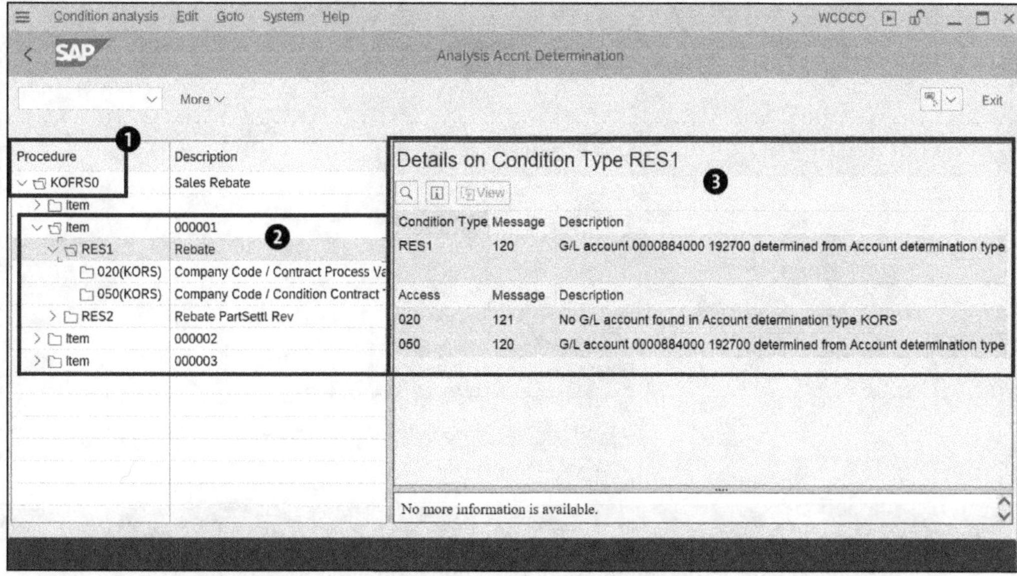

Figure 2.87 Rebate Settlement Account Determination Analysis: Screen 1

On this screen, the first step is to check the account determination procedure, as shown in ❶. Here the procedure is **KOFRSO** which is correct based on our configuration. If the procedure is wrong, then we have to check the assignment of the procedure to the settlement document type in the last subsection when we assigned an account determination procedure to a settlement document type. If the procedure is correct, then we can continue our analysis.

As shown in Figure 2.87, on the left ❷, you'll see the different items included in the settlement document. Under each item, we can have several conditions. For example, for item **0000001**, we have the conditions **RES1** (**Rebate**) and **RES2** (**Rebate PartSettll Rev**, rebate partial settlement reversal). Each of these conditions will impact the accounting entry. This settlement document will reverse the partial settlement posted before and post the final settlement. Under each condition, notice the different condition tables that SAP checked for the account determination. By clicking on condition **RES1**, as shown in Figure 2.87, on the right ❸, the result of the account determination for this condition. Notice how the system has determined account **884000** for rebate expenses and determined account **192700** for rebate accruals. From the messages in ❸, note also that no account was found in the first condition table that was assigned to the access sequence **020**. Therefore, the system checked the second table that has the sequence **050**, where it was able to find an account assigned there.

To review more details on how the system found this account inside the condition table, click on the condition table with the access sequence **050** on the left side of the screen 2, and you'll see its details, as shown in Figure 2.88.

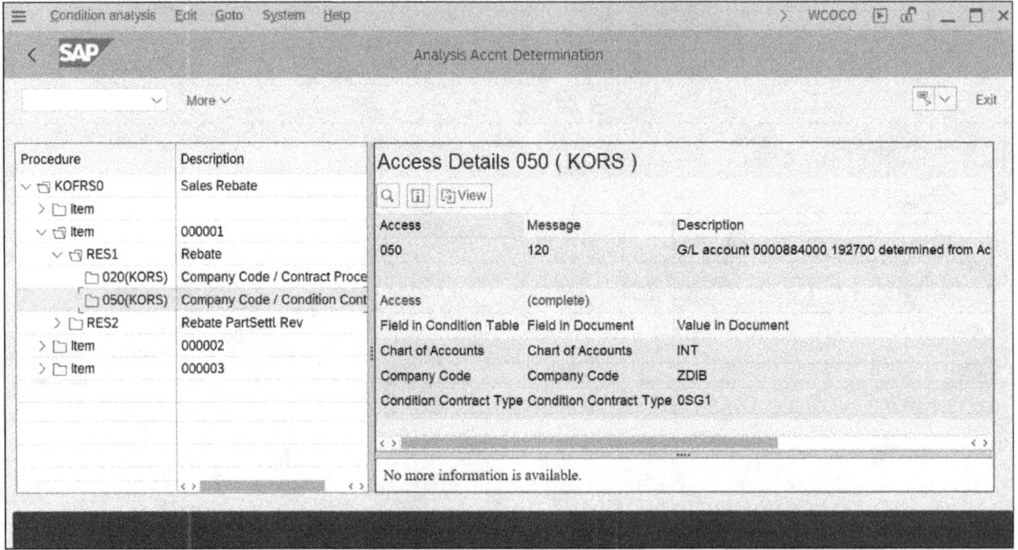

Figure 2.88 Rebate Settlement Account Determination Analysis: Screen 2

In our example, the system has found the account in the condition table assigned to the **Chart of Accounts INT**, **Company Code ZDIB**, and **Condition Contract Type 0SG1**.

Now, you know how to configure the account determination for the sales rebates process and how to analyze the account determination of the rebate settlement document.

2.10 Summary

This chapter provided a comprehensive overview of various sales processes within SAP S/4HANA, emphasizing the critical role of account determination in aligning financial transactions with business operations. We systematically examined different sales scenarios, including sales returns, third-party sales, cash sales, and FoC sales, each presenting unique challenges in account determination and process flow.

Key takeaways include the importance of precise account determination in ensuring financial accuracy across diverse sales scenarios. SAP S/4HANA's flexibility is evident in its ability to handle complex sales structures, such as third-party sales logistics and rebate management, while maintaining detailed financial records. We delved into the nuances of configuring SAP S/4HANA to accommodate various sales types, highlighting the system's adaptability in mapping sales activities to specific general ledger accounts.

The insights from this chapter underline SAP S/4HANA's capability to cater to a wide range of business needs, ensuring that every sale, regardless of its complexity, is accurately reflected in the financial statements. The configuration steps and examples provided serve as a practical guide for businesses to optimize their sales and distribution management using SAP S/4HANA.

Moving forward, the book will continue to explore other facets of SAP S/4HANA, deepening the understanding of its extensive functionalities. By aligning business processes with financial reporting, SAP S/4HANA proves to be an invaluable tool for modern businesses aiming for operational efficiency and financial integrity.

Chapter 3
Production Planning and Product Costing

The production planning module in SAP is responsible for managing manufacturing processes, which are among the most complex areas in any business. In this chapter, we'll focus on the common make-to-stock discrete manufacturing process to analyze different accounting entries and show you how to configure the determination of the correct accounts. The concepts explained in this process will also be applicable to other manufacturing processes.

Production planning is closely integrated with product costing controlling (CO-PC), a vital component of the Controlling module in SAP, closely integrated with the production planning process. Product cost controlling comprises three components: product cost planning (CO-PC-PCP), cost object controlling (CO-PC-OBJ), and actual costing/ material ledger (CO-PC-ACT).

Product cost planning aims to estimate the cost of products you plan to produce, based on various cost components such as raw materials and manufacturing activities.

Cost object controlling is tasked with comparing planned product costs against the actual costs of production, analyzing variances between planned and actual, and reporting all types of discrepancies.

The material ledger is responsible for capturing and recording the actual costs of materials during production and procurement. It tracks the real expenses incurred, including material prices, production costs, and overheads. With the material ledger, you can track your inventory's value in multiple currencies and valuation views. One feature of material ledger is actual costing, with which you can revaluate inventory values for the period at period end, after you've reported all your actual costs. Revaluating the inventory values means posting an accounting entry, which requires account determination configuration. Actual costing is an optional feature in SAP S/4HANA.

In this chapter, we'll dive into the process flow, accounting entries, and account determination configuration for product cost planning and cost object controlling. Let's begin our exploration with product cost planning, the initial step in the make-to-stock manufacturing sequence.

Since actual costing is an advanced optional feature in SAP S/4HANA, we won't include this topic in this book so we can focus on the mandatory components in product costing.

3.1 Product Cost Planning

As we embark on our journey through account determination in production planning, it's essential we start with product cost planning—the initial and foundational step any company does before starting production.

Product cost planning serves as the backbone for effective cost management, enabling businesses to forecast the costs associated with producing each product. This process involves a detailed evaluation of various cost components, mainly materials, manufacturing activities, and manufacturing overheads. By accurately estimating these costs, your company can be better equipped to make informed decisions, budget effectively, and maintain a competitive edge in pricing its products.

The end result of product cost planning in an SAP system is a product cost estimate with detailed cost components ❶ and cost itemization ❷, as shown in Figure 3.1.

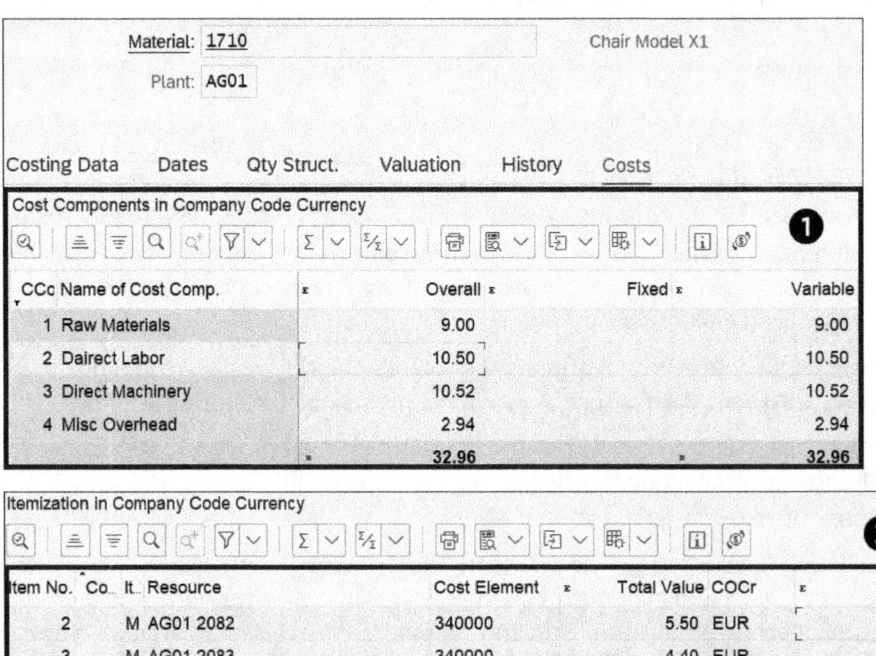

Figure 3.1 Cost Components and Cost Itemization

The cost elements are displayed in the cost itemization, shown in Figure 3.1 ❷. These cost elements are mapped to the different **Cost Components**. For example, as shown in Figure 3.1, the **Cost Element 340000** is mapped to the **Cost Component 1 Raw Materials**, and **921000** is mapped to **Direct Labor**.

A cost element is another name for a general ledger account that we use within the controlling or costing areas in SAP, In the next sections, you'll learn how these general ledger accounts are determined in the standard cost estimate and how they are mapped to the different cost components.

Cost Elements in SAP S/4HANA versus SAP ERP

You can use the terms "general ledger account" and "cost element" interchangeably in SAP S/4HANA because both are fully integrated for both secondary and primary cost elements. However, for older versions of SAP ERP, secondary cost elements were created separately in the controlling area without integration with a general ledger account.

The cost components and their cost itemizations can be split into three main components:

1. **Direct materials**
   These components are used in the production process, as determined through the bill of materials (BoM). As shown in Figure 3.1, this cost component is 1 **Raw Materials.**
2. **Direct manufacturing activities**
   These activities are needed to convert the material components into the finished product, and they are determined based on the routing or the recipe. As shown in Figure 3.1, these cost components are 2 **Direct Labor** and 3 **Direct Machinery.**
3. **Manufacturing overhead**
   These manufacturing expenses are not directly related to the manufacturing process. These costs are assigned as a percentage of the direct activity or as direct material costs through the use of an overhead costing sheet.

Let's start with the first step of the product cost planning process: the determination of direct material costs and their related general ledger accounts.

3.1.1 Cost Elements: Direct Material Costs

When you perform a product standard cost estimate in SAP, you'll start by inserting a costing variant—the first step in the determination of the direct material costs, as shown in Figure 3.2.

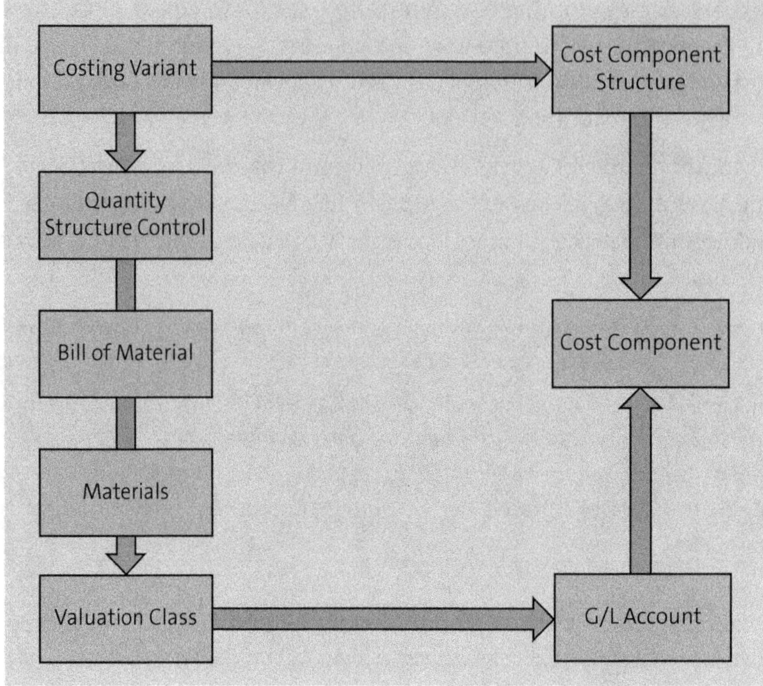

Figure 3.2 Direct Materials Account and Cost Component Determination

The configuration of the costing variant can be performed in Transaction OKKN, which can be found by following the menu path **Controlling • Product Cost Controlling • Product Cost Planning • Material Cost Estimate with Quantity Structure • Define Costing Variants**. The configuration of costing variant **PPC1** is shown in Figure 3.3.

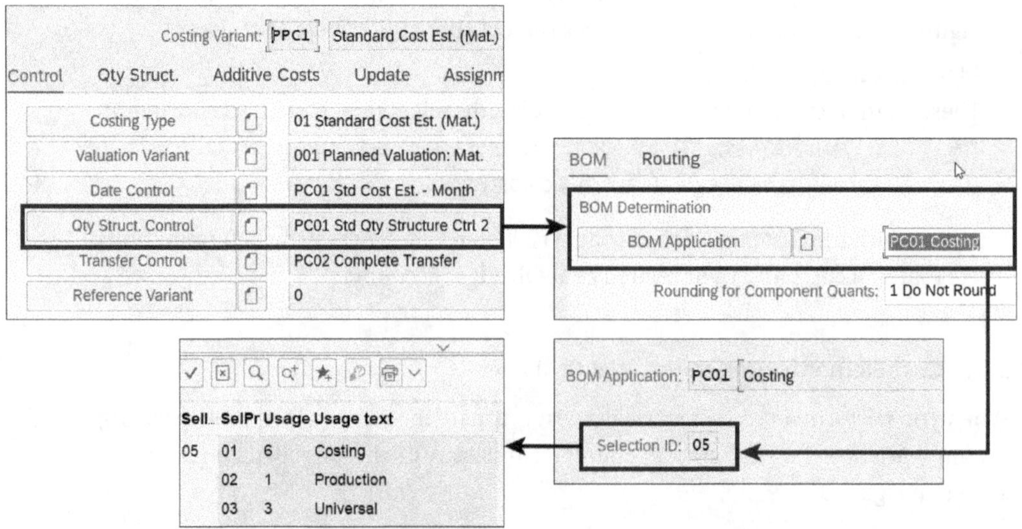

Figure 3.3 BoM Usage Determination

In the costing variant, select a *quantity structure control* (**PC01**), as shown in Figure 3.3. In the configuration of the quantity structure control, assign a BOM application. In the configuration of the BOM application, we assign a selection ID. The selection ID determines how the BoM is selected. As shown in Figure 3.3, the BoM will be selected based on the following priority: Usage 6 (Costing), then Usage 1 (Production), and finally Usage 3 (Universal).

When creating any BoM in SAP, selecting a usage is mandatory, and the usage can be displayed later in the BoM header in Transaction CS03 (Display Bill of Material), as shown in Figure 3.4.

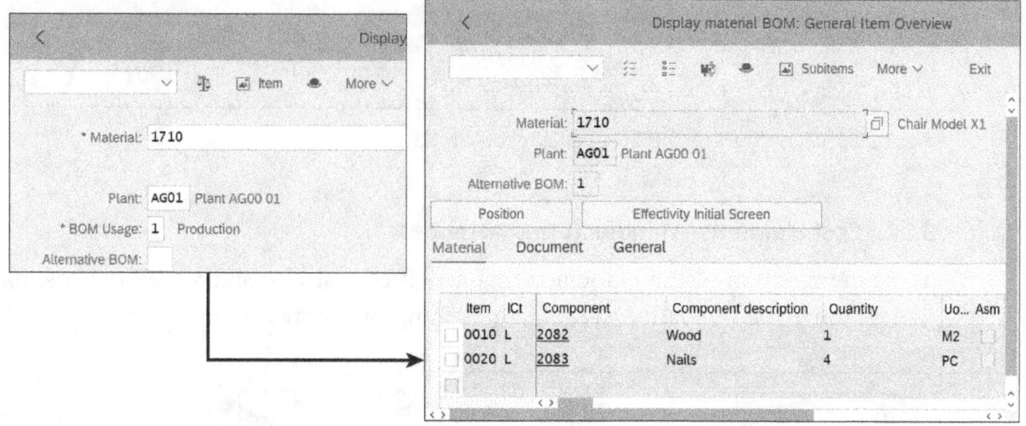

Figure 3.4 Display Material BoM

On the **Display material BOM General Item Overview** screen, you can insert different material components and their quantities. For example, as shown in Figure 3.4, to produce 1 unit of material **1710**, we need **1** M2 of material **2082** and **4** PCs of material **2083**. Each component has a valuation class in the material master data definition, and this class is used for the account determination of the consumption account, as described in Chapter 1, Section 1.1.2. This consumption account is shown in the cost itemization of a standard cost estimate (shown earlier in Figure 3.1). The account determination for this account is configured in the Transaction OBYC (Maintain Automatic Posting). The account is assigned to the valuation class of the material, the transaction key **GBB**, and the general modification **VBR**, as shown in Figure 3.5.

You can then map this general ledger account or cost element to a cost component in the cost component structure configuration. This step is common for all the different cost components and is explained in Section 3.1.4.

Now, you understand the account determination for the direct materials consumption account and how this is determined in as standard cost estimate. Next, let's move to the second cost component shown in Figure 3.1—manufacturing activities.

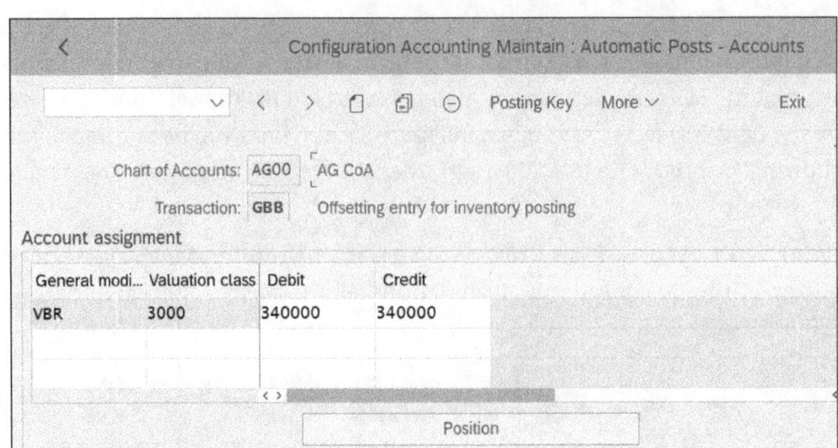

Figure 3.5 Account Determination of Direct Materials

3.1.2 Cost Elements: Manufacturing Activities

The determination of the manufacturing activities cost also starts with the costing variant use in the standard cost estimate, as shown in Figure 3.6.

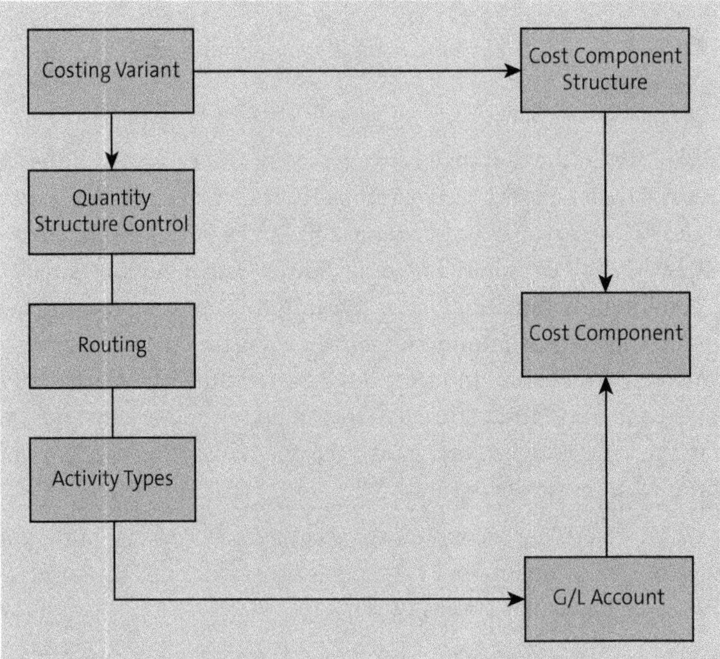

Figure 3.6 Manufacturing Activities Accounts and Cost Component Determination

The manufacturing activities are maintained in the routing master data. The routing is where we maintain all the manufacturing activities needed to convert the components into the final product. Figure 3.7 shows the configuration of the costing variant and how the routing is determined in the standard cost estimate.

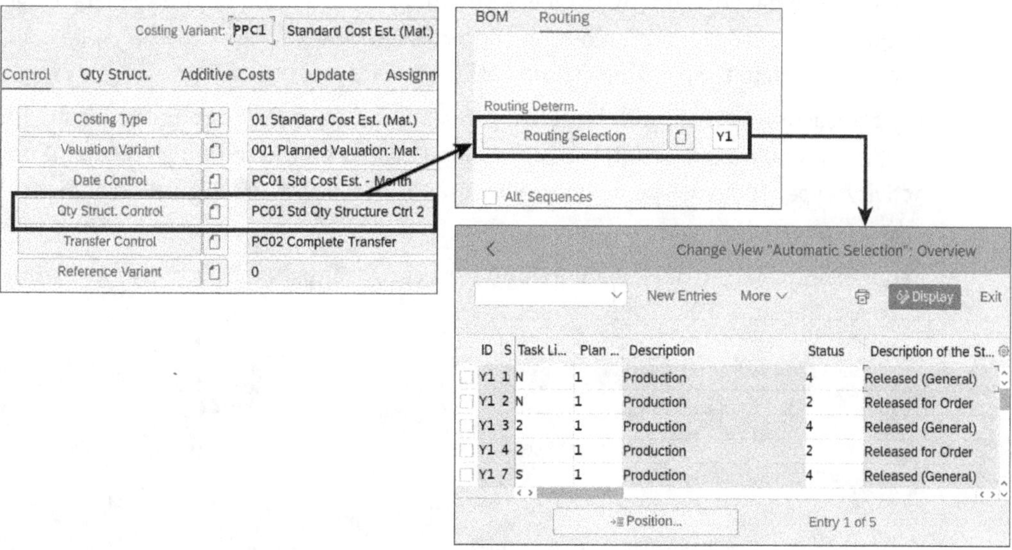

Figure 3.7 Routing Determination

In the **Costing Variant PPC1**, we've selected the **Quantity Structure Control PC01** and there we have selected the **Routing Selection Y1**. In the configuration of the routing selection Y1, we specified the priorities for automatically selecting a routing in the cost estimate. For example, as shown in Figure 3.7, the priority number 1 is for **Task List Type N (Routing), Plan Usage 1 (Production), and Status 4 (Released general).**

The routing is defined by the production department and can be displayed using Transaction CA01 (Display Routing). Figure 3.8 shows an example of routing master data.

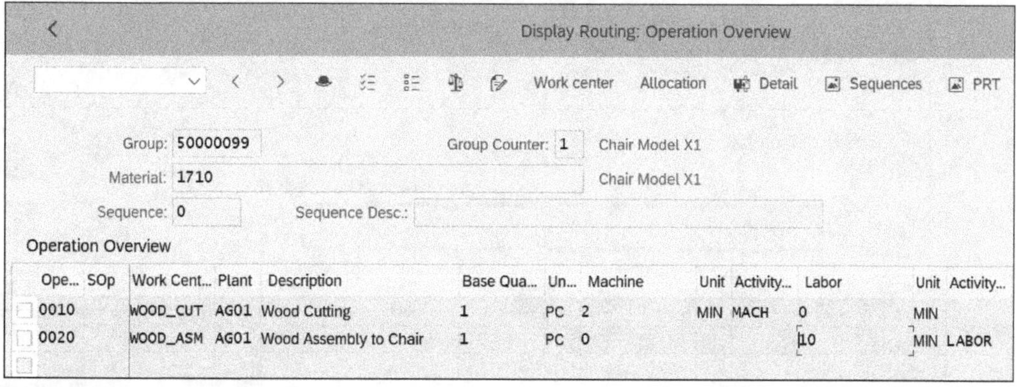

Figure 3.8 Routing Operations Overview

As shown in Figure 3.8, to produce **1 PC** of the final product **1710**, we need two operations. The first one is **Wood Cutting** that requires **2 MIN** of machinery, for which the **Activity Type** is **MACH** and the second operation is **Wood Assembly** that requires **10 MIN** of Labor with the **Activity Type LABOR**.

In the master data of the activity type, we maintain the cost element (the general ledger account) that's used for any accounting postings done by through this activity type and also that's seen in the cost itemization in the standard cost estimate.

The transaction to maintain the activity type master data can be found following the SAP user menu path **Accounting • Controlling • Cost Center Accounting • Master Data • Activity Type**. To create an activity type, use Transaction KL01. Figure 3.9 shows the activity type master data.

In our example, the cost element **921000** has been assigned to the activity type **LABOR**, as shown in Figure 3.9.

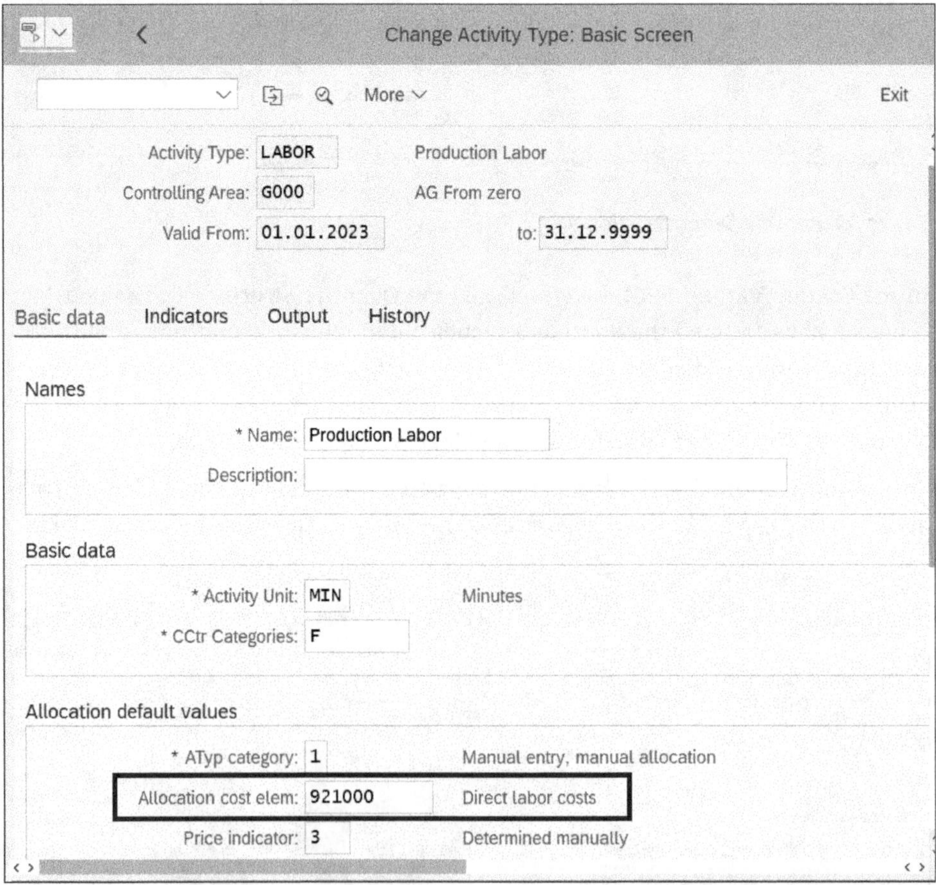

Figure 3.9 Activity Type Master Data: Cost Element

Note that this cost element must be a secondary cost element of type **43** (**Internal activity allocation**). To check this value in the cost element master data, use Transaction FS00 (Display G/L Account Centrally), as shown in Figure 3.10.

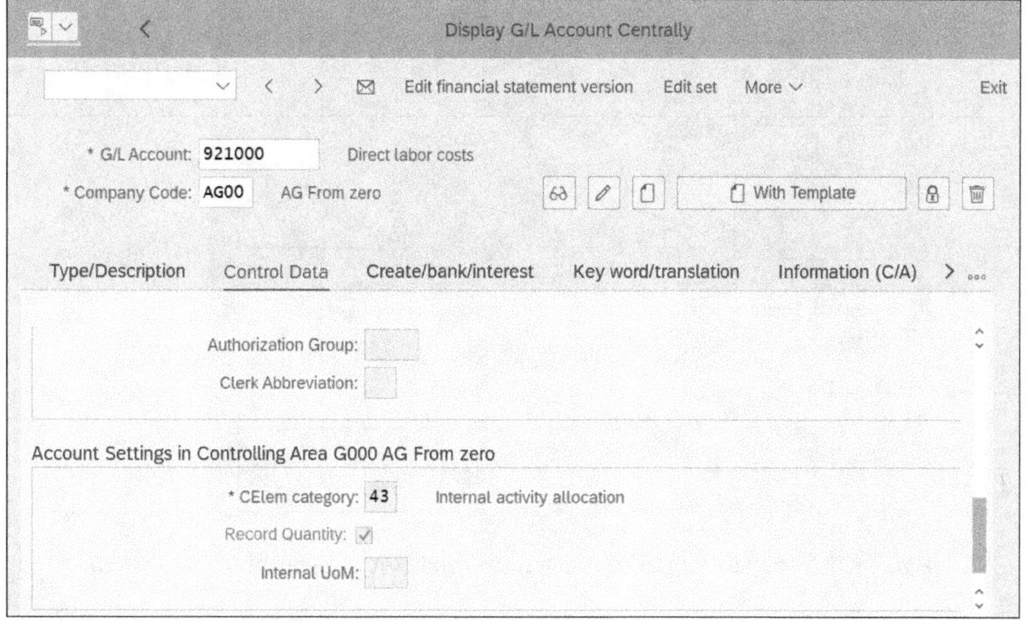

Figure 3.10 Cost Element Category for Activity Types

Now, you know how activity types are determined in a standard cost estimate and how they are connected to general ledger accounts. Next, let's look into the third type of cost component: manufacturing overhead costs.

3.1.3 Cost Elements: Manufacturing Overhead

The determination of manufacturing overhead costs and their accounts also starts with a costing variant in the standard cost estimate, as shown in Figure 3.11.

The overhead percentages and bases are maintained in costing sheets. You can assign a costing sheet to the costing variant used in the standard cost estimate. Figure 3.12 shows the assignment of the costing sheet **G0** in the configuration of the costing variant **PPC1** in Transaction OKKN.

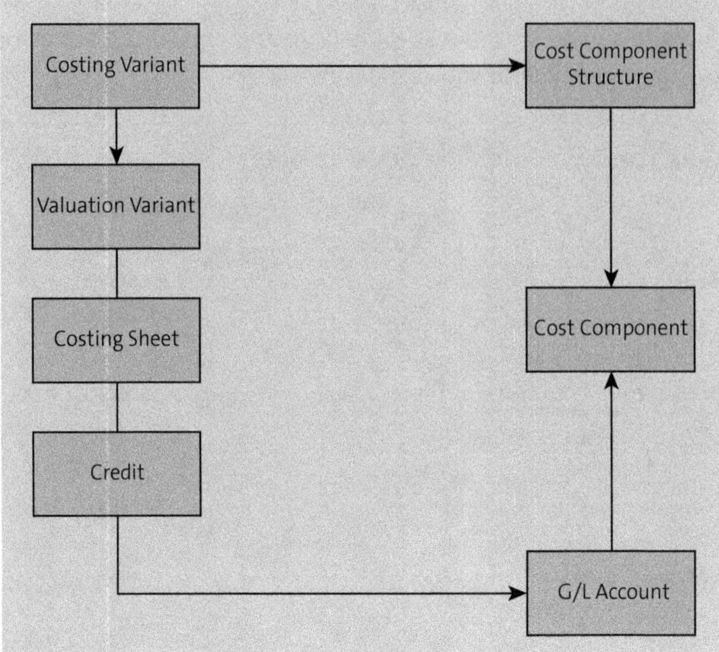

Figure 3.11 Manufacturing Overhead Accounts and Cost Component Determination

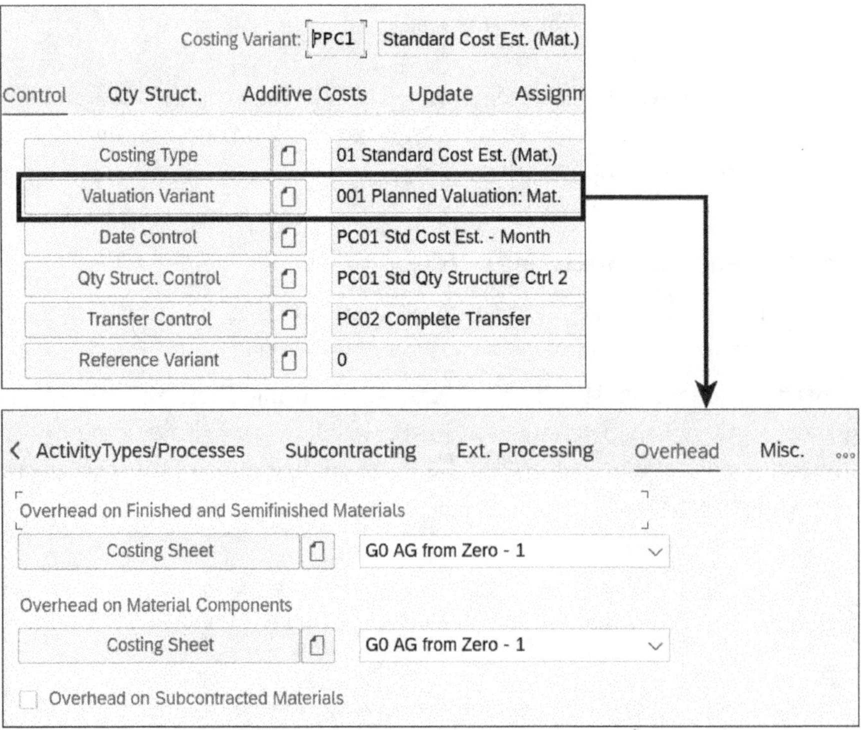

Figure 3.12 Costing Sheet Determination

The costing sheet is a configuration object where you control how the manufacturing overhead is calculated. The costing sheet can be configured in Transaction KZS2, which can be found by following the menu path **Controlling • Product Cost Controlling • Product Cost Planning • Basic Settings for Material Costing • Overhead • Define Costing Sheets**. Figure 3.13 shows the configuration of costing sheet **GO**.

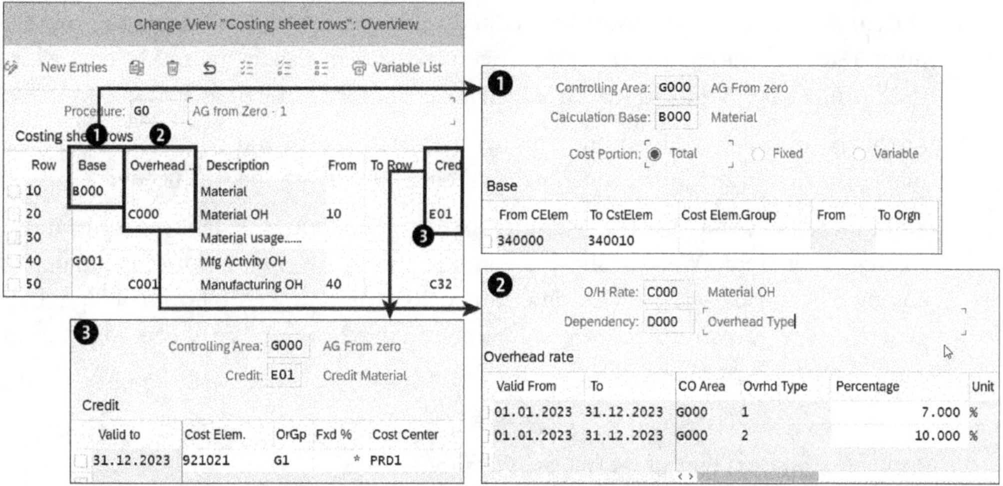

Figure 3.13 Configuration of Manufacturing Overhead Calculation and Cost Element in Costing Sheets

The first step in costing sheet configuration is to define the base of the overhead calculation. As shown in Figure 3.13 ❶, the definition of the calculation base **B000** is for the total costs posted to the cost elements from **340000** to **340010**, which are the material consumption accounts. In this way, we're saying that the overhead will be a percentage of the direct material costs posted to these general ledger accounts.

After defining the base, you can define the **Overhead (O/H) Rate**, as shown in Figure 3.13 ❷. Note that the overhead rate will be multiplied by the base to calculate the overhead cost. **Overhead Type 1** is for actual overhead (for actual production order costs) and **2** is for planned overhead (for cost estimates and planned production order costs). As shown in Figure 3.13, we have a planned overhead of **10%** and an actual overhead of **7%** for the period from **01.01.2023** until **31.12.2023**. These percentages can be changed directly in the production system and are often changed at each month-end closing.

On the **Costing Sheet Rows Overview** screen, shown in Figure 3.13, notice how **row 20** includes the **Overhead Rate C000** in the **From** column, with a value of **10**. This value means the overhead rate should be multiplied by the **Base** in row **10**, which is **B000**, which includes the material consumption account. In this way, the following formula calculates our planned overhead:

```
10% * Total costs of the accounts from 340000 to 340010
```

The third and final step in the costing sheet is to define the **Credit**, as shown in Figure 3.13 ❸. This step is the most important part because it includes the **Cost Element** that will be used for the overhead assessment and that is mapped to the cost component in the standard cost estimate. In our example, the overhead assessment cost element is **921021**. Note also the **cost center** from which the overhead will be allocated to the production order, which is **PRD1** in our example. For the credit **E01**, we're saying that, until **31.12.2023**, this overhead should be credited from cost center **PRD1** using the cost element **921021**. In the costing sheet rows overview, shown also in Figure 3.13, notice how this credit is maintained in row **20**, which is calculating on the base of row **10**.

At this point, we've defined our first overhead rate, base, and credit. Simply repeat these steps for other kinds of overheads in the same costing sheet as needed.

You now understand how the overhead assessment account is determined from the costing sheet credit. We also showed you how the cost elements for direct materials and manufacturing activities are determined. Next, let's look into how to assign these general ledger accounts to cost components and how the cost components are assigned to the costing variant.

3.1.4 Assign Cost Elements to Cost Components

In this step, we'll assign the cost elements to the cost components in the cost component configuration, as shown in Figure 3.14.

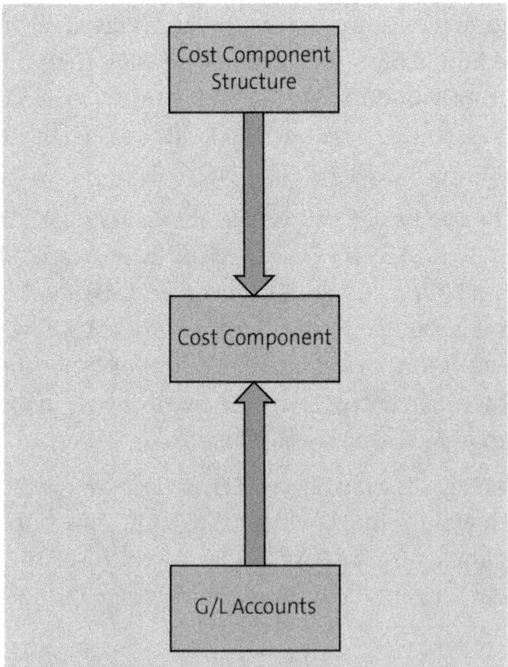

Figure 3.14 Assign General Ledger Accounts to Cost Components

The cost component structure configuration can be accessed via Transaction OKTZ or by following the menu path **Controlling • Product Costing • Product Cost Planning • Basic Settings for Material Costing • Define Cost Component Structure**. The configuration of the cost component structure **G0** is shown in Figure 3.15.

Cost Comp. Str.	Cost Compo...	Name of Cost Comp.
G0	1	Raw Materials
G0	2	Dairect Labor
G0	3	Direct Machinery
G0	4	Misc Overhead
G0	8	External Activity
G0	9	Machine Setup
G0	10	Miscellaneous

Cost Comp. Str.	Chart of Accts	From cost el.	Origin Group	To cost elem.	Cost Compo...	Name of Cost Comp.
G0	AG00	340000		340010	1	Raw Materials
G0	AG00	921000		921010	2	Dairect Labor
G0	AG00	921011		921020	3	Direct Machinery
G0	AG00	921021		921030	4	Misc Overhead

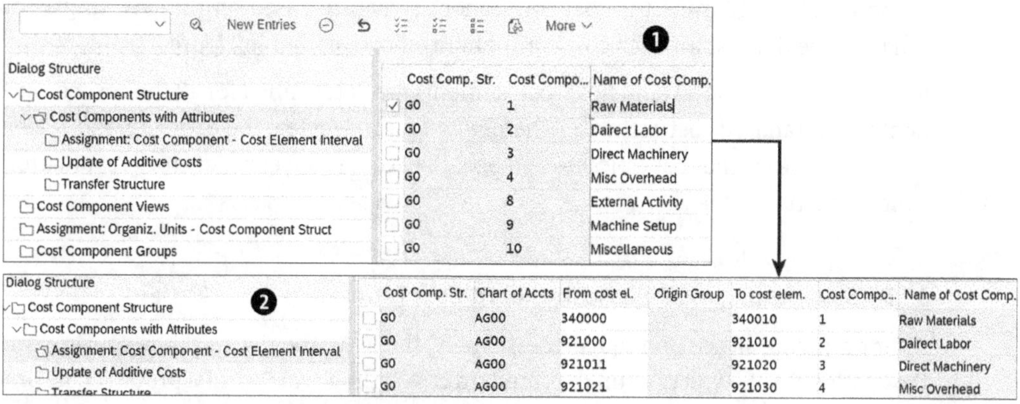

Figure 3.15 Configure Cost Component Structure: Assign Cost Elements

First, we'll define the different cost components from the **Dialog Structure** menu, by selecting the **Cost Components with Attributes** option, as shown in Figure 3.15 ❶. Multiple options are available in the cost component definition that are not related to account determination.

Second, you can assign each cost component to an interval of cost elements. From the **Dialog Structure** menu, select the **Assignment: Cost Component – Cost Element Interval** option ❷. In our example, the **cost component 1** (**Raw Materials**) is assigned to the cost elements from **340000 to 340010**. These accounts are configured as the material consumption accounts as explained in the Section 3.1.1 section. The same applies to the other cost elements for manufacturing activities and manufacturing overhead.

The final step is to assign the cost component structure to company codes. From the **Dialog Structure** menu, select the **Assignment: Organiz. Units – Cost Component Struct** option, as shown in Figure 3.16.

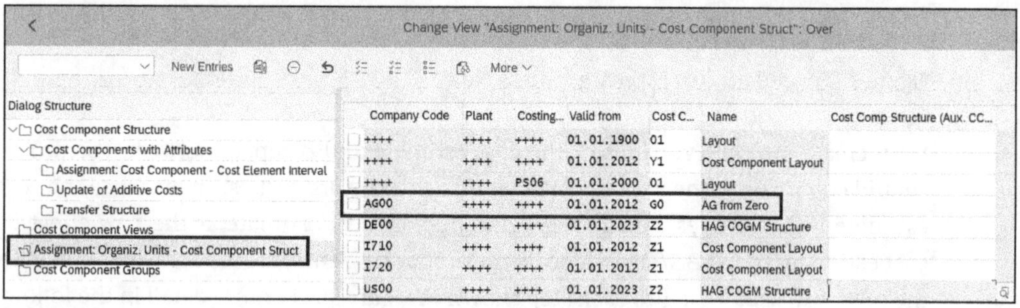

Company Code	Plant	Costing...	Valid from	Cost C...	Name	Cost Comp Structure (Aux. CC...
++++	++++	++++	01.01.1900	01	Layout	
++++	++++	++++	01.01.2012	Y1	Cost Component Layout	
++++	++++	PS06	01.01.2000	01	Layout	
AG00	++++	++++	01.01.2012	G0	AG from Zero	
DE00	++++	++++	01.01.2023	Z2	HAG COGM Structure	
I710	++++	++++	01.01.2012	Z1	Cost Component Layout	
I720	++++	++++	01.01.2012	Z1	Cost Component Layout	
US00	++++	++++	01.01.2023	Z2	HAG COGM Structure	

Figure 3.16 Assign Cost Component Structure to Company Codes

The combination of company code **AG00** with plant **++++** and costing variant **++++** is assigned to the cost component structure **G0**. The value ++++ means all plants and costing variants for company code AG00.

When you run a standard cost estimate and choose a costing variant, the system will automatically determine the cost component structure based on these settings. You can then see the cost for each cost component (shown earlier in Figure 3.1).

Now, you should understand how cost elements are determined for all the cost components in a standard cost estimate. The next step is to review the estimate and if all is okay then release the cost estimate and post the accounting entry of stock revaluation, which we'll turn to next.

3.1.5 Save, Mark, and Release Standard Cost Estimates

After preparing all the prerequisites—namely, the BoM, the material, the routing, and the costing sheet—you can run the standard cost estimate of the final product. The process overview is shown in Figure 3.17.

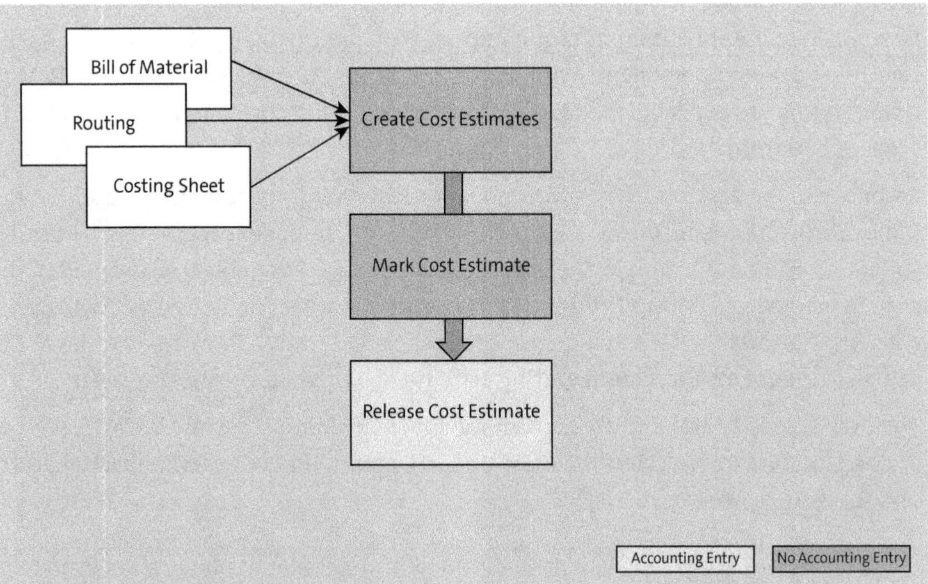

Figure 3.17 Standard Cost Estimate Process Flow

First, create multiple versions of the cost estimate and compare between them. You should choose only one version and mark it as the future cost estimate. Once the cost estimate period arrives, you'll release the marked cost estimate as the new effective cost estimate, and SAP S/4HANA posts the accounting entry for stock revaluation with the new cost. Let's look into each step shown in Figure 3.17 in more detail in the following sections.

Save Cost Estimate

You can save as many versions as you want of the cost estimate for the same material and the same period. Every time you run a cost estimate, you must insert a costing variant and costing version. Cost estimates can be created for each material individually using Transaction CK11n (Create Material Cost Estimate with Quantity Structure) or for multiple items at once using Transaction CK40n (Edit Costing Run).

Each cost estimate has a validity period that must take end after the end of the month. You can create a cost estimate that's valid for multiple months.

When you run the cost estimate, an important step is that you confirm that the BoM, routing, and costing sheet have been determined correctly from the costing variant. Confirm these details in Transaction CK11n, as shown in Figure 3.18.

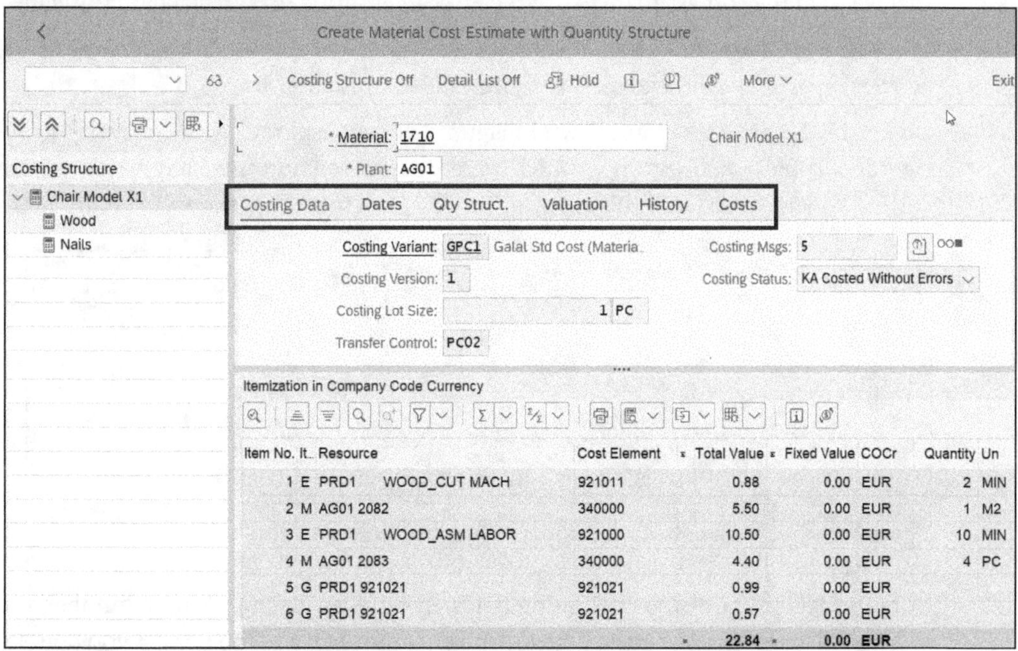

Figure 3.18 Create Material Cost Estimate with Quantity Structure: Main Screen

The costing header tabs highlighted on the screen shown in Figure 3.18 include the details we want. Under the **Costing Data** tab, notice the costing variant **GPC1** that we've selected for this cost estimate. Note also see the **Costing Version 1**.

Under the **Dates** tab, you'll see details related to the dates, as shown in Figure 3.19.

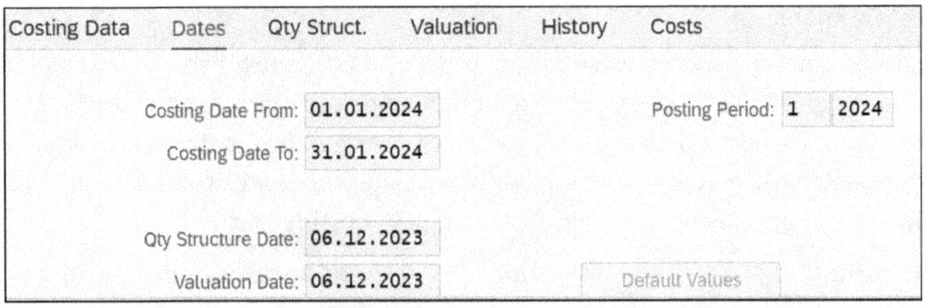

Costing Data | Dates | Qty Struct. | Valuation | History | Costs

Costing Date From: 01.01.2024 Posting Period: 1 2024
Costing Date To: 31.01.2024
Qty Structure Date: 06.12.2023
Valuation Date: 06.12.2023 Default Values

Figure 3.19 Create Material Cost Estimate: Dates

In our example, this cost estimate is valid from **01.01.2024** until the end of the period **31.01.2024**. Note also that the **Quantity Structure Date** is **06.12.2023**. This date means that we want the cost estimate to use the BoM and routing that are valid on this date.

The **Valuation Date 06.12.2023** means we want to use the value of the components, activity types, and overhead rates that are valid on this date.

Under the **Qty Structure** tab, shown in Figure 3.20, you can review the quantity structure details (BoM and routing), which were determined automatically based on the costing variant, as so far in this chapter. The **Qty Structure** tab is shown in Figure 3.20.

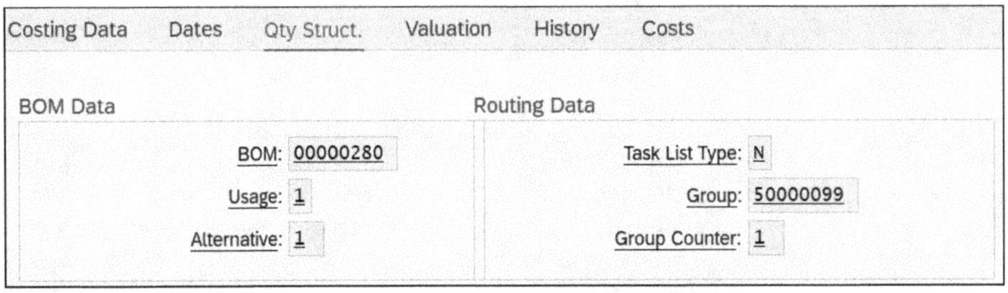

Costing Data | Dates | Qty Struct. | Valuation | History | Costs

BOM Data
BOM: 00000280
Usage: 1
Alternative: 1

Routing Data
Task List Type: N
Group: 50000099
Group Counter: 1

Figure 3.20 Create Material Cost Estimate: Quantity Structure

Note the data in the **BOM** and **Routing** fields. By double-clicking the BoM number, the system will open another transaction displaying the details of the BoM. If you double-click the routing group number in the **Group** field, then the system will display the details about the routing.

Now, we only have the costing sheet determination remaining, which you can see under the **Valuation** tab, as shown in Figure 3.21.

Note the costing sheet determined is **GO**. By double-clicking on **GO**, you can display the details of the costing sheet.

Now, you know how to analyze the different details in your cost estimates to understand where specific cost components and cost itemization numbers are coming from. Once you're satisfied with the results, save the cost estimate version. Then, you can create and save other versions as needed to simulate different BoM, routing, or quantity structure dates.

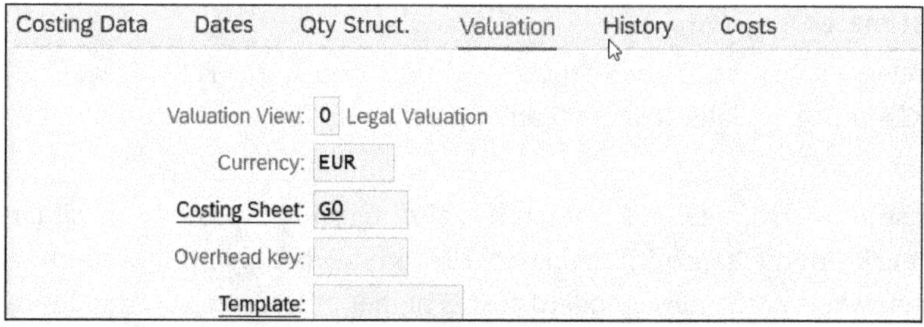

Figure 3.21 Create Material Cost Estimate: Valuation

Once you have an acceptable cost estimate version, you can proceed to the next step—marking the cost estimate for release.

Mark Cost Estimate

Only one cost estimate version can be marked for release. Use either Transaction CK24 (Price Update: Mark Standard Price) for single items or Transaction CK40n (Edit Costing Run) for mass costing.

Once you mark a standard cost estimate, this estimate will be displayed in the material master data, under the **Costing 2** tab, as the future material cost. You can view these details in Transaction MM03 (Display Material Master Data), as shown in Figure 3.22.

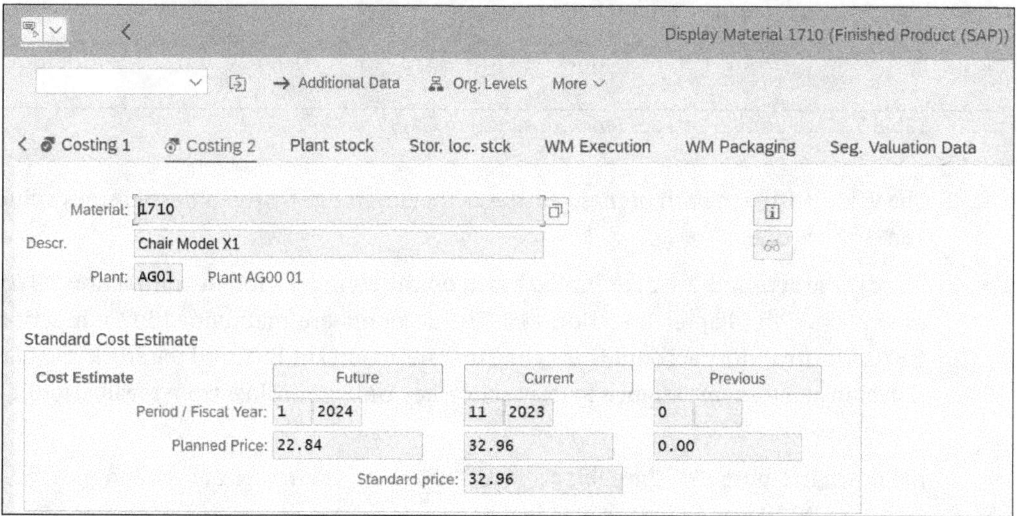

Figure 3.22 Display Material Marked Cost Estimate

So far, no accounting entries have been posted for the stock revaluation because the cost hasn't been released yet. Once the marked cost estimate's validity period starts, you can release the cost estimate.

Release Cost Estimate

You can only release a cost estimate after its validity period starts. For this task, use Transaction CK24 (Price Update: Release Standard Price) or Transaction CK40n (Edit Costing Run).

When you release a cost estimate, an automatic accounting entry is posted to revaluate all the stock available of the item with the new cost estimate. The resulting accounting entry depends on whether the new standard cost is higher or lower than the old standard cost. If the new cost is higher, the accounting entry shown in Table 3.1 will be posted.

Debit	Credit	Debit Amount ($)	Credit Amount ($)
Inventory		1000	
	Stock revaluation gain		1000

Table 3.1 Accounting Entry of Stock Revaluation Gain

If the new cost is lower than the old cost, then the accounting entry shown in Table 3.2 will be posted.

Debit	Credit	Debit Amount ($)	Credit Amount ($)
Stock Revaluation Loss		1000	
	Inventory		1000

Table 3.2 Accounting Entry of Stock Revaluation Gain

The value posted in both of these entries is the difference between the old stock value and the new one.

All of these accounts are determined based on the configuration in Transaction OBYC, as explained in Chapter 1, Section 1.1.2. The accounts are maintained in Transaction OBYC. The inventory account is assigned to transaction key BSX, and the stock revaluation gain and loss are assigned to transaction key UMB (Gain/loss from revaluation), as shown in Figure 3.23.

As shown in Figure 3.23, the debit account **330001** is used in case of loss, and the credit account **411000** is used in case of gain.

With releasing the standard cost estimate, we're done with the first part of the process: product cost planning. Now, we're ready to create production orders and start the actual production process and the cost object controlling.

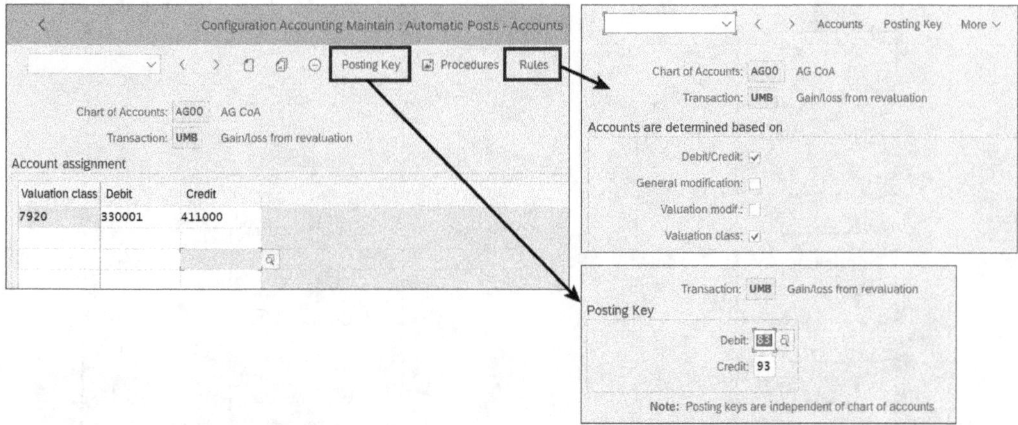

Figure 3.23 Assign Accounts, Rules, and Posting Keys: Stock Revaluation

3.2 Cost Object Controlling

Having delved into the details of product cost planning, our next focus in the SAP financial and controlling area is *cost object controlling*. This crucial component serves as a bridge between planning and actual production, ensuring that the financial aspects of manufacturing align with the initial cost estimates.

Cost object controlling is central to monitoring and managing the actual production costs. It involves tracking and comparing the planned costs, as established in the product cost planning phase, with the actual costs incurred during production. This comparison is vital for identifying variances—the differences between what was expected and what actually happened in terms of spending.

In this section, we'll explore how SAP S/4HANA facilitates this critical comparison. We'll cover how costs are collected, tracked, and allocated to production orders through the various accounting entries that are posted throughout the production process.

Multiple approaches exist for cost object controlling: For example, you can track the actual costs per financial period or per production order. This choice has no impact on account determination, so for our examples, we'll use product costing by order.

The production process is one of the most complex and variable processes in any organization. In this section, we'll focus solely on the main parts of production that are integrated with finance and controlling.

3.2.1 Cost Object Controlling Process Overview

Figure 3.24 shows an overview of the production order process, which is closely integrated with cost object controlling.

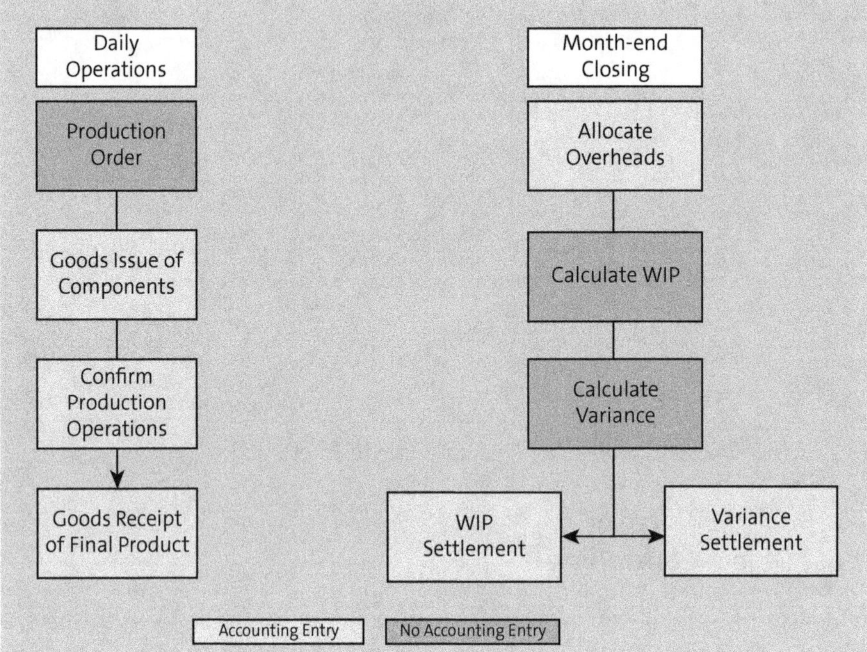

Figure 3.24 Cost Object Controlling Process Overview

In the following sections, we'll explore each step shown in Figure 3.24 and their related accounting entries.

Production Order

A production order is the main cost object we use in product costing by order. Each order has two costing variants assigned to it, which are determined automatically based on our configuration. These costing variants can be accessed via Transaction CO03 (Display Production Order), as shown in Figure 3.25.

In this example, the value **PPP1** in the **PlnCstgVar** field is the planning costing variant that's used to determine the preliminary (planned) cost of the order when the order is saved or released. The value **PPP2** in the **ActCstgVar** PPP2 field is the actual costing variant used to determine the production order's actual costs. The determination of the actual production costs during the production process is called *simultaneous costing*.

Both of these costing variants work the same way as the costing variant used in the standard cost estimate. Planned production order costs are determined using the planning costing variant, and actual costs are determined using the actual costing variant.

The creation of a production order doesn't post any accounting entries. After the order is saved and released, you can move to the following step—issuing components to start the production process.

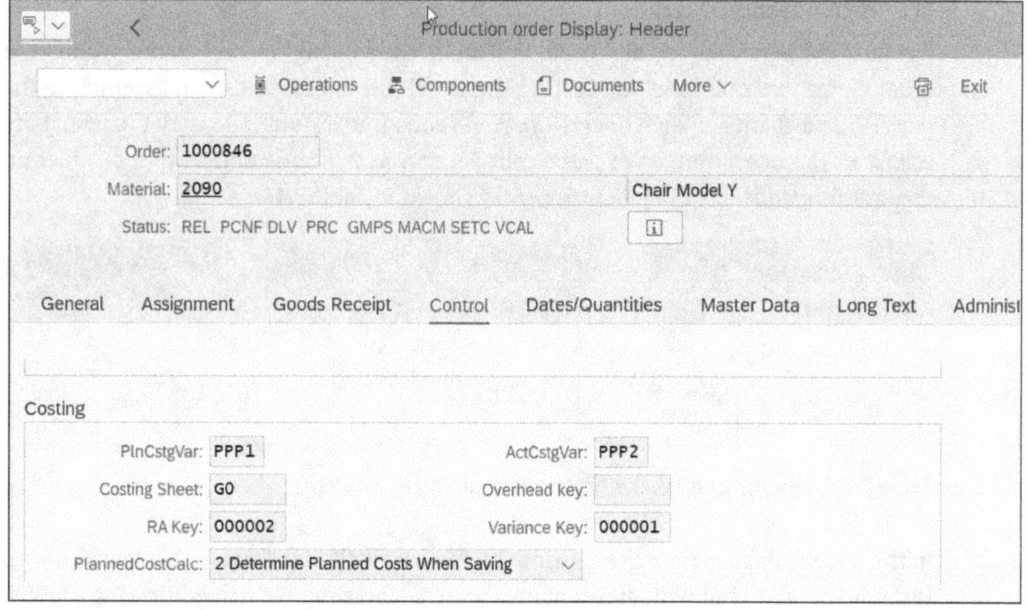

Figure 3.25 Production Order Costing Variants

Good Issue of Components to the Production Order

In this step, you'll issue the components needed for the production process from free stock to the production order. Ideally, we should issue the same components as what is maintained in the BoM, but in real life, it's normal to issue different components based on specific production requirements.

Issuing material components to a production order results in the accounting entry shown in Table 3.3.

Debit	Credit	Cost Object	Debit Amount ($)	Credit Amount ($)
Material consumption		Production order	1000	
	Inventory	N/A (balance sheet)		1000

Table 3.3 Accounting Entry: Issue Material Components to Production Order

The value posted in this accounting entry is equal to the component quantity multiplied by the component cost per unit as per the component master data and whether we are using standard costing or moving average costing for this component. This entry will increase the cost allocated to the production order by the amount posted.

We covered both the material consumption account and the inventory account earlier in Section 3.1.1. After issuing the material components, you move to the next step and perform some production operations.

Confirm Production Operations

Production operations are determined through the routing selected in the production order. Some examples include cutting the wood into pieces and then assembling the pieces into a chair. For each operation, you'll consume different activity types that are linked to the operation, as explained in Section 3.1.2. Confirmation of a production operations results in the accounting entry shown in Table 3.4.

Debit	Credit	Cost Object	Debit Amount ($)	Credit Amount ($)
Activity cost element		Production order	100	
	Activity cost element	Cost center		100

Table 3.4 Accounting Entry: Confirming Production Activities

In this accounting entry, we're posting the same cost element (general ledger account) in debit and credit but with different cost objects: This time, we're crediting the production cost center and debiting the production order.

This internal cost allocation in cost accounting has no impact on financial statements. Thus, the same account is used for both debit and credit. The net accounting impact is ultimately zero, but from a costing point of view, the cost has been allocated from the cost center to the production order.

The activity cost element is determined from the activity type master data, as explained in Section 3.1.2. The value posted in this accounting entry equals the activity quantity multiplied by the activity price. The activity price is determined based on the planning costing variant maintained in the production order.

After confirming the production operation, you can issue more components and perform further production operations as needed until the final product is ready. Then, you can move to the next step—receiving the final product.

Goods Receipt of the Final Product

Once the final product is ready, you can post a goods receipt for the final product from the production order to your stock. This results in the accounting entry shown in Table 3.5.

Debit	Credit	Cost Object	Debit Amount ($)	Credit Amount ($)
Inventory			2000	
	Cost of goods manufactured (COGM)	Production order		2000

Table 3.5 Accounting Entry: Goods Receipt (GR) of Final Product from a Production Order

The amount posted in this accounting entry is equal to the standard cost of the final product multiplied by the quantity received. Note that the standard cost of the final product is the one that was released for the period, as you can see in the material master data. Thus, this value is not related to the actual costs of the production order. Any difference between this standard cost and the actual costs spent will be reported as a production variance. This accounting entry will also remove the cost from the production order and capitalize it into your stock account.

The inventory and COGM accounts used in this accounting entry are determined based on the valuation class of the final product, as explained in Chapter 1, Section 1.1.2. The inventory account is determined from the transaction key BSX, while the COGM is determined based on the transaction key GBB and the general modification AUF, as shown in Figure 3.26, which displays Transaction OBYC (Maintain Account Determination).

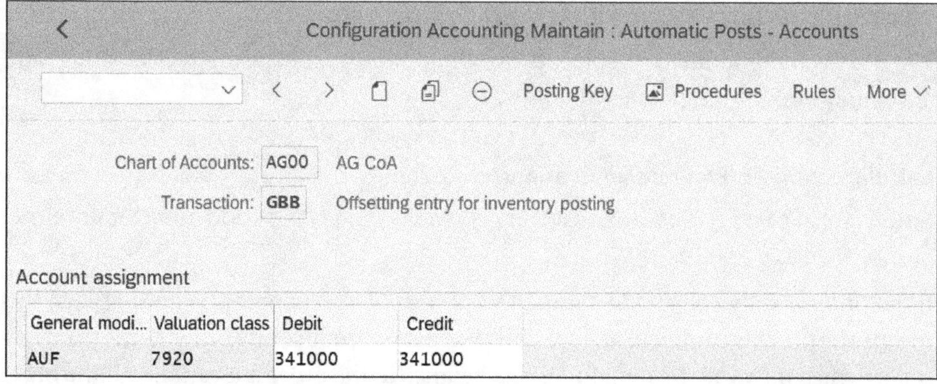

Figure 3.26 Account Determination of COGM Account

The COGM account must be created as a primary cost element with the cost category 1 (Primary costs/cost-reducing revenues). At this point, you've seen the daily operation that impact the cost object controlling. Let's look into month-end activities next.

Allocate Overhead

The overhead costs that are calculated using costing sheets are not posted automatically to a production order. To allocate the overhead cost to production orders, use Transaction KGI2 (Actual Overhead Calculation) for a single order, or for multiple orders, use Transaction CO43. Both transactions calculate the overhead based on the costing sheet assigned to the actual costing variant in the production order (shown earlier in Figure 3.25) and allocates the overheads to production orders through posting the accounting entry shown in Table 3.6.

Debit	Credit	Cost Object	Debit Amount ($)	Credit Amount ($)
Overhead cost element		Production order	150	
	Overhead cost element	Cost center		150

Table 3.6 Accounting Entry: Allocate Actual Overhead to Production Orders

In this accounting entry, both the debit and credit are posted to the same overhead allocation cost element (general ledger account). This entry will move the costs from the overhead expenses cost center to the production order.

The value posted in this accounting entry is calculated based on the costing sheet allocation bases and allocation rates, and the overhead cost element is determined through the costing sheet credits as explained in Section 3.1.3.

The next activity in the cost object controlling month end closing is calculating variances and work-in-progress (WIP).

Calculate Work-in-Progress and Variance

At the end of every period, any cost balance available in production orders must either be classified as variances or WIP.

If the order is fully delivered (status DLV) or technically completed (status TECO) this means the order is complete and we don't expect to produce any other items from this order, which mean all the remaining cost balance in the order is a variance. If the order has any other status, this means the order is still open and still in process, so any cost balance at the end of the period is considered WIP.

To calculate the production order variance, we use the transaction Variance Calculation individual processing (Transaction KKS2) or collective processing (Transaction KKS1). To calculate the WIP we use the transaction WIP calculation individual processing (Transaction KKAX) or collective processing (Transaction KKAO). The calculation doesn't post any accounting entry or impact the order cost, it's a prerequisite for the settlement step where the entries will be posted.

Both the variances and WIP for the different orders are posted using the same transaction: Order Actual Settlement individual processing (Transaction KO88) or collective processing (Transaction CO88). The transaction will post the WIP or variance calculated in the calculation transaction.

Variance Settlement

In this step, we settle the production orders that have calculated variances from the previous step. This transaction will take all the remaining actual cost balance in the production order and post it to variance accounts allocated to margin analysis. The accounting entry is shown in Table 3.7.

Debit	Credit	Cost Object	Debit Amount ($)	Credit Amount ($)
	COGM	Production order		1,000
Price difference		Profitability segment	1,000	
	Price difference	Profitability segment		1,000
Variance category 1			200	
Variance category 2			500	
Variance category 3			300	

Table 3.7 Accounting Entry: Order Variance Settlement with Splitting

This accounting entry removes the cost remaining in the order using the COGM account in the first line and posts this amount as a price difference in margin analysis (profitability analysis) in the second line. This entry is the main accounting entry; the other lines are optional.

In SAP S/4HANA, you usually use accounting-based profitability analysis or margin analysis, in which case you should split the price difference account based on the variance category to be able to understand where the variance is coming from. A similar concept applies to a COGS split, as described in Chapter 2, Section 1.1.3. This feature is called *production variances splitting*. If you configure this feature, then the entry in Table 3.7 will include the additional lines from line 3 onwards. The price difference in the third line will set off the price difference in the second line, and this price difference value is split between different variance accounts as per our configuration. We'll explain this configuration in detail after the business process overview.

Now, let's look at the accounts that form the accounting entry shown in Table 3.7:

- **First line: COGM account**
  - Configure this account as described earlier in this section when we covered the GR of the final product.
  - The value posted is the actual cost balance of the production order being settled.
  - This line will credit all the cost from the production order.
  - This cost element is a primary cost element of category 1.

- **Second line: Price difference account**
  - Configure this account as described earlier in this section when we covered releasing the cost estimate.
  - The value posted is also equal to the actual cost balance of the production order being settled.
  - This line will debit all the variances to a profitability segment (margin analysis).
  - This cost element is a primary cost element of category 1.
- **Third line: Price difference**
  - This line offsets the second line, so it uses the same account, the same value, and the same profitability segment.
- **The remaining lines: Variance accounts**
  - You can assign an account for each production variance category in the configuration (explained in the next section).
  - The value posted to each variance account is equal to the variance value related to this variance category as calculated by the system.
  - These variance accounts will debit the variances to a profitability segment (same segment as the second and third posting lines).
  - These cost elements are primary cost elements of category.

Now that you understand the accounting impact of variance settlements, let's look at WIP settlement.

Work-in-Progress Settlement

In this step, we'll settle the orders that have WIP calculated two steps ago when we calculated the WIP and the variance. These orders have specific statuses showing that the order is not delivered, which means that the expenses posted to the order are related to WIP. The settlement will credit your profit and loss (P&L) account and capitalize the WIP value into the balance sheet. The accounting entry is shown in Table 3.8.

Debit	Credit	Cost Object	Debit Amount ($)	Credit Amount ($)
WIP inventory – element 1			1000	
	Inventory change WIP – element 1	Production order		1000
WIP inventory – element xx			200	

Table 3.8 Accounting Entry: Order WIP Settlement with Splitting

Debit	Credit	Cost Object	Debit Amount ($)	Credit Amount ($)
	Inventory change WIP – element xx	Production order		200

Table 3.8 Accounting Entry: Order WIP Settlement with Splitting (Cont.)

In this entry, the WIP inventory is posted to the balance sheet account, and the inventory change WIP is posted to the P&L account. In our configuration, we decided whether to post all WIP to the same WIP inventory and WIP change accounts, or we can specify specific accounts based on the source of the WIP. For example, you can have different WIP accounts for raw materials, machine cost, labor cost, overheads, and so on, or you can group them as needed depending on the level of detail you need in financial accounting. We'll explain this configuration in detail in the next section after the business process.

In every period when we run order settlement, the system checks the order status and the WIP posted in previous periods and then proceeds according to the following conditions:

- If the order is subject to WIP (not yet delivered or technically completed), then any changes in the WIP value is settled.
- If the order is no longer subject to WIP, then the previously posted WIP is reversed.

Now, let's look at the nature of these accounts and the values posted earlier in Table 3.8:

- **WIP inventory**
  - This account is a balance sheet account.
  - The value posted is equal to the order's WIP capitalizable value. Note that we're not obligated to capitalize all the production order's balance as WIP. You can choose, in the configuration, which costs are capitalizable. We'll return to this topic in Section 3.2.2.
  - If you're splitting the WIP accounts based on the source (e.g., material, activity, etc.), then you'll have several lines with different WIP inventory accounts as configured.
- **Inventory change WIP**
  - This account is a P&L account that should be created as a non-operating expense or as a revenue account because the WIP value shouldn't be credited from the order cost. The objective of WIP capitalization is to correctly represent these values in your financial statements in accordance with financial standards, but these values shouldn't impact your own internal cost reports.
  - The value posted is equal to the value posted to the WIP inventory line.
  - You can also split the inventory change WIP account based on the WIP source, or you can post all the WIP to a single inventory change account.

At this point, we've covered all the process steps shown earlier in Figure 3.24 and we've seen the accounts and the accounting entries posted in the various steps. Now, let's switch to the configuration side and learn how to configure the account determinations for these accounts.

We've already explained the account determination configurations for most of the accounts mentioned in this section. For example, the accounts determined for material consumption, manufacturing activities, and overhead were all covered in earlier processes. The remaining accounts to explain are the settlement accounts for production variances and for WIP.

3.2.2 Work-in-Progress Settlement Accounts

We introduced you to the WIP settlement accounts in the previous section, as shown earlier in Table 3.8. All the configuration steps related to the determination of these accounts can be found in the SAP configuration menu shown in Figure 3.27.

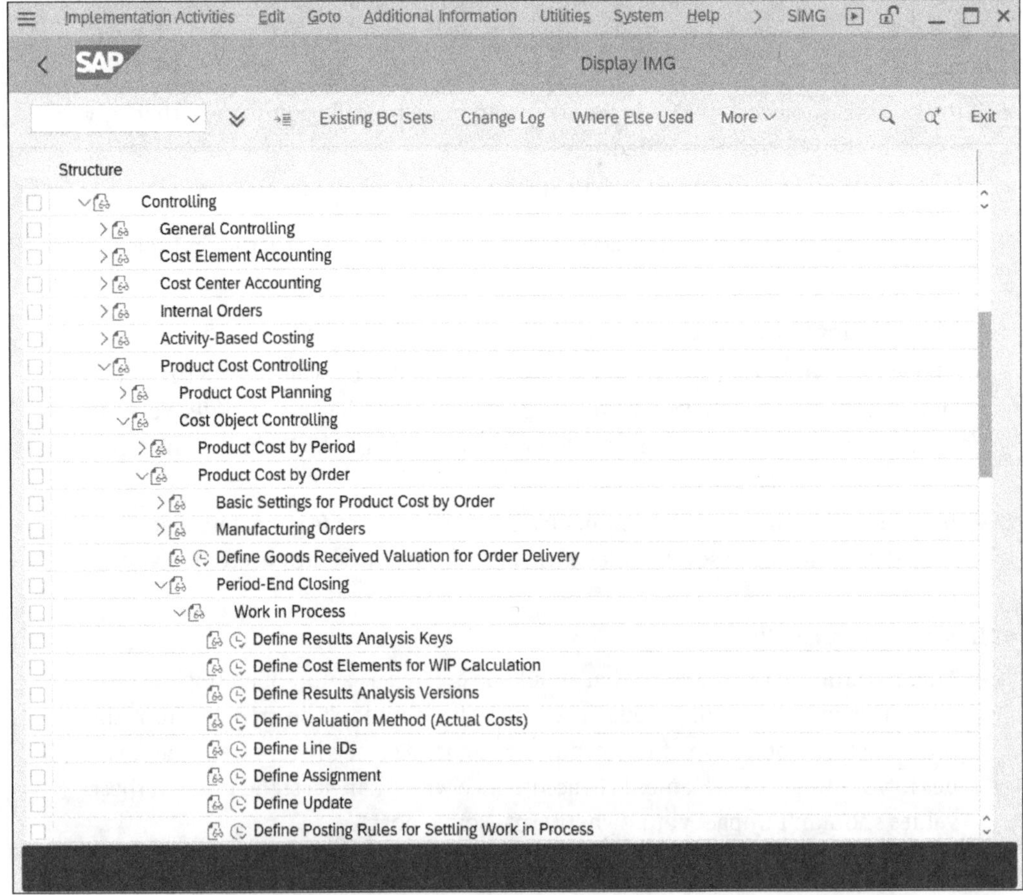

Figure 3.27 WIP Settlement Accounts Configuration Menu

Now, let's walk through each of these configuration steps to learn how WIP settlement accounts are determined.

Note that results analysis (RA) is one of the most complicated tasks in product costing and includes many options to fit specific industries and business scenarios. Our focus will be on options that impact the account determination of the WIP settlement accounts shown earlier in Table 3.8.

Define Results Analysis Keys

The RA key is used in the following configuration steps and is maintained in the header of any production order that we want to include in WIP calculation. This key is the link between the production orders and our WIP configuration.

The definition of the RA key is just a key and a description, as shown in Figure 3.28.

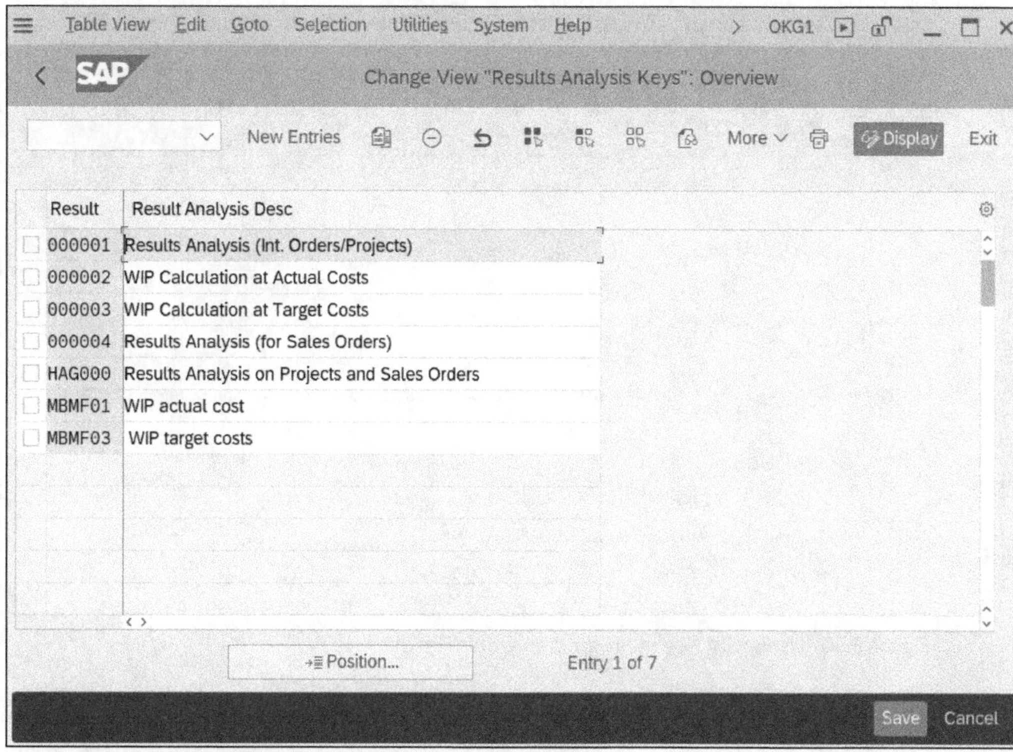

Figure 3.28 WIP Settlement: Define Results Analysis Key

The RA key can be determined by default when a production order is created. This determination is based on the order type and the plant. You can maintain this setting by following the menu path **IMG • Controlling • Product Cost Controlling • Cost Object Controlling • Product Cost By Order • Manufacturing Orders • Define Cost-Accounting-Relevant Default Values for Order Types and Plants,** as shown in Figure 3.29.

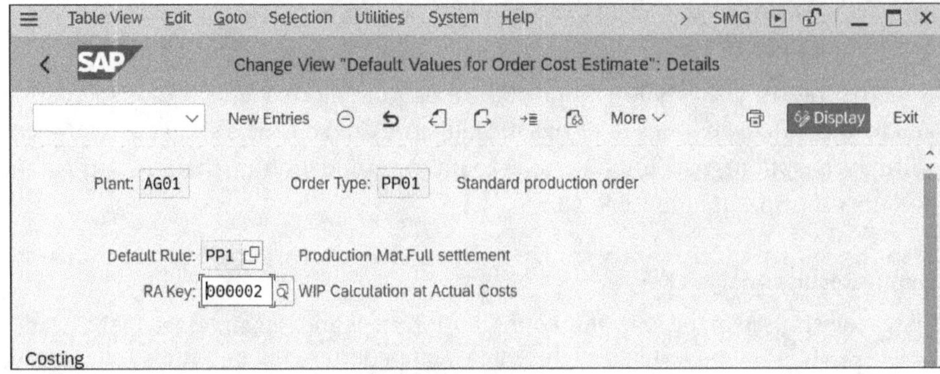

Figure 3.29 Default RA Key in Plant and Order Type

This RA key can be reviewed in the production order header, which can be displayed through Transaction CO03 (Production Order Display), as shown in Figure 3.30.

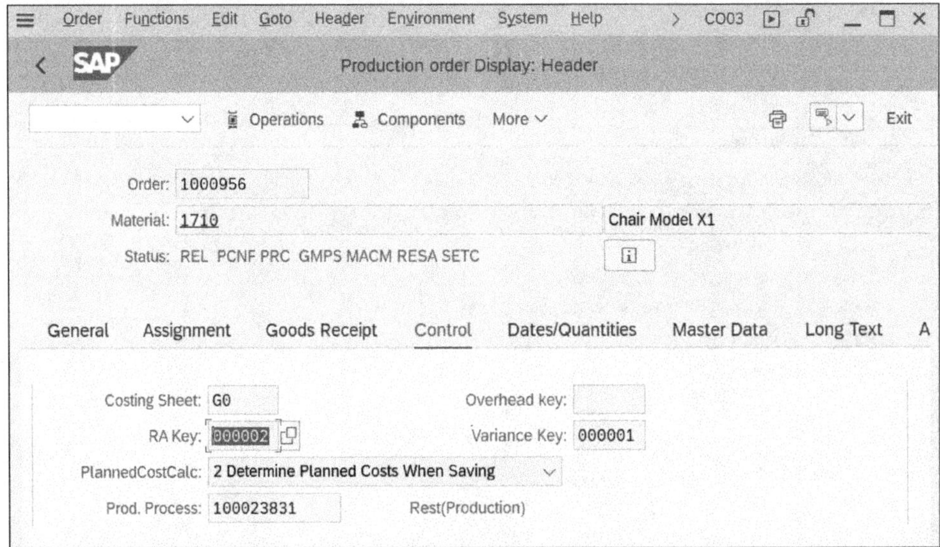

Figure 3.30 Display RA Key in Production Order Header

After creating the RA key, proceed with the next configuration step, as shown earlier in Figure 3.27.

Define Results Analysis Versions

RA versions are defined at the level of each controlling area and include extremely important parameters for WIP calculation and settlement. You can define multiple RA versions and have a different WIP calculation method for each RA version. You can also determine, on the level of each RA version, which cost elements should be capitalized in WIP. You can also run WIP calculations for multiple versions simultaneously.

If you want a specific RA version to be settled to financial accounting (with postings of accounting entries), then you must select the **Transfer to Financial Accounting** checkbox, as shown in Figure 3.31.

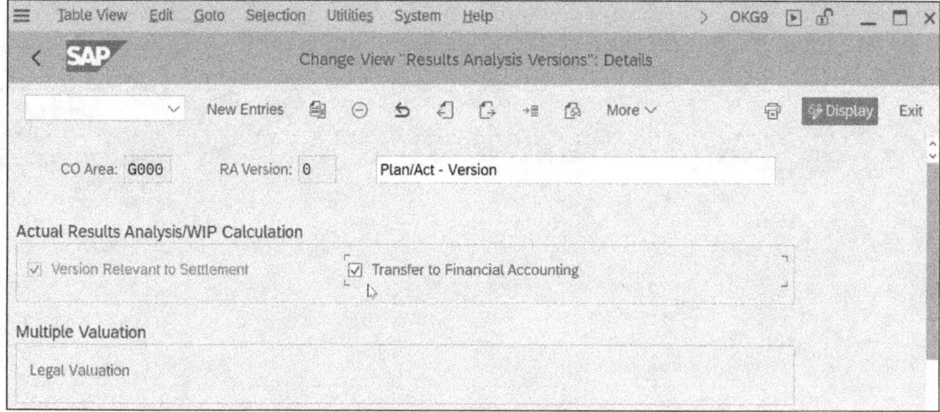

Figure 3.31 WIP Settlement: Define Results Analysis Version

Now that we've defined some RA versions, let's proceed with the next step.

Define Valuation Method (Actual Costs)

In this step, we'll create a link between the controlling area, the RA key, the RA version, and the production order system status, as shown in Figure 3.32.

Table View Edit Goto Selection Utilities System Help — OKGC

Change View "Valuation Method for Work in Process": Overview

New Entries More Display Exit

CO Area	RA Version	RA Key	Status	Status Number	RA Type
G000	0	000002	REL	2	WIP Calculation on Basis of Actual Costs
G000	0	000002	DLV	3	Cancel Data of WIP Calculation and Results Ana
G000	0	000002	PREL	1	WIP Calculation on Basis of Actual Costs
G000	0	000002	TECO	4	Cancel Data of WIP Calculation and Results Ana

Figure 3.32 WIP Settlement: Define Valuation Method

Now, you'll maintain the **RA Type** field for each order status. For example, WIP is calculated for orders with status REL, and when the order is in status DLV, the WIP is reversed.

After maintaining the valuation method, you can proceed to the next step.

Define Line IDs

Through line IDs, you can group the WIP and reserves for unrealized costs according to the requirements of financial accounting. In these steps, you'll assign the source cost elements to each line ID.

The definition of a line ID is just an ID and a description, as shown in Figure 3.33.

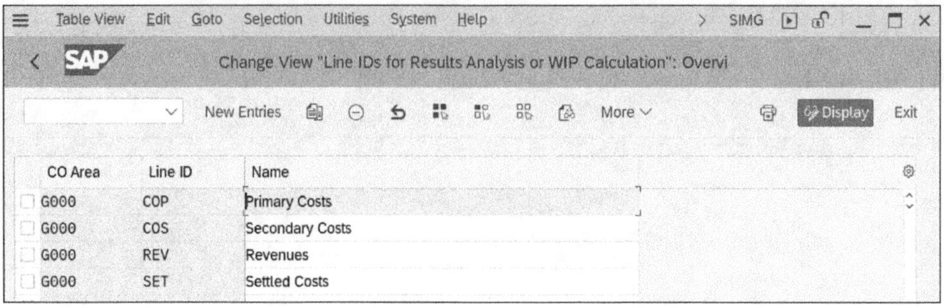

CO Area	Line ID	Name
G000	COP	Primary Costs
G000	COS	Secondary Costs
G000	REV	Revenues
G000	SET	Settled Costs

Figure 3.33 WIP Settlement: Define Line IDs

After defining your line IDs, proceed to the next step.

Define Assignment

In this step, you'll assign source cost elements to line IDs and specify which cost elements must be capitalized, are optionally capitalized, or prohibited from capitalization entirely, as shown in Figure 3.34.

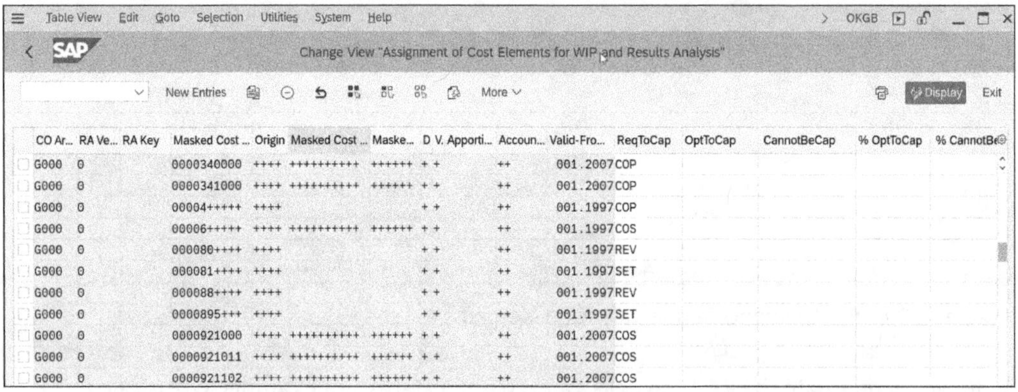

CO Ar...	RA Ve...	RA Key	Masked Cost ...	Origin	Masked Cost ...	Maske...	D	V.	Apporti...	Accoun...	Valid-Fro...	ReqToCap	OptToCap	CannotBeCap	% OptToCap	% CannotB…
G000	0		0000340000	++++	++++++++++	++++++	+	+		++	001.2007	COP				
G000	0		0000341000	++++	++++++++++	++++++	+	+		++	001.2007	COP				
G000	0		00004+++++	++++			+	+		++	001.1997	COP				
G000	0		00006+++++	++++	++++++++++	++++++	+	+		++	001.1997	COS				
G000	0		000080++++	++++			+	+		++	001.1997	REV				
G000	0		000081++++	++++			+	+		++	001.1997	SET				
G000	0		000088++++	++++			+	+		++	001.1997	REV				
G000	0		0000895+++	++++			+	+		++	001.1997	SET				
G000	0		0000921000	++++	++++++++++	++++++	+	+		++	001.2007	COS				
G000	0		0000921011	++++	++++++++++	++++++	+	+		++	001.2007	COS				
G000	0		0000921102	++++	++++++++++	++++++	+	+		++	001.2007	COS				

Figure 3.34 WIP Settlement: Define Assignment

Any field you see containing **++++** is masked, which means that this assignment applies to all the values. You can also use the masking to insert all the cost elements starting with "6" by entering "6++++."

The cost elements we insert now are the source cost elements that post costs to the production orders, such as the material consumption cost elements and the activity type cost elements. You must make sure that all the source cost elements used in production are assigned on this screen.

For each line in Figure 3.34, insert a line ID in one of the fields: either **ReqToCap**, **OptToCap**, or **CannotBeCap**. You can also maintain the percentage that should be capitalized. If the **% OptToCap** field is left empty, 100% of the value will be capitalized.

Now that you've mapped source cost elements to line IDs, let's proceed to the next step.

Define Update

Now, we'll maintain the technical WIP settlement cost elements for each line ID and cost element category, as shown in Figure 3.35.

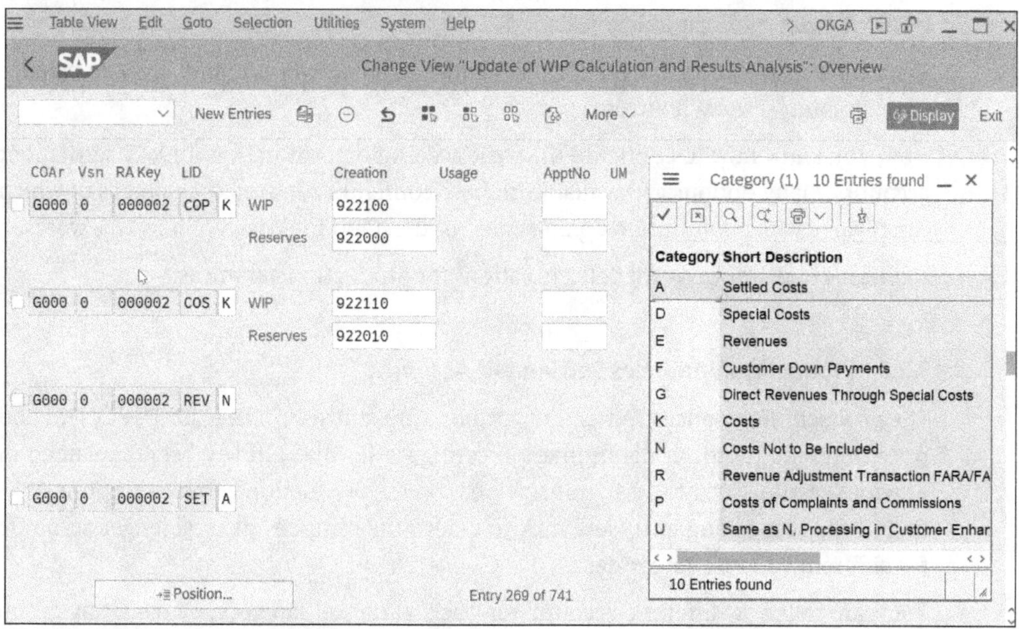

Figure 3.35 WIP Settlement: Define Update

In our example, notice that, for controlling area **G000**, RA version **0**, RA key **000002**, line ID **COP**, and category **K**, we've assigned the cost element 922100 for WIP and 922000 for reserves. These cost elements are secondary cost elements with the cost element category 31. In the next step, we'll map these cost elements to the WIP settlement general ledger accounts.

Define Posting Rules for Settling Work-in-Progress

In this step, you'll assign the WIP settlement general ledger accounts, as shown in Figure 3.36.

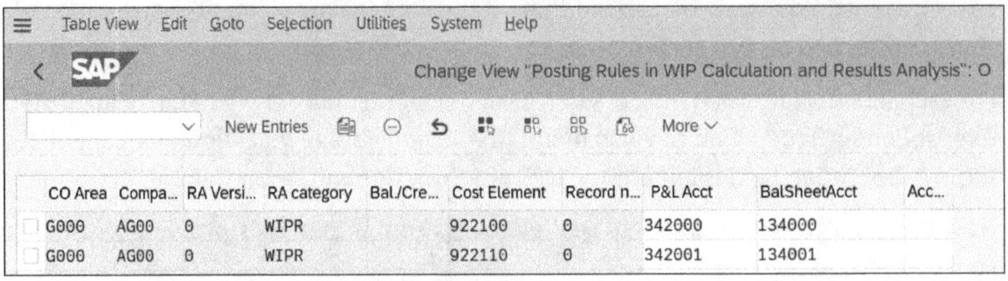

CO Area	Compa...	RA Versi...	RA category	Bal./Cre...	Cost Element	Record n...	P&L Acct	BalSheetAcct	Acc...
G000	AG00	0	WIPR		922100	0	342000	134000	
G000	AG00	0	WIPR		922110	0	342001	134001	

Figure 3.36 WIP Settlement: Define Posting Rules

In this step, you'll assign the WIP settlement general ledger accounts to a combination of controlling area, company code, RA version, RA category, and technical WIP settlement cost elements (which we configured in the previous step). The RA category WIPR is for WIP that must capitalize costs.

The P&L account you see here is the inventory change WIP account, and the balance sheet account is the WIP inventory account.

Now, you know how to configure the basic account determination for WIP settlement accounts. Many options are available in these configuration steps, the details of which can fill an entire book, but now, you should understand the basics of how this works.

Next, let's look into account determinations for production variances.

3.2.3 Production Variances Settlement Accounts

The production variances settlement accounts are shown in Table 3.7. The COGM and price difference account determination configuration has already been explained in previous sections. Our focus in this section is on the production variance account splitting in the accounting entry and how to determine different general ledger accounts for different variance categories.

You can assign a different account for each variance category, or you can group together multiple categories with the same account. Follow the menu path **Financial Accounting • General Ledger Accounting • Periodic Processing • Integration • Materials Management • Define Accounts for Splitting Price Differences.**

We'll start by defining a price difference splitting profile. Inside the profile, you'll assign different accounts to variance categories, as shown in Figure 3.37.

In this example, as shown in Figure 3.37, we've assigned a different account for each variance category. A mandatory step is to choose a default account assignment that the system will use whenever it can't determine a splitting account for any reason. Indicate the default assignment by selecting the **Default** checkbox. In our example, the default assignment is line **0010**.

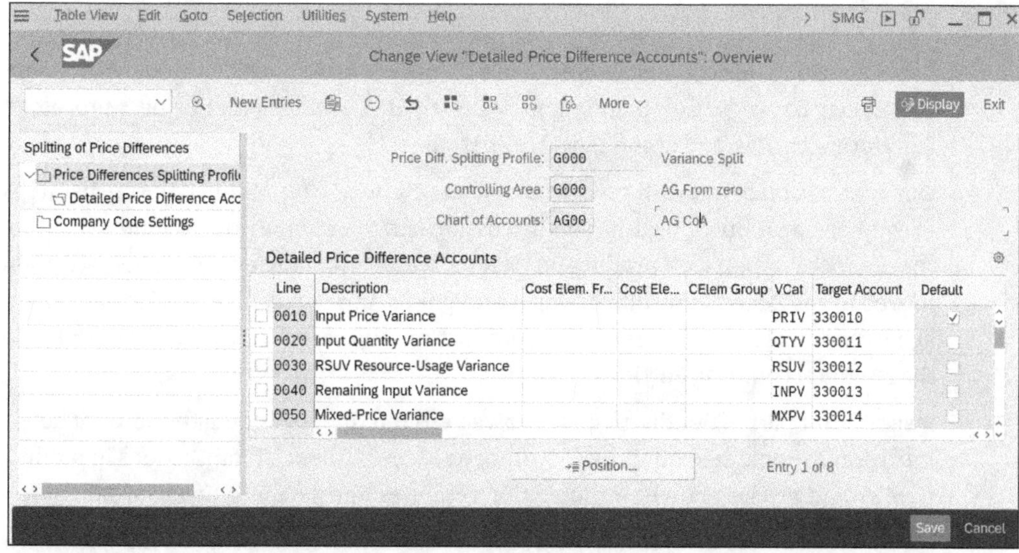

Figure 3.37 Price Difference Splitting Profile Account Assignment to Variance Categories

You can also determine the splitting accounts based on cost elements. In this case, maintain the **Cost Element From** and **Cost Element To** fields or the **Celem Group** field. The cost elements that should be inserted in these fields are the original cost elements that post expenses the production order, such as the material consumption, activity, and overhead cost elements.

After defining the splitting profile, you can assign it to one or more company codes, as shown in Figure 3.38.

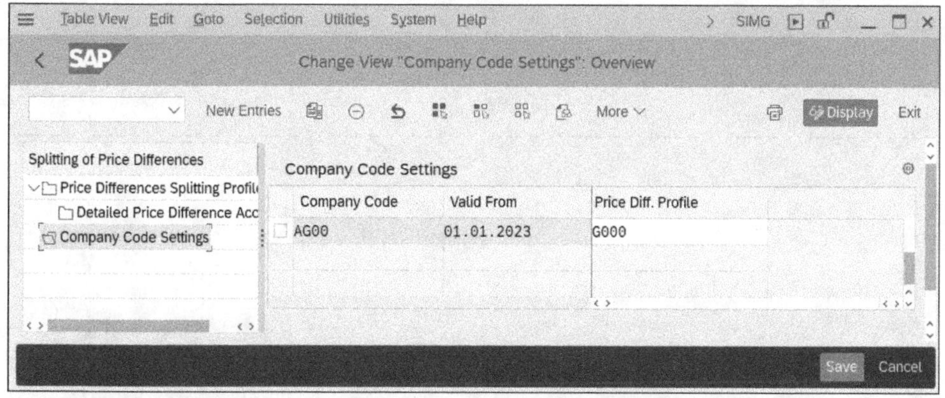

Figure 3.38 Price Difference Splitting Profile Company Code Assignment

Now, you understand how to configure SAP to split the price difference from production variances to different accounts based on the variance category.

3.3 Summary

As we conclude this chapter, we've taken a comprehensive journey through the essential aspects of production planning in SAP S/4HANA, focusing on business processes, accounting entries, and account determination.

Our exploration began with product cost planning, where we outlined the crucial steps of estimating production costs. This discussion laid the groundwork for understanding the financial aspects of production before actual manufacturing begins. We then moved to cost object controlling, which brought us into the realm of tracking and managing the actual costs incurred during production, and we provided a detailed look into the related accounting entries.

A significant part of our discussion revolved around understanding the account determination. We analyzed how SAP allocates and records financial transactions in production, providing clarity on how production activities translate into financial data.

We covered complexities of handling production variances, focusing on the settlement of these variances and the configuration of variance accounts. Our discussion included a detailed analysis of how to configure the system to manage different general ledger accounts for various variance categories, thereby ensuring accurate financial reporting and compliance.

The objective of this chapter was to equip you with an understanding of the business processes within SAP S/4HANA's production planning and product costing, along with the necessary knowledge to navigate its accounting entries and account determination. Our hope is that this detailed exploration will enhance your ability to effectively manage and integrate production planning with financial operations in SAP S/4HANA.

Chapter 4

Accounts Receivable and Accounts Payable

In this chapter, we'll explain the main business processes in the accounts receivable and accounts payable areas that have an impact on accounting. We'll discuss important fields in the general ledger account master data and provide instructions for configuring automatic determination for all accounting entries.

Within financial accounting, every company commonly utilizes two primary subledgers: accounts receivable (AR) and accounts payable (AP). AR handles all customer transactions, facilitating sales and revenue generation. Conversely, AP manages transactions with vendors and suppliers, contributing to cost of goods sold (COGS) and other operational expenses. In SAP S/4HANA, both subledgers are not only seamlessly integrated with the general ledger but also intricately linked with operational modules like materials management and sales and distribution, as discussed in Chapters 1 and 2.

Since the switch from SAP ERP to SAP S/4HANA, the technical concept of AR and AP has changed. SAP S/4HANA has subsumed AP and AR under the *business partner* category. The business partner is a central concept used to manage information about individuals and organizations that are involved in business transactions. SAP provides a comprehensive way to store and manage data about customers, vendors, suppliers, contacts, and other entities in a unified manner through the business partner concept.

In this chapter, we investigate the integration of business partner transactions into the general ledger via reconciliation accounts. We examine configuring value-added taxes (VATs) in sales and purchase transactions, explore both manual and automatic payment processing methods, discuss the unique SAP-specific goods receipt/invoice receipt (GR/IR) account, and cover additional topics like withholding taxes and interest calculations.

4.1 Reconciliation Accounts

The *reconciliation account* is an important concept used to manage and monitor financial transactions between a main account and its associated subledger accounts. In theoretic accounting, it is also commonly referred to as a *control account*.

A reconciliation account is a balance sheet general ledger account that serves as a summary account or control account. It represents a category of transactions, such as AR or AP, at the general ledger level. Underneath the reconciliation account, there are subledger accounts that hold detailed transaction data (see Figure 4.1).

In the subsequent sections, we will uncover the mechanics of business partner integration and the foundational principles of its configuration. We also will scrutinize how reconciliation accounts are assigned to business partners.

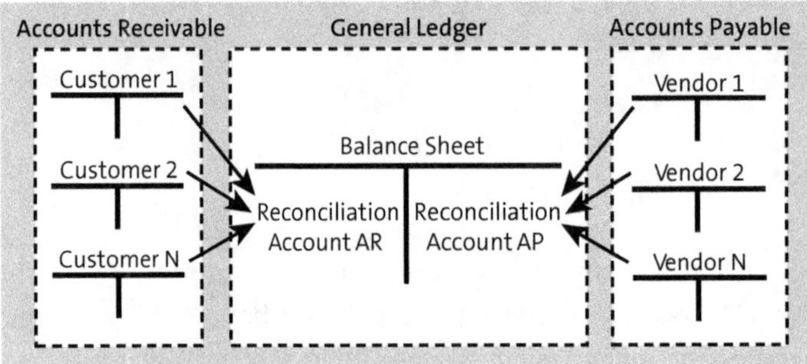

Figure 4.1 Simplified Visualized Reconciliation Account Concept

4.1.1 Concept Overview

Typically, each subledger requires more than one reconciliation account. For instance, for AR, you probably would set up separate reconciliation accounts to distinguish receivables from external customers and internal, affiliated companies. Even for external customers, you may want to separate receivables from domestic customers and receivables from foreign customers.

Prior to initiating any accounting transactions within AR or AP associated with a customer or vendor account, you must assign a reconciliation account in the respective master data of the customer or vendor first.

4.1.2 Create Reconciliation Accounts

You can convert any regular general ledger balance sheet account to a reconciliation account by using Transaction FS00 or following SAP menu path **Accounting • Financial Accounting • General Ledger • Master Records • G/L Accounts • Individual Processing • Centrally.** You first go to the **Type/Description** tab, where you categorize the account as a **Balance Sheet Account** under the **G/L Account Type** dropdown, for account group **Recon.account AP/AR**, selected under the **Account Group** dropdown, as shown in Figure 4.2.

Then you need to specify in the **Control Data** tab what type of reconciliation account it will be (under the **Recon. Account for Acct Type** dropdown). You can choose **Assets**, **Customer**, **Vendors**, or **Contract Accounts Receivable**, as shown in Figure 4.3. Specifying the account type and the reconciliation account type is highly recommended as doing so contributes to achieving consistent settings for and documentation of reconciliation accounts.

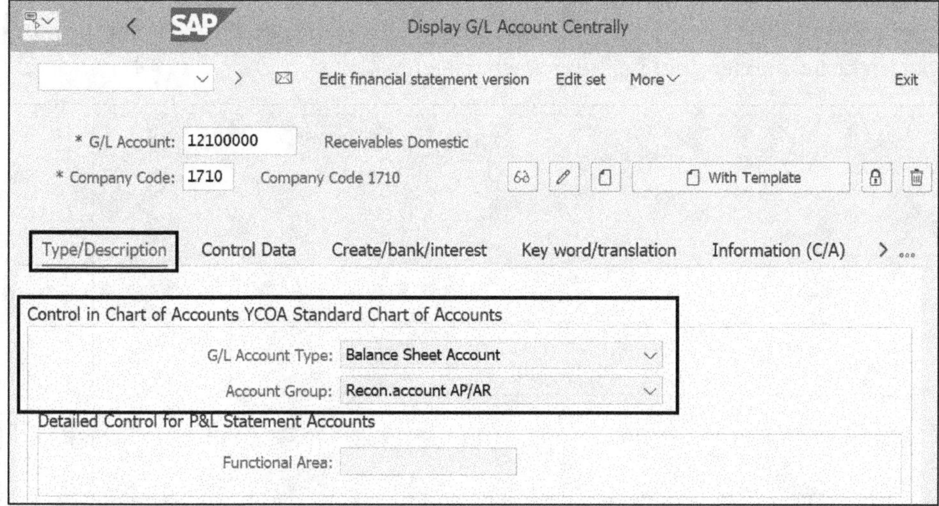

Figure 4.2 Type/Description of Reconciliation Account

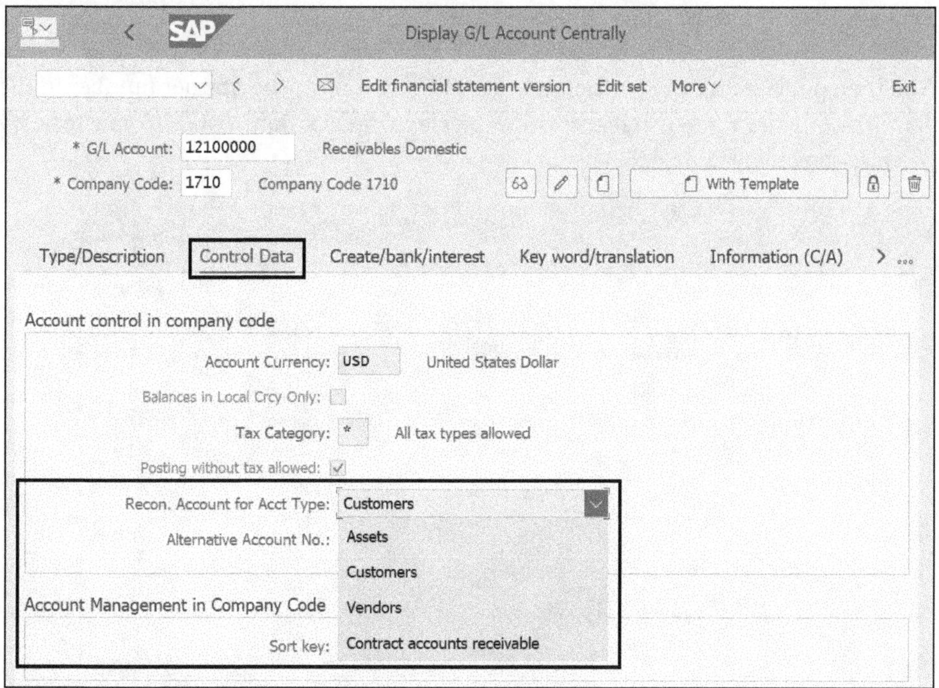

Figure 4.3 Control Data to Specify Reconciliation Account Type

4.1.3 Assign Reconciliation Accounts to Business Partners

Once you have created the necessary reconciliation account, the next step is to assign it to a business partner in the role of a customer or a vendor account. To do so, you first have to access the relevant business partner using Transaction BP or through the following path in the SAP menu: **Logistics • Sales and Distribution • Credit Management • Master Data • Business Partner Master Data.** If you know the number of the business partner who will want to access, select **Number** in the **By** dropdown and enter the number in the **BusinessPartner** field, as shown in Figure 4.4.

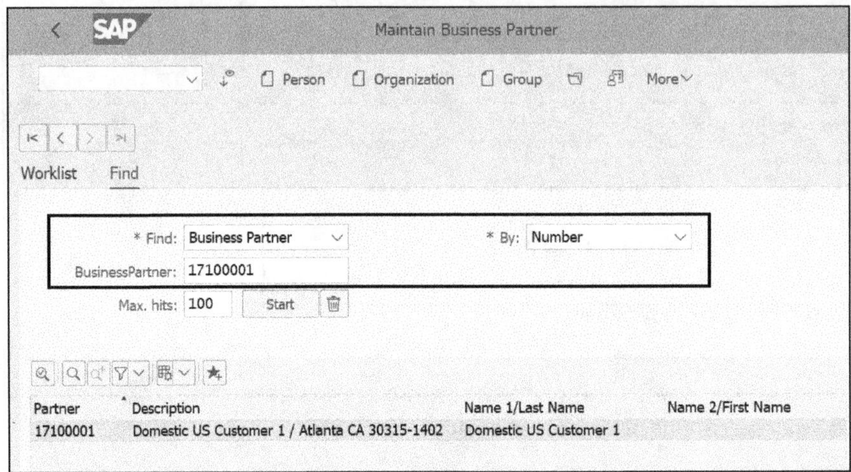

Figure 4.4 Access and Maintain Business Partner

You then double-click the partner number or in the line of the partner number in the lower half of the screen, and access the **Display Organization: 17100001** master data view, as shown in Figure 4.5.

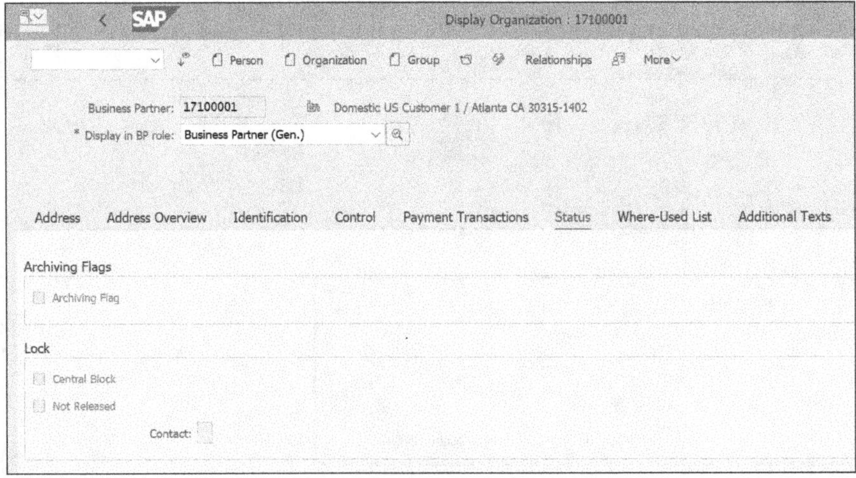

Figure 4.5 Business Partner Display Organization

Next, you change the role in **Display in BP Role** to **FI Customer**, as shown in Figure 4.6. You can click **Company Code** to see all company code–relevant data, such as the reconciliation account, as shown in Figure 4.7.

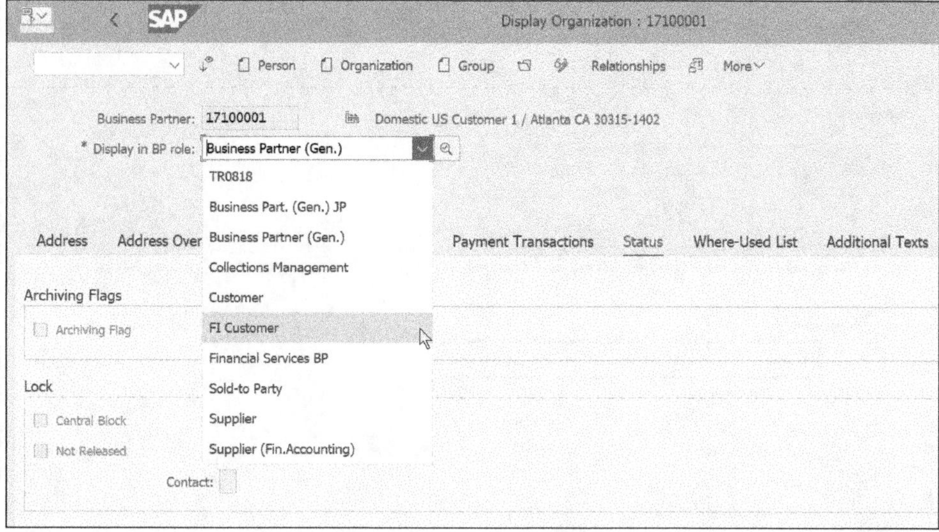

Figure 4.6 Select FI Customer to Access Customer-Specific Data for Relevant Business Partner

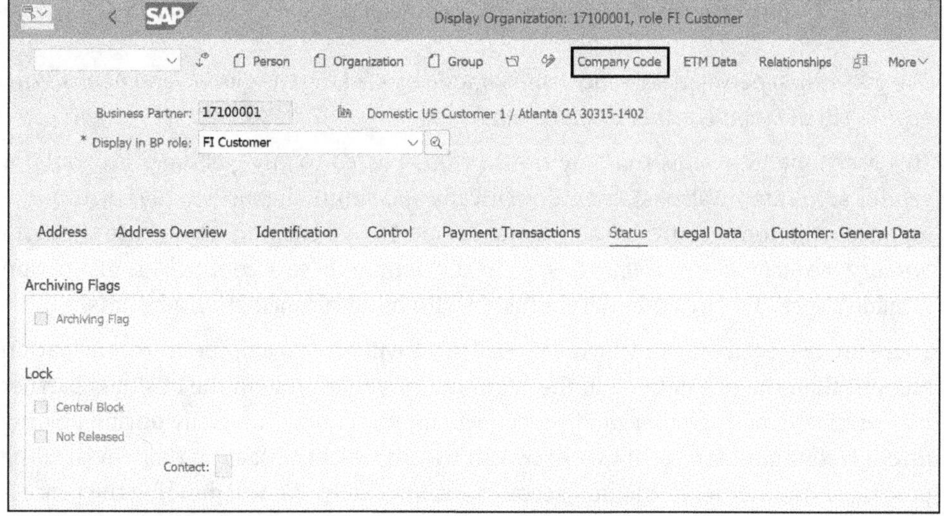

Figure 4.7 Display Organization: 1700001, Role FI Customer

After clicking **Company Code**, you will see the assigned reconciliation account for business partner 17110001 in the role of a **Customer** in the **Customer: Account Management** tab, and the **Reconciliation Acct** field (see Figure 4.8).

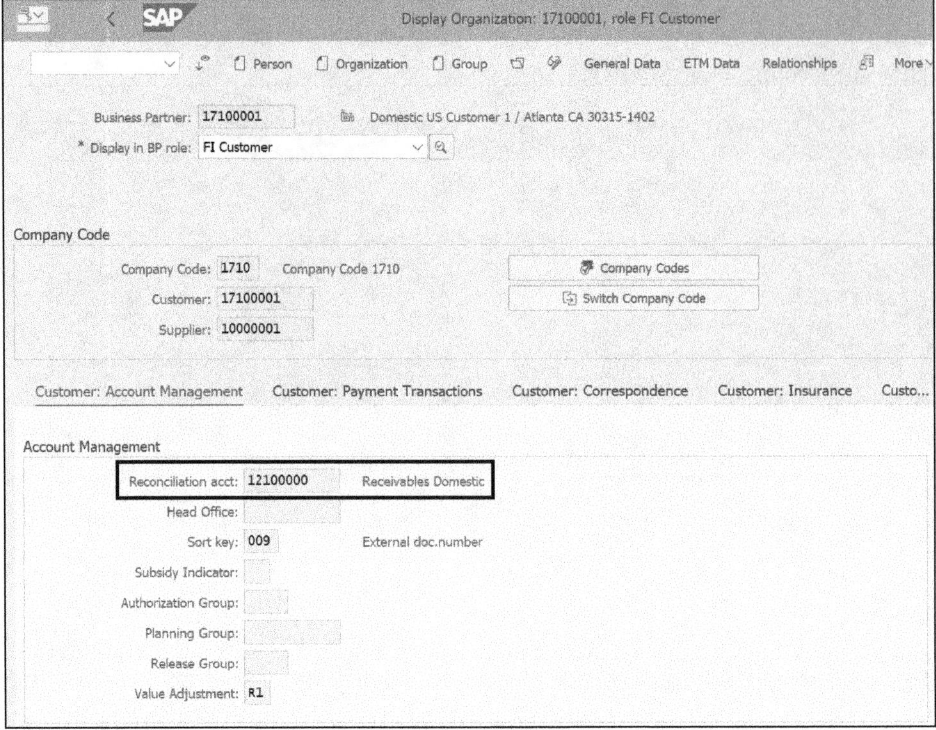

Figure 4.8 Display Reconciliation Account of Customer

You can switch between edit and display mode by clicking the glasses and pencil symbol ( ) in the menu header of the screen.

This assignment ensures that any transactions posted to any customer accounts or vendor accounts are also posted automatically and simultaneously to the corresponding reconciliation account in the general ledger. In this chapter, we will use the term *primary reconciliation accounts* to specifically denote these accounts. This distinction is made to prevent any potential confusion with other reconciliation accounts.

There are two scenarios in which SAP S/4HANA will post to a different reconciliation account than the one defined in the customer or vendor role in the business partner master data. You can either modify reconciliation accounts manually during posting through configuration settings, or you can use a special transaction that deviates the posting to another reconciliation account automatically. We will describe the process for setting up an alternative reconciliation account in the very common case of handling down payments in the following section.

4.2 Down Payments and Down Payment Requests

Across various sectors, making a preliminary payment when placing an order is a prevalent procedure. In these situations, preliminary payments are called *down payments*.

These payments. This could entail your firm making payments to suppliers, or you could be handling down payments received from customers. In both situations, you have the option to execute down payment procedures within SAP S/4HANA, choosing between the net method and the gross method.

When utilizing the *net method* for down payment processing, the posting to the business partner's account, whether it's a customer or a vendor, does not factor in the tax amount. Conversely, if down payments are managed through the *gross method*, the posting to the business partner's account does include the tax amount. This implies that when employing the gross method for posting down payments—where taxes are integrated into the down payment posting to the business partner—the system necessitates an additional offsetting or clearing account for the tax entries.

Now let's look at the streamlined process in action before delving into a discussion of its associated account determination.

4.2.1 Business Process Overview

Ideally, the process is initiated by generating a sales or purchase order, precisely detailing the desired goods or services. From this order, you proceed to establish a down payment request, intricately tied to the assigned order number. Subsequently, this down payment request is forwarded to the respective business partner.

You or the business partner fulfills the requested payment, and the down payment executed effectively counterbalances the initially raised down payment request. For VAT reasons, you must set up a down payment invoice the moment you collect the money. As soon as all performance obligations (e.g., shipments and services) have been performed, a regular invoice is submitted for the goods and services.

The difference between the received or sent down payments and a regular (final) invoice—the billing for the goods and services delivered—amounts to the outstanding balance, which ultimately needs to be settled by your company (in case of a previous purchase order) or the business partner (in case of a previous sales order). The whole process is outlined in Figure 4.9.

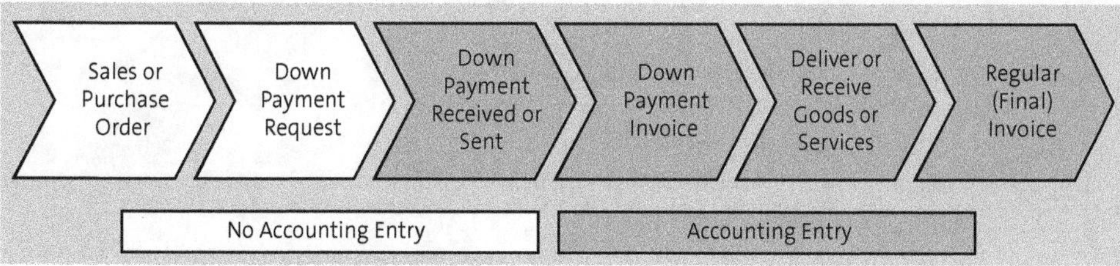

Figure 4.9 Down Payment Process

4.2.2 Assign Down Payment Accounts

You will find the configuration activities to assign down payment accounts under the configuration menus in the IMG for AR and AP business transactions. You assign special reconciliation accounts via menu path **Tools • Customizing • Execute Project**: there, click **SAP Reference IMG** and follow the configuration path **Financial Accounting • Accounts Receivable and Accounts Payable • Business Transactions • Postings with Alternative Recon Account • Other Special G/L Transactions**, then **Define Alternative Recon Accounts for Customers** or **Define Alternative Recon Accounts for Vendors.**

Let's walk through how to maintain down payment accounts for customers. If you follow the menu path provided in the previous paragraph through **Alternative Recon Accounts for Customers**, you will reach the screen shown in Figure 4.10.

Maintain Accounting Configuration : Special G/L - List

More Exit

Acct type	Sp.G/LInd.	Name	Description
D	D	Dbt rec	Doubtful Receivables
D	E	IVA	Individual Value Adjustment
D	H	Securit	Security Deposit
D	I	DP, IA	Down Payments, Intang. Assets
D	P	PmntReq	Payment Request
D	Z	Int.Rec	Interest Receivable

Figure 4.10 Alternative Reconciliation Accounts for Special General Ledger Transactions

Here you will see various situations for which different types of alternative reconciliation accounts are set up. Because we deviate from standard general ledger accounts for specific transactions, the configuration screen is called **Special G/L—List**. Within the down payment process, we will focus on the payment request and on down payments.

The configuration depends on the chart of accounts.There when selecting **Down Payments, Intag. Assets**, for instance, the system prompts you for the relevant chart of accounts (see Figure 4.11).

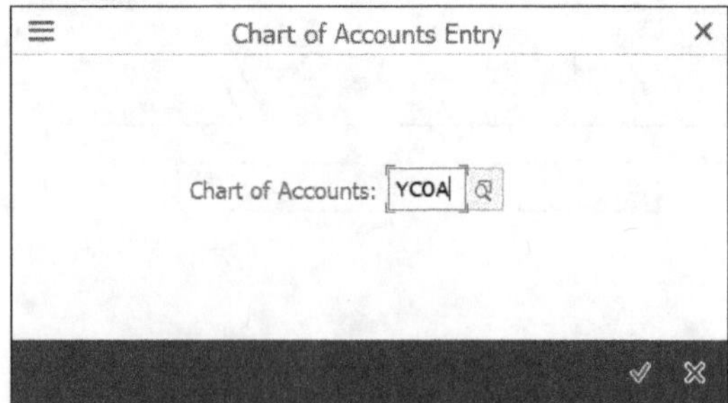

Figure 4.11 Enter Relevant Chart of Accounts before Maintaining Alternative Reconciliation Accounts

After you've selected the chart of accounts and clicked **Payment Request**, you will reach the configuration level at which you can enter the alternative (deviating) reconciliation account (here denoted as **Special G/L account**), such as 21186000, to the standard AR reconciliation account, **121000000** (see Figure 4.12).

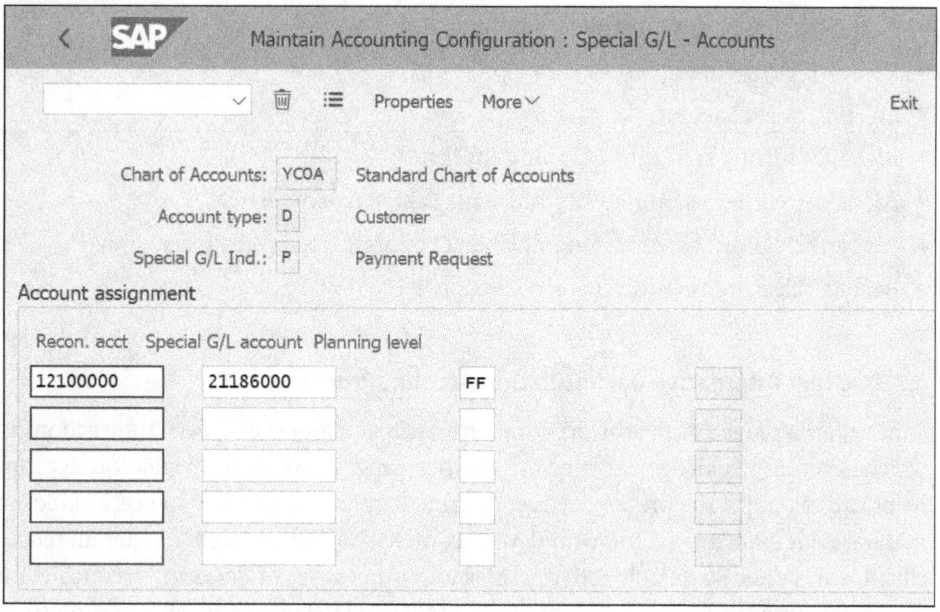

Maintain Accounting Configuration : Special G/L - Accounts

Properties More Exit

Chart of Accounts: YCOA Standard Chart of Accounts

Account type: D Customer

Special G/L Ind.: P Payment Request

Account assignment

Recon. acct	Special G/L account	Planning level
12100000	21186000	FF

Figure 4.12 Maintain Alternative Reconciliation Accounts for Down Payment Requests

In SAP S/4HANA, there is one field more to configure for the alternative reconciliation account: the planning level. The planning level is used to control displays in cash management for cash forecast reasons.

The following levels are possible:

- Only noted items in the cash position
- Only noted items in the liquidity forecast
- Purchase orders
- Sales orders
- Notified incoming payments
- Financial accounting payment settlements (incoming checks, outgoing checks, bills of exchange [BoEs], payments)
- Financial accounting bank transactions (bank statement postings, cashed checks, BoE payments)

You can freely choose the planning levels, but you should follow a clear naming convention. For the best display of the cash position and liquidity forecast, you should reserve levels that start with *F* or *B* for automatically updating data during posting.

F levels should be used for bank accounts, customers, and vendors, and B levels for bank clearing accounts.

Some examples of these names are as follows:

- F0: bank accounts
- F1: purchasing/sales (customers/vendors)
- FF: down payment requests
- FW: bills of exchange
- B1: bank clearing accounts, outgoing checks
- B2: bank clearing accounts, outgoing bank transfers (domestic)
- B3: bank clearing accounts, outgoing bank transfers (foreign country)
- B4: bank clearing accounts, bank collection, etc.

4.2.3 Other Alternative Reconciliation Account Process

Other alternative reconciliation accounts are necessary for other special transactions in which recording subledger entries on another general ledger reconciliation account helps to distinguish and present different business transactions. For instance, all kinds of manual value adjustments for AR and AP document should be posted to different reconciliation accounts. By default, SAP classifies four other cases of deviating reconciliation accounts within the AR process (compare the following to Figure 4.10, shown earlier):

- **Doubtful receivables**
  Doubtful receivables, also known as doubtful accounts or bad debts, are the portions of a company's AR that are uncertain or unlikely to be collected from customers. These are accounts where there is significant doubt about the customer's ability or willingness to pay the outstanding amount, often due to financial instability, disputes, bankruptcy, or other factors.
- **Individual value adjustments**
  Individual value adjustments, such as in AR, are specific adjustments made to the outstanding balances of individual customer accounts. These adjustments are usually driven by circumstances that affect payment obligations. Individual value adjustments are often necessary to accurately reflect the financial reality of the accounts and maintain accurate financial statements. They might be reasons to adjust values in case of settlement agreements, disputes, product returns, or other such situations.
- **Security deposit**
  A security deposit is a sum of money paid by one party, usually a tenant or a customer, to another party, typically a landlord or a service provider, as a form of assurance or protection against potential future financial losses, damages, or noncompliance with agreements.

- **Interest receivable**
  Interest receivable for AR refers to the interest that a company expects to receive on outstanding customer invoices or balances. When customers do not pay their invoices promptly, companies may include terms in their sales agreements that allow them to charge interest on overdue payments. This interest amount is considered interest receivable.

4.3 Value-Added Taxes

The taxation of services and goods depends on the jurisdiction. Broadly, we can categorize the two main systems as follows: the sales and use tax regime, which is used in the US, and the VAT regime, which is used mainly in the European Union and certainly in other parts of the world. Therefore, SAP S/4HANA provides tax calculation procedure templates for several countries. In this section, we focus on the VAT system, covering the business process overview and input and output tax accounts.

Value-added tax, often abbreviated as VAT, is a prevalent type of consumption tax imposed at each stage of a product's journey, starting from the acquisition of raw materials and ending at the ultimate purchase by a consumer. This taxation method is used by over 170 countries globally, encompassing all European Union member states, for levying taxes on both goods and services.

4.3.1 Business Process Overview

VAT is designed to ensure that tax is paid throughout the supply chain, from the initial production of raw materials to the final sale to the end consumer. Let's consider a detailed explanation of how the VAT regime works.

Say that you have a bakery in a country with a 10% VAT rate. Here's how the VAT regime works for your bakery:

- You purchase flour, sugar, and other ingredients from a supplier for $1,000. You pay $100 in VAT on these purchases.
- You bake and sell cakes to customers for $1,200. You charge them $120 in VAT on the sales invoice.
- In your VAT return, you report $120 in output VAT (collected from customers) and $100 in input VAT (paid on your purchases).
- You owe the government $20 ($120 - $100) in VAT.
- You pay the $20 to the tax authorities when you submit your VAT return.

In this example, the VAT of $20 represents the value added to the ingredients through your bakery's production process, and it is ultimately paid by the end consumer when they purchase the cake. The VAT system ensures that the tax is collected and remitted at each stage of production and distribution.

The general VAT business process is illustrated in Figure 4.13.

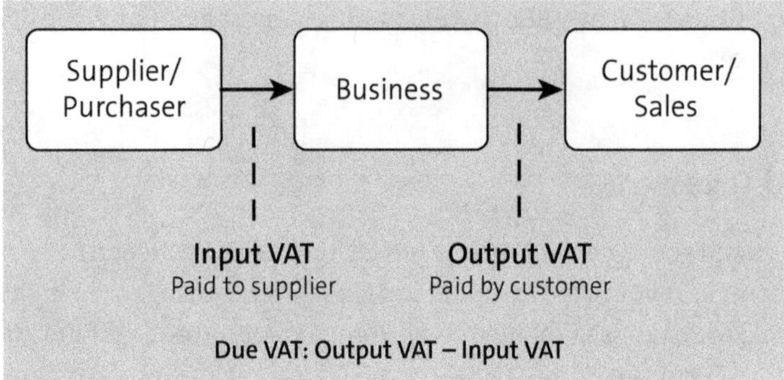

Figure 4.13 General VAT Business Process

The input taxes from purchases have to be paid to the suppliers, the output taxes are charged to and paid by the customers. The VAT amounts for both input and output taxes must be shown on the invoice in a separate line or field. The difference between input and output taxes must be paid to the fiscal authority on a monthly basis. This is accomplished by filing VAT returns. To enable this process, a taxable entity must undergo registration within the jurisdiction where it conducts its business activities.

In Europe, VAT rates typically range between 17% and 27%. However, the specific rates can vary from one European country to another. Here's a breakdown of the common VAT rate categories in Europe:

- **Standard rate**
  This is the most common VAT rate and usually falls within the range of 17% to 27%. Many European countries apply a standard VAT rate of around 20% or 21%.
- **Reduced rate**
  Some goods and services are subject to a reduced VAT rate, which is lower than the standard rate. The reduced rate can vary from around 5% to 15% and is often applied to essential items like food, books, pharmaceuticals, and certain services.
- **Super-reduced rate**
  A few European countries have an even lower super-reduced VAT rate, typically around 2% to 5%. This rate is often reserved for necessities like basic food items, water, and medical supplies.
- **Zero rate**
  In some cases, certain goods and services are subject to a 0% VAT rate. This means that no VAT is charged, but businesses can still reclaim the VAT they paid on their inputs. Zero rates are often applied to exports, international services, and certain financial services.

- **Exempt**
  Some goods and services may be exempt from VAT altogether. Unlike zero-rated items, businesses cannot recover the VAT they paid on inputs for exempt goods and services. Exemptions are commonly applied to healthcare, education, and some financial services.

It's important to note that each European country can set its own VAT rates and exemptions within the framework provided by European Union regulations. As a result, there can be significant variation in VAT rates from one country to another. In addition, VAT rates may change over time due to economic and legislative factors, so it's advisable to check with local tax authorities for the most up-to-date information.

4.3.2 Input and Output Tax Accounts

The tax determination rules are ideally set up in material management for purchases and in sales and distribution for sales. However, you can also maintain all sales/purchase tax-relevant accounts using the configuration menu path **Tools • Customizing • Execute Project**. There, click **SAP Reference IMG** and follow the configuration path **Financial Accounting • Financial Accounting Global Settings • Tax on Sales/Purchases • Posting • Define Tax Accounts**. Figure 4.14 shows the maintenance of general ledger accounts for different transaction keys.

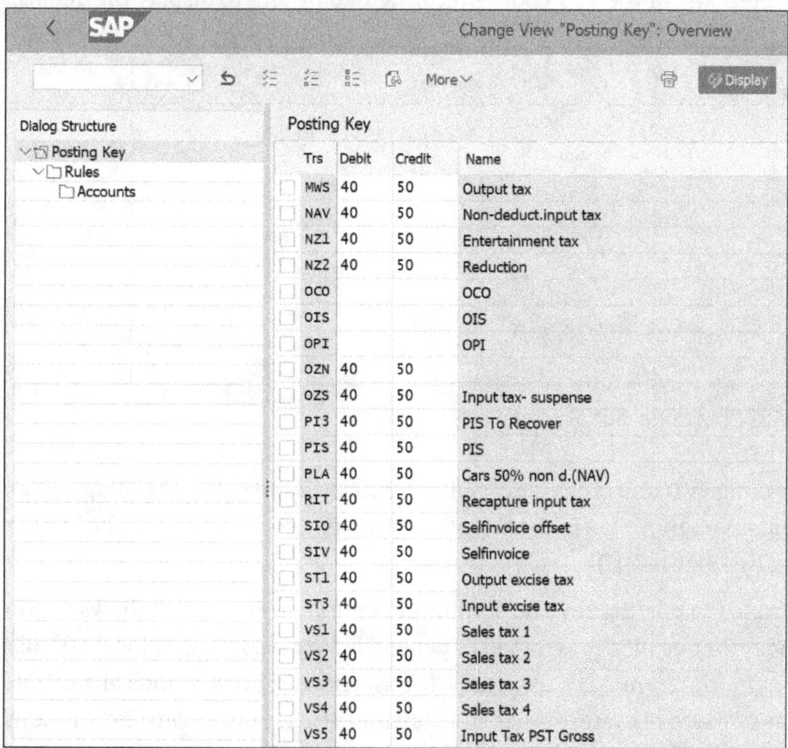

Trs	Debit	Credit	Name
MWS	40	50	Output tax
NAV	40	50	Non-deduct.input tax
NZ1	40	50	Entertainment tax
NZ2	40	50	Reduction
OCO			OCO
OIS			OIS
OPI			OPI
OZN	40	50	
OZS	40	50	Input tax- suspense
PI3	40	50	PIS To Recover
PIS	40	50	PIS
PLA	40	50	Cars 50% non d.(NAV)
RIT	40	50	Recapture input tax
SIO	40	50	Selfinvoice offset
SIV	40	50	Selfinvoice
ST1	40	50	Output excise tax
ST3	40	50	Input excise tax
VS1	40	50	Sales tax 1
VS2	40	50	Sales tax 2
VS3	40	50	Sales tax 3
VS4	40	50	Sales tax 4
VS5	40	50	Input Tax PST Gross

Figure 4.14 Transaction Keys for Automatic Tax Postings

MWS is the standard key for automatic postings of output tax. *VST* is the standard key for automatic postings of input tax. These keys both have to be assigned to the respective chart of accounts.

When implementing a new SAP system, tax codes, **Tax Codes** and content are delivered by SAP for each country with the country template from SAP. If you want to check the settings, you can, for instance, enter Transaction FTXP, where you will be prompted to enter your region (see Figure 4.15). In our example, we enter "DE" for Germany.

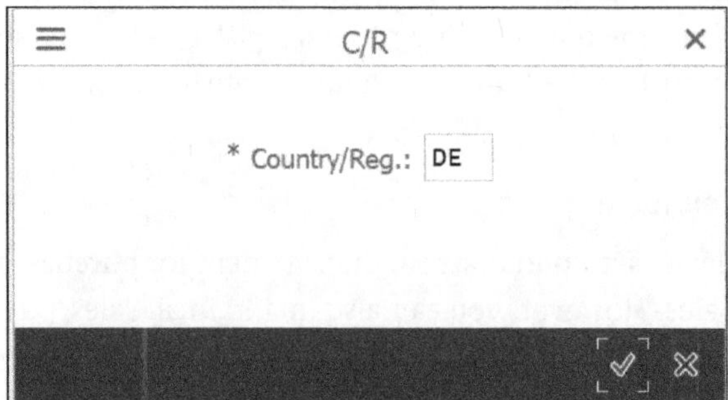

Figure 4.15 Transaction FTXP to Display Tax Code Details

After clicking the green checkmark, you are prompted to enter a **Tax Code** (see Figure 4.16). You can select any of the VAT codes preconfigured by SAP to display the details.

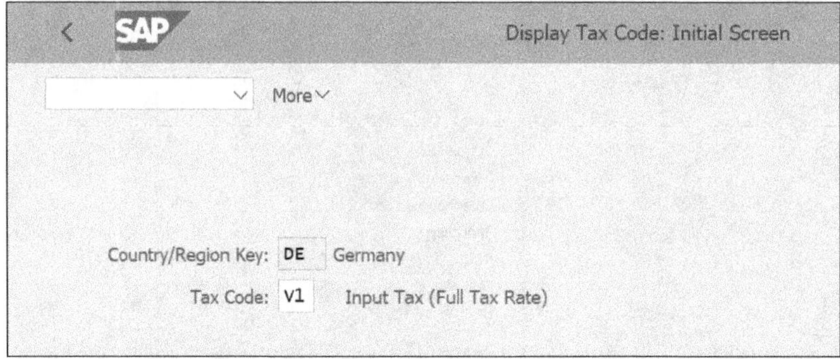

Figure 4.16 Display Tax Code: Initial Screen

In our case, we enter "V1" and press [Enter]. The results will show the percentage of taxation for a regular VAT input tax in Germany. The screen shows 19.000% in the **Tax Percent. Rate** field (see Figure 4.17).

Now let's return to the posting key and accounts assignment. Figure 4.18 shows all configured tax codes for output tax assigned to general ledger accounts. As local tax rules can lead to many different taxation cases (transactions with a standard tax rate, reduced tax rate, or zero tax rate; domestic acquisitions; EU transactions; exemptions; etc.), many general ledger accounts are needed to group the different transactions.

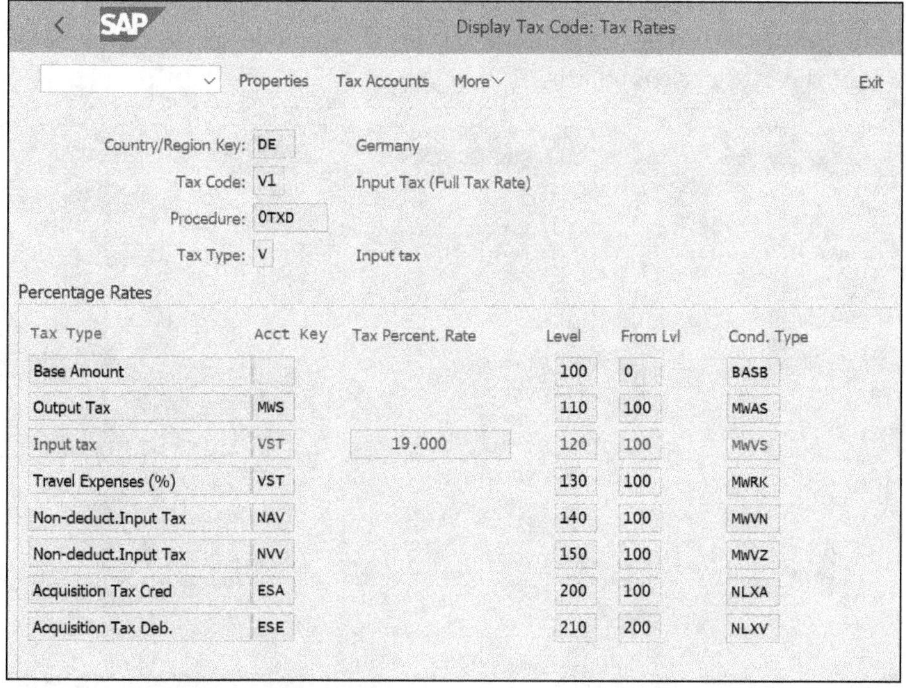

Figure 4.17 Display Tax Code: Tax Rates

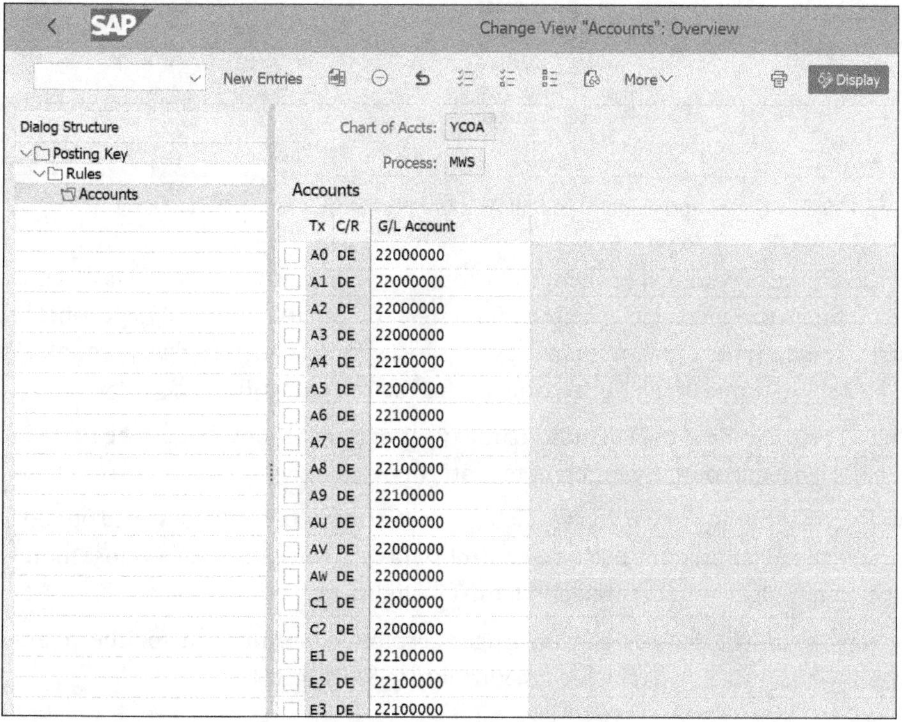

Figure 4.18 Tax Codes and General Ledger Accounts for Standard Output Tax (MWS)

Figure 4.19 shows all configured tax codes for the output tax; there can be many, depending on the local rules (e.g., transactions with the standard tax rate, a reduced tax rate, a zero tax rate, or exemptions).

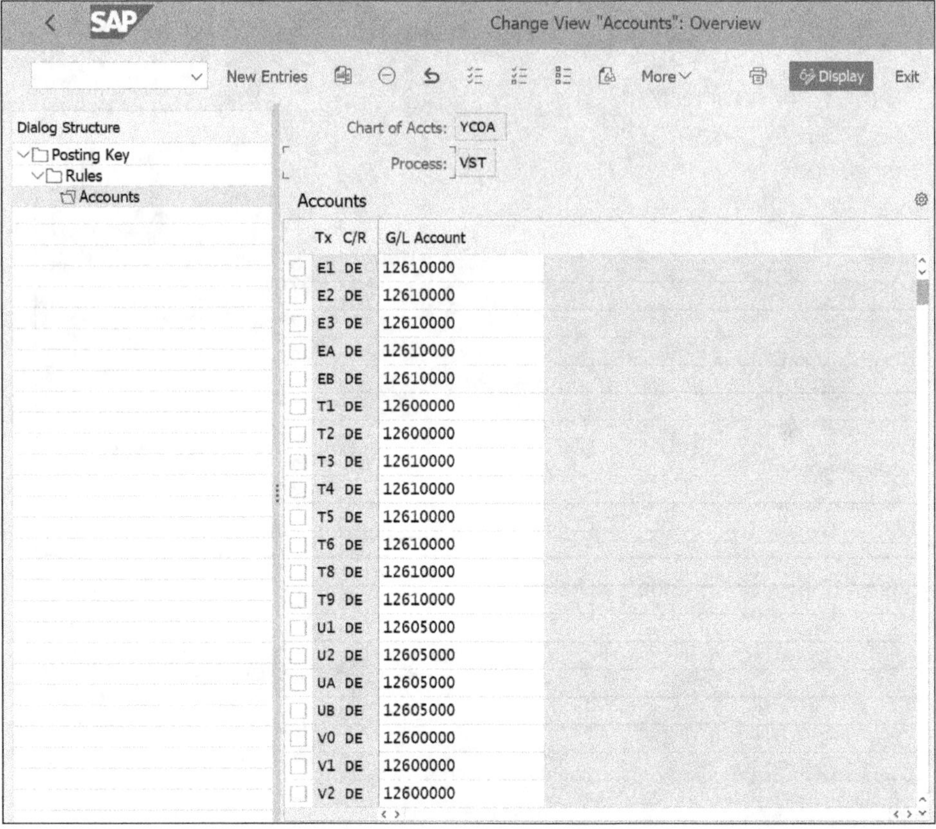

Figure 4.19 Tax Codes and General Ledger Accounts for Standard Input Tax (VST)

Tax accounts can be added or changed in both views. You also can control tax transactions through the general ledger account master. You can use the **Tax Category** field to restrict certain types of tax-relevant postings to a specific general ledger account. Figure 4.20 shows some of the values available for the **Tax Category** field.

Table 4.1 provides brief explanations of the types of transactions that can be posted to a general ledger account for different tax category values.

It's worth mentioning that if you've enabled the **Posting without Tax Allowed** option in the general ledger account master's **Control Data** tab, you can make a transaction to that general ledger account even if it doesn't include a tax code.

The next section discusses general ledger account determination for the manual incoming and outgoing payments functionality in SAP S/4HANA.

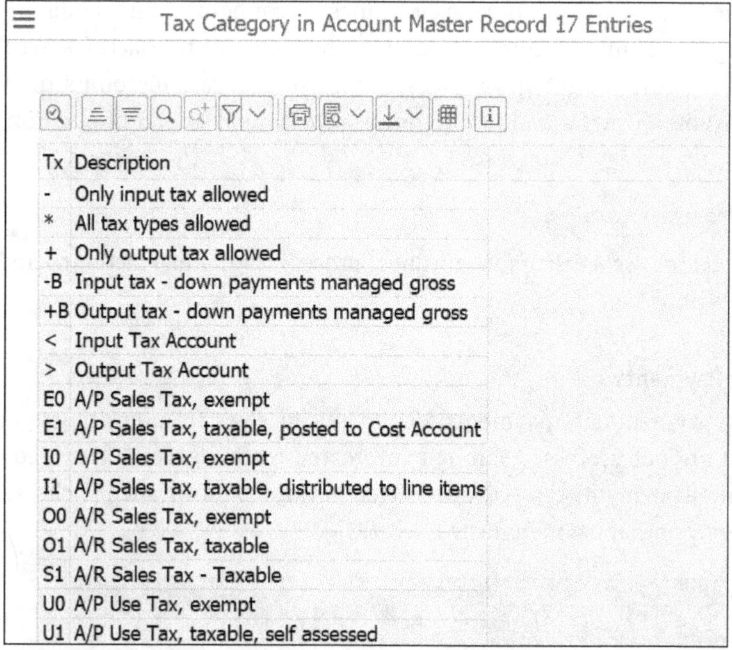

Tax Category in Account Master Record 17 Entries

Tx	Description
-	Only input tax allowed
*	All tax types allowed
+	Only output tax allowed
-B	Input tax - down payments managed gross
+B	Output tax - down payments managed gross
<	Input Tax Account
>	Output Tax Account
E0	A/P Sales Tax, exempt
E1	A/P Sales Tax, taxable, posted to Cost Account
I0	A/P Sales Tax, exempt
I1	A/P Sales Tax, taxable, distributed to line items
O0	A/R Sales Tax, exempt
O1	A/R Sales Tax, taxable
S1	A/R Sales Tax - Taxable
U0	A/P Use Tax, exempt
U1	A/P Use Tax, taxable, self assessed

Figure 4.20 Tax Category in General Ledger Accounts

Tax Category	Transactions Allowed for the General Ledger Account
*	You can post transactions containing input tax or output tax.
-	You can only post transactions that contain input tax.
+	You can only post transactions that contain output tax.
Individual tax codes	You can only post transactions that contain the specific tax code.
<	The general ledger account is an input tax account.
>	The general ledger account is an output tax account.

Table 4.1 Tax Category Values

4.4 Manual Incoming and Outgoing Payments

In SAP, the manual incoming and outgoing payments processes record and manage financial transactions in which payments are received from customers or made to vendors or other entities manually, without involving automated payment processes such as electronic funds transfer (EFT) or automatic payment program (APP). These manual payment processes are typically used for various reasons, such as handling exceptional cases, one-time payments, or situations where automated payments are not feasible or preferred.

In the following sections, we will explore the guidelines for configuring and modifying the posting settings for manual incoming and outgoing payment transactions. We'll also discuss how to handle and set up various scenarios such as cash discounts, overpayments, underpayments, and rounding differences within the system configuration.

4.4.1 Business Process Overview

In the following sections, we'll discuss the business process for manual incoming and outgoing payments.

Manual Incoming Payments

Manual incoming payments are payments a company receives from customers or other entities that are not processed through automated methods like EFT or credit card payments. These payments can also be received in various forms, such as checks, cash, bank transfers, or other payment instruments.

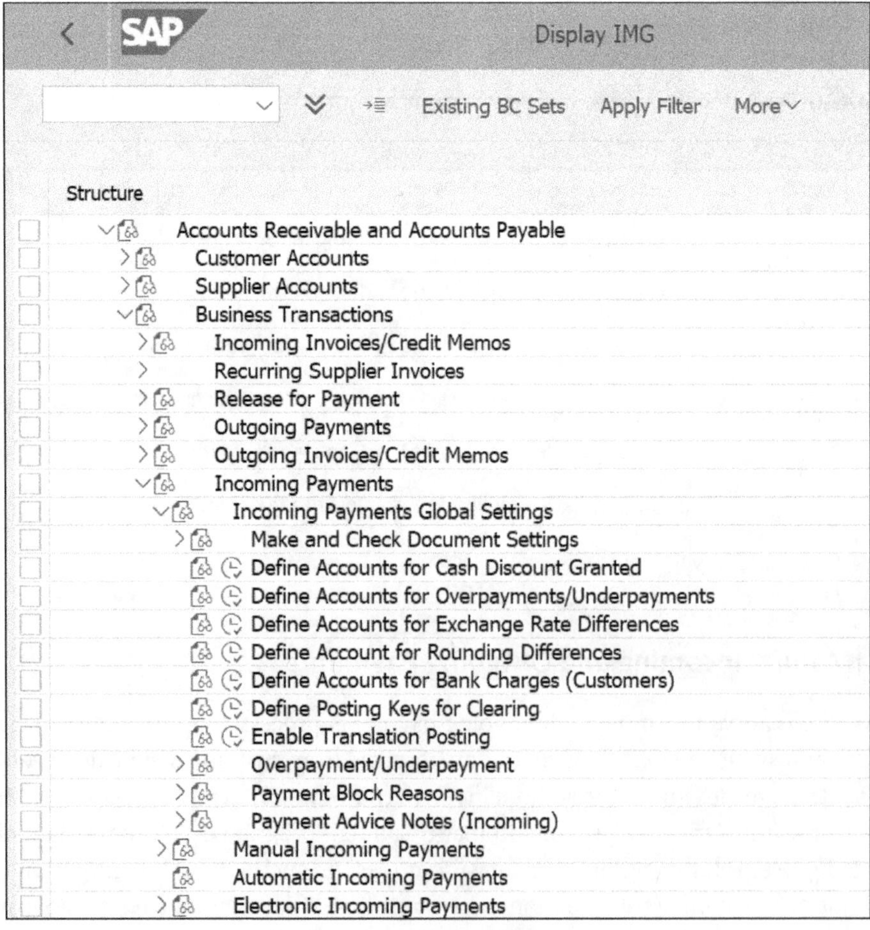

Figure 4.21 Incoming Payment Transactions Configuration

To record a manual incoming payment in SAP, you typically use transactions like Transaction F-28 for incoming payments or Transaction F-26 for posting incoming payments manually. You'll need to enter details such as the customer's information, payment amount, payment method, and relevant accounting data.

SAP then updates the relevant accounts, such as customer accounts and cash accounts, and records the transaction in the financial accounting module. The configuration for all incoming payment transactions is found at following path: **IMG • Financial Accounting • Accounts Receivable and Accounts Payable • Business Transactions • Incoming Payments**, as shown in Figure 4.21.

Manual Outgoing Payments

Manual outgoing payments are payments made by a company to external parties, such as vendors, suppliers, or service providers, without using the automated payment functionality. These payments can be made using various methods, including checks, bank transfers, cash, or other payment instruments.

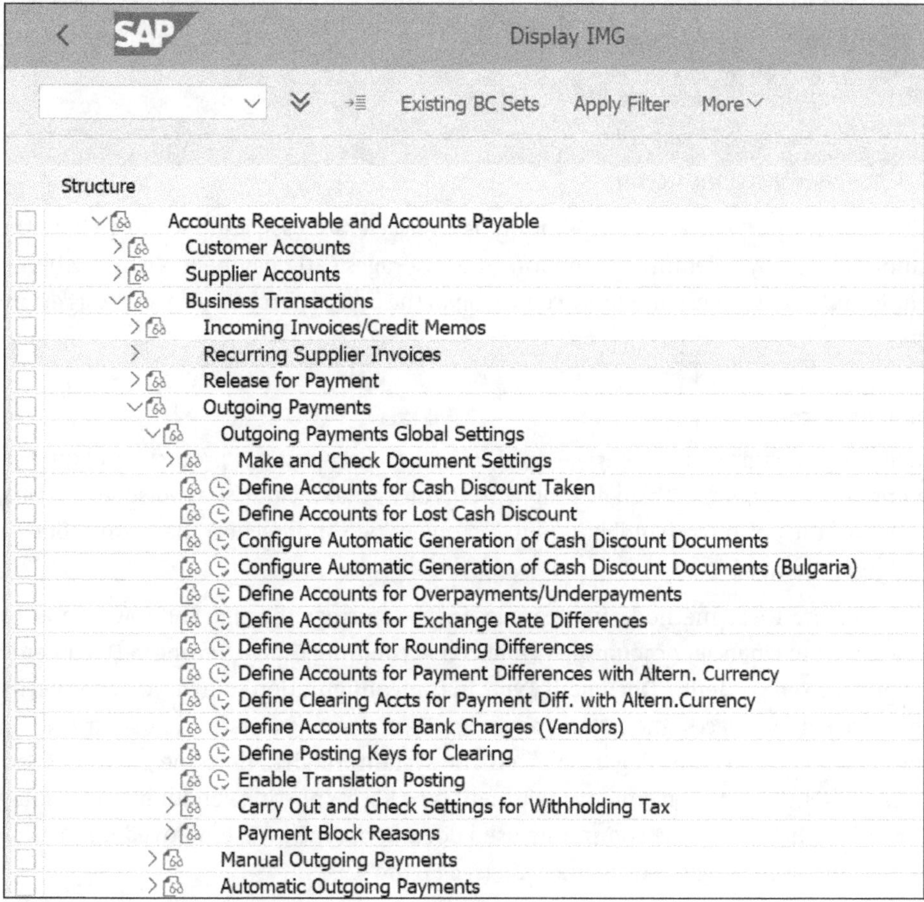

Figure 4.22 Outgoing Payment Transactions Configuration

To record a manual outgoing payment in SAP, you typically use a transaction like Transaction F-53 for a post or Transaction F-58 for a payment order. You'll need to enter the necessary payment details, including the vendor's or payee's information, the payment amount, the payment method, and any related accounting information.

After recording the payment, SAP updates the relevant accounts and records the transaction in the financial accounting module.

The configuration for all outgoing payment transactions is under the following path: **IMG • Financial Accounting • Accounts Receivable and Accounts Payable • Business Transactions • Outgoing Payments**, as shown in Figure 4.22.

Manual Payment Transactions

Manual incoming and outgoing payments are useful when dealing with exceptional cases, situations with unique payment requirements, or when there is no integration with electronic payment systems. However, it's essential to ensure that all manual payment transactions are accurately recorded and reconciled to maintain accurate financial records and comply with accounting standards and regulations. SAP also provides robust tools and reporting features to help manage and reconcile manual payment transactions effectively.

4.4.2 Cash Discount Accounts

The primary purpose of cash discounts is to accelerate cash inflow and reduce the amount owed by a debtor. For vendors, it encourages early payment, ensuring better liquidity. For customers, it offers a reduction in the cost of purchases if paid by a certain date.

Vendor Discounts

Many suppliers provide incentives in the form of discounts for prompt or early payment of their invoices. When dealing with invoices that include these discount terms, you have the option to handle them using one of two approaches: the net method or the gross method.

To configure these methods, you can access the settings through the IMG menu by navigating to **Financial Accounting • Financial Accounting Global Settings • Document • Document Types • Define Document Type**. In this configuration screen, you can specify the method by selecting the **Net Document Type** flag to indicate the use of the net method. If you leave this flag unselected, the system will default to the gross method. In SAP S/4HANA, you can employ document type KN for posting vendor invoices using the net method, while the commonly used document type KR is employed for posting vendor invoices using the gross method (see Figure 4.23).

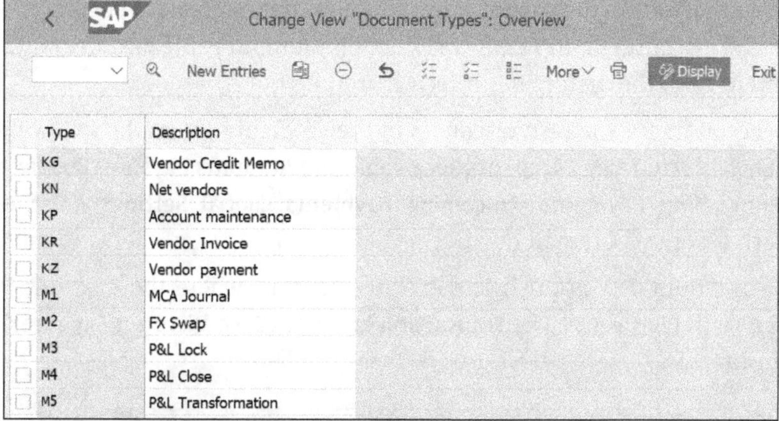

Figure 4.23 Document Types KN and KR

Only for the gross method of posting invoices, a general ledger account determination is required to be maintained. Under this method, any discount is calculated and posted only at the time when actual payment is made against the invoice. Any discount given on an outgoing payment is posted to the general ledger account determined using this configuration. You set up the accounts for the cash discounts in the configuration by following the menu path **IMG • Financial Accounting • Accounts Receivable and Accounts Payable • Business Transactions • Outgoing Payments • Outgoing Payments Global Settings • Define Accounts for Cash Discounts Taken.**

In our example, account 70040000 is the revenue account where the discount is posted to, as shown in Figure 4.24. The **Transaction** key **SKE** for **Cash Discount Received** (or earned) cannot be changed.

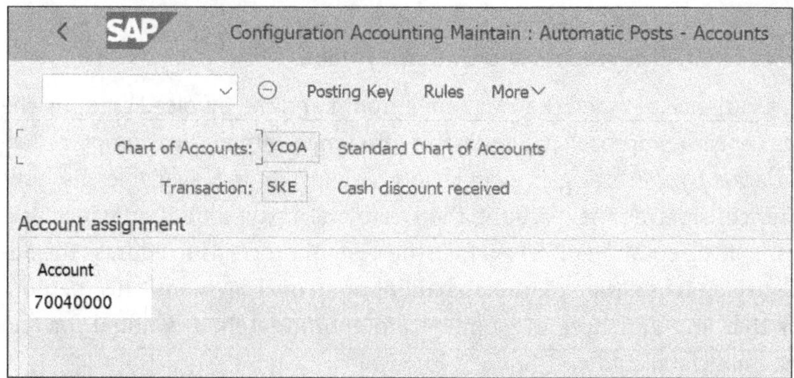

Figure 4.24 Account for Vendor Cash Discount (Gross Method)

Customer Discounts

Just like how vendors provide discounts in the accounts payable process, your company might offer discounts to customers for paying their invoices early or on time. It's important to note that, unlike the vendor side, there is no net method for posting customer invoices and discounts. In this case, customer discounts are only recorded when the actual payment from the customer is received.

You set up the accounts for the cash discounts in the configuration at the following path: **IMG • Financial Accounting • Accounts Receivable and Accounts Payable • Business Transactions • Incoming Payments • Incoming Payments Global Settings • Define Accounts for Cash Discounts Granted**.

In our example, account 71050000 is the expense account where the discount is posted to, as shown in Figure 4.25. The **Transaction** key, **SKT** (**Cash Discount Expenses**), cannot be changed.

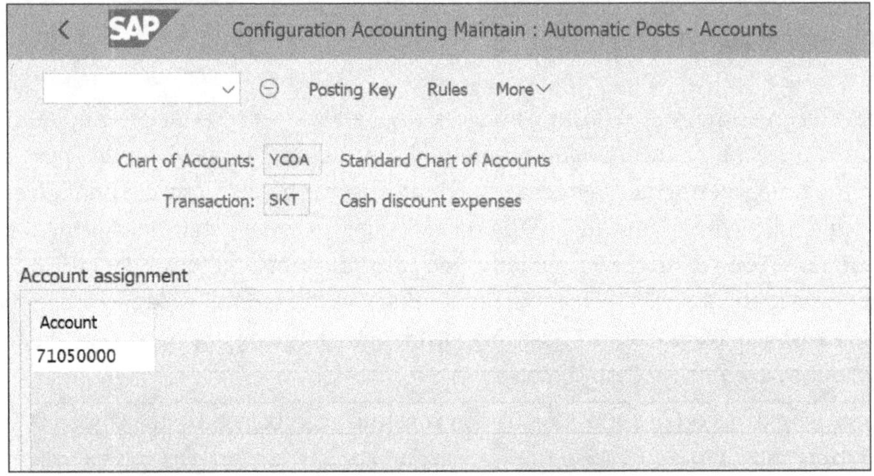

Figure 4.25 Account for Customer Cash Discount (Gross Method)

4.4.3 Accounts for Overpayments and Underpayments

When there is a variance between the payment amount and the invoiced amount, the system follows a specific approach to handle this difference. Initially, it attempts to reconcile the variance by offsetting it against any applicable and available discount amount. If, after considering the discount, there remains a payment discrepancy or if there is no available discount to offset against, the system proceeds to address the difference by classifying it as either an underpayment or an overpayment. This methodology ensures that any variances in payment are appropriately accounted for and resolved in the financial records.

The system employs an automated process to designate the appropriate general ledger accounts for underpayment or overpayment, depending on whether the payment amount falls short of or exceeds the corresponding invoice amount. The system identifies these general ledger accounts by following the configuration settings found in the IMG path outlined in Figure 4.26: **Financial Accounting • Accounts Receivable and**

Accounts Payable • Business Transactions • Outgoing Payments • Outgoing Payments Global Settings • Define Accounts for Overpayments/Underpayments.

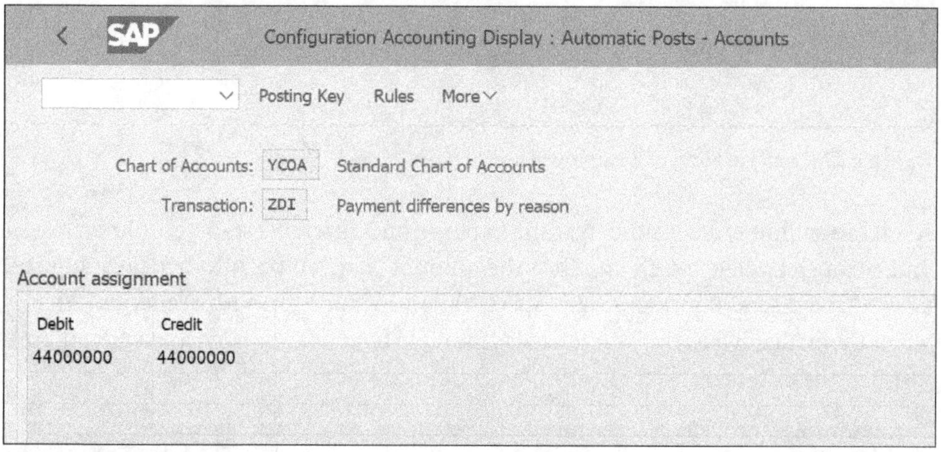

Figure 4.26 Account for Overpayments/Underpayments

In our example, account 4400000 is used for both underpayments and overpayments.

Interestingly, whether it's an overpayment or an underpayment happening in customer payments or vendor payments, the system employs the same transaction key for general ledger account determination. When you're reviewing these transactions involving overpayments and underpayments, it's crucial to have a good understanding of the concept of tolerance limits in SAP S/4HANA.

Tolerance limits play a crucial role in governing the acceptable level of payment differences for both overpayments and underpayments in transactions involving vendors and customers. To establish and manage these limits effectively, you follow a two-step process. First you configure tolerance groups to define specific tolerance criteria. Then you associate these tolerance groups with individual customer and vendor accounts. In scenarios in which a customer or vendor account lacks an assigned tolerance group, you also have the flexibility to create a company-wide tolerance group at the company code level to ensure consistency.

These tolerance groups encompass various parameters, and Figure 4.27 illustrates the key parameters that are particularly pertinent to the discussion in this section. These parameters determine the boundaries within which payment discrepancies are considered acceptable, providing essential control over financial transactions. The tolerance groups can be configured following IMG path **Financial Accounting • Accounts Receivable and Accounts Payable • Business Transactions • Incoming Payments • Manual Incoming Payments • Define Tolerance Groups for Employees.**

Permitted payment differences

	Amount	Percent		Cash discnt adj.to
Revenue:	5.00	2.0	%	1.0
Expense:	5.00	2.0	%	1.0

Figure 4.27 Configuration of Tolerance Limit

A tolerance limit specifies the maximum permitted difference as an absolute amount and as a percentage. It also specifies the amount that can be automatically adjusted against the discount amount. This specification is made for both scenarios—that is, when there is gain (i.e., the difference is favorable to your company) and when there is loss (i.e., the difference is not favorable to your company).

Note

The system evaluates the overpayment or underpayment variance by comparing it to the tolerance limits that have been assigned to the specific customer, vendor, and employee (user ID) responsible for the transaction. It considers the tolerance limits from these different sources, and the most restrictive (lowest) of these limits is applied to the transaction. This ensures that the system adheres to the most stringent tolerance limit when addressing payment discrepancies.

The payment difference that cannot be adjusted against discount amounts and underpayment/overpayment tolerance amounts is posted back to the customer account or vendor account.

Let's now discuss general ledger account determination for rounding differences.

4.4.4 Rounding Difference Accounts

The general ledger accounts for rounding differences utilized to record the discrepancies arising from inaccuracies in the automatic postings' calculations. This situation can be particularly prevalent in foreign currency transactions, where varying numbers of decimal places and exchange rates defined with up to four or five decimal points come into play. To shed more light on this, let's delve into the configuration, found at the following IMG path: **Financial Accounting • Accounts Receivable and Accounts Payable • Business Transactions • Incoming Payments • Incoming Payments Global Settings • Define Account for Rounding Differences.**

In the example given, depending on debit or credit postings, the rounding differences go to account 7202000 or 7253000, as shown in Figure 4.28.

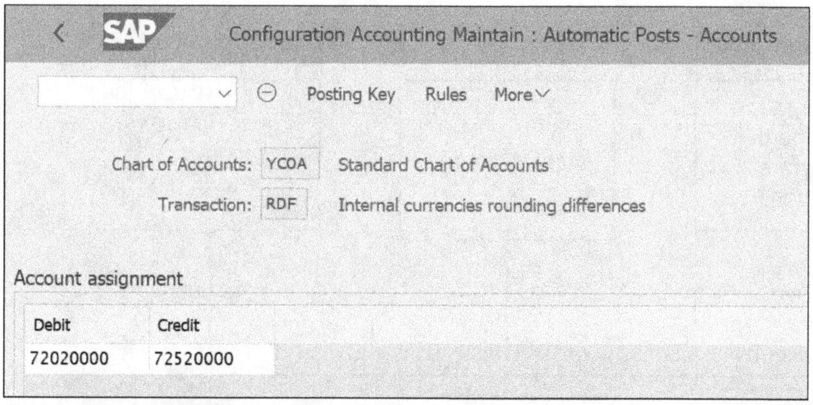

Configuration Accounting Maintain : Automatic Posts - Accounts

Posting Key Rules More

Chart of Accounts: YCOA Standard Chart of Accounts

Transaction: RDF Internal currencies rounding differences

Account assignment

Debit	Credit
72020000	72520000

Figure 4.28 Configuration of Accounts for Rounding Differences

In scenarios involving transactions in foreign currencies where distinct exchange rates apply to various line items, any total difference is bifurcated into two components.

First, there is a portion attributed to the variance caused by fluctuations in exchange rates. This portion is posted to general ledger accounts determined as explained in Chapter 2 (presumably, these are accounts designated for exchange rate variance).

Second, the remaining segment of the difference pertains to calculation or rounding disparities. Only this part of the difference is recorded in the general ledger accounts under discussion in this section. These accounts are designed specifically to account for these calculations and rounding discrepancies.

4.5 Automatic Incoming and Outgoing Payments

In the context of SAP, automatic incoming and outgoing payments use automated processes for handling financial transactions involving payments made to vendors or received from customers. These automated payment processes are designed to streamline and expedite the payment and collection of funds, improving efficiency and reducing manual intervention in financial operations.

In the following sections, we will sketch the business process of automatic payments and will describe the automated payments process for vendors in detail. For the customer side, see Chapter 5.

4.5.1 Business Process Overview

In the following sections, we'll discuss the business process of automatic incoming and outgoing payments. In general, there are two major tools to process payments automatically: electronic bank statements and the automatic payment program. Figure 4.29 shows both customer and vendor automatic payments, which we explain ahead.

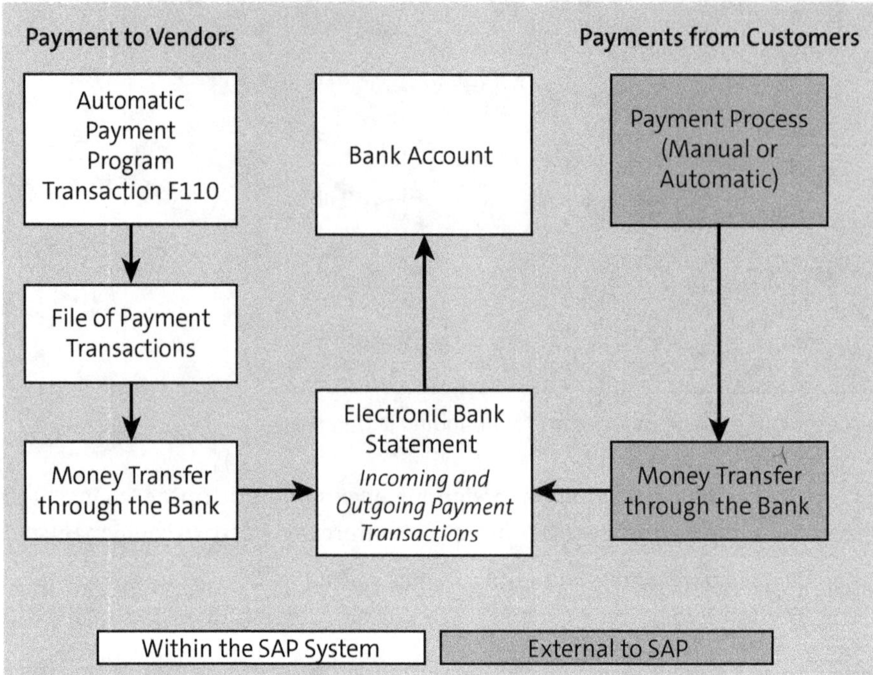

Figure 4.29 Automatic Payment Process

Automatic Incoming Payments (Customer Payments)

Automatic incoming payments in SAP are primarily related to the accounts receivable (AR) process, where a company expects to receive payments from its customers or clients.

SAP S/4HANA provides features for automating the receipt and allocation of customer payments. Automatic cash application is one such functionality, which matches incoming payments with open customer invoices.

When customers make payments, whether through EFT, lockbox processing, or other means, SAP S/4HANA can automatically identify and apply these payments to the corresponding customer invoices, reducing manual effort and minimizing errors.

Typically, transactions are recorded through the house bank of the company. The house bank provides bank statements on a daily basis. For many years, banks have provided electronic bank statements to their customers. SAP can process these statements automatically and can generate accounting entries to record the payment and clear the open invoices, ensuring accurate and up-to-date financial records. Here, we refer to Chapter 5, Section 3, where we explain the configuration in detail.

Automatic Outgoing Payments (Vendor Payments)

Automatic outgoing payments in SAP are primarily associated with the AP process, where a company needs to make payments to its vendors, suppliers, or service providers.

SAP offers various tools and functionalities for automating vendor payments, such as the automatic payment program. APP enables a company to configure payment criteria, including payment methods, payment terms, and due dates.

Once configured, APP automatically generates payment proposals based on open vendor invoices and their due dates, taking into account any available cash discounts.

Users can review and approve these payment proposals, and upon approval SAP generates payment files that can be sent to banks for processing, or it can generate checks and payment advice documents.

4.5.2 Automatic Payment Accounts

For customers, incoming automatic payment processing is usually handled through the configuration of automatic bank statement processing. The account for automatic payment processing is a bank account, and the counter account is the customer account. In Chapter 5, we will discuss automatic payments further and show the configuration in detail.

For vendors, the automatic payment processed is triggered by the company. Ahead, we will walk through the complete process, including the configuration required.

Automatic Payment Program

To execute the automated payment program, enter Transaction F110, or follow the application menu path **Accounting • Financial Accounting • Accounts Payable • Periodic Processing • Payments**. The **Automatic Payment Transactions: Status** screen appears, as shown in Figure 4.30.

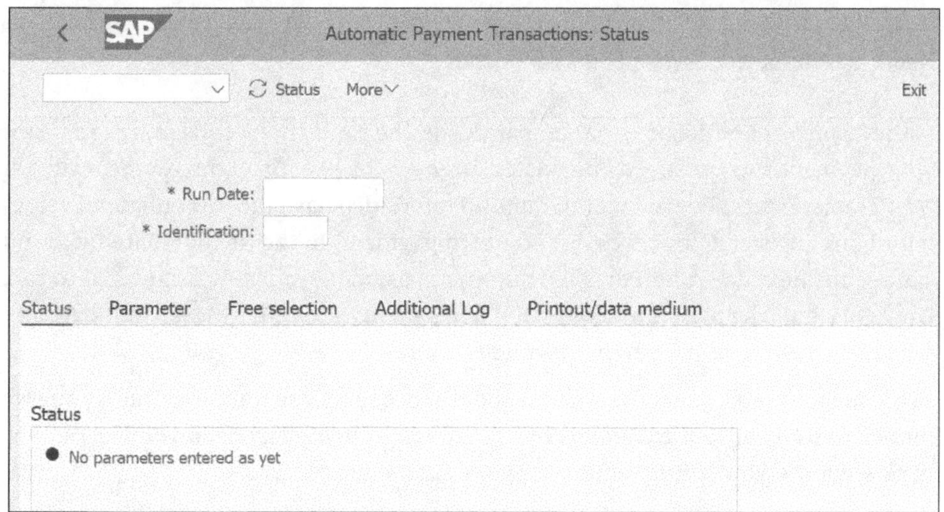

Figure 4.30 Automatic Payment Transactions: Status Screen

Fill out the required fields: **Run Date** (usually the current date) and **Identification** (any character string with five characters). You can later search for the identification to find the performed payment run quicker.

Fill in the parameters for your payment run on the **Parameter** tab, as shown in Figure 4.31. The **Posting Date** and **Docs Entered Up To** fields are prepopulated with the current date.

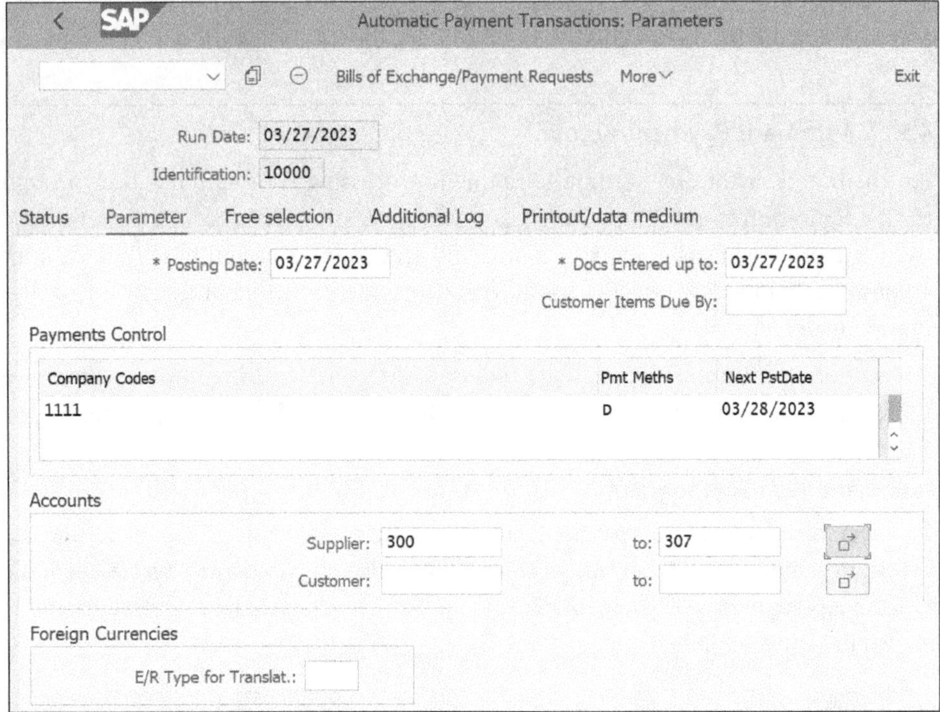

Figure 4.31 Automatic Payment Run: Parameters

In the **Payments Control** box, the company code, the payment method, and the date for your payment run are already populated after you have defined the payment run on the **Parameter** tab. The payment methods (**Pmt Meths**) have the preconfigured values **B** for bank transfer, **C** for check, or **N** for card payment. In the **Next PstDate** field, you can see the next date when the next automatic payment program is expected to run. Until this date, the program finds all the due date invoices after the last run date and schedules them to be paid in this payment run.

In the **Accounts** box, select the range of **Supplier** accounts you want to settle. If you owe money to some of your customers or have the right to directly debit your customers' bank accounts, you can also enter a range of **Customer** accounts.

Save the entries you've made so far by clicking the **Save Parameters** button. You'll receive a message that the details have been saved.

In the **Free selection** tab, define your exception to exclude certain open items that you don't want to pay. Then, in the **Additional Log** tab, define what details should be logged in the payment log. Finally, to print the payment transaction, select the printer and media in the **Printout/Data Medium** tab.

Go back to the **Status** tab, and you can see that it has turned green, as shown in Figure 4.32.

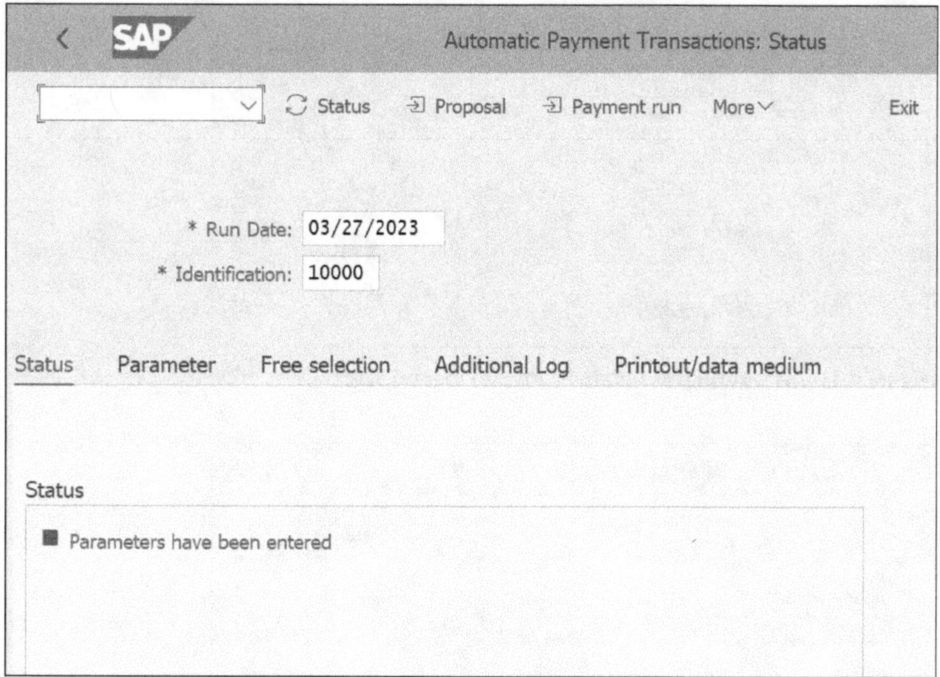

Figure 4.32 Automatic Payment Run: Green Status

Generate a payment proposal run by clicking the **Proposal** menu item, or directly start the payment run with the **Payment Run** menu item. Before executing the payment program, we recommend that you double-check the settings configuration.

You can configure the payment program by choosing **More • Environment • Maintain Configuration** from the menu bar. In the screen that opens (see Figure 4.33), settings for the automatic payment program are categorized into several groups in which you can configure the relevant settings and options:

- **All company codes**
  Click **All company codes**, and you'll arrive at the screen shown in Figure 4.34. Here you can select one of the company codes and double-click it to see more details.

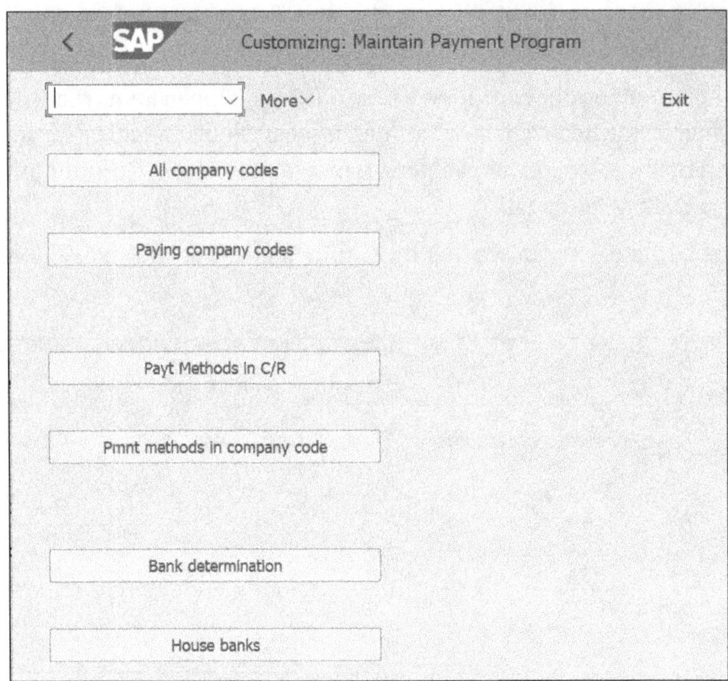

Figure 4.33 Customizing: Maintain Payment Program Screen

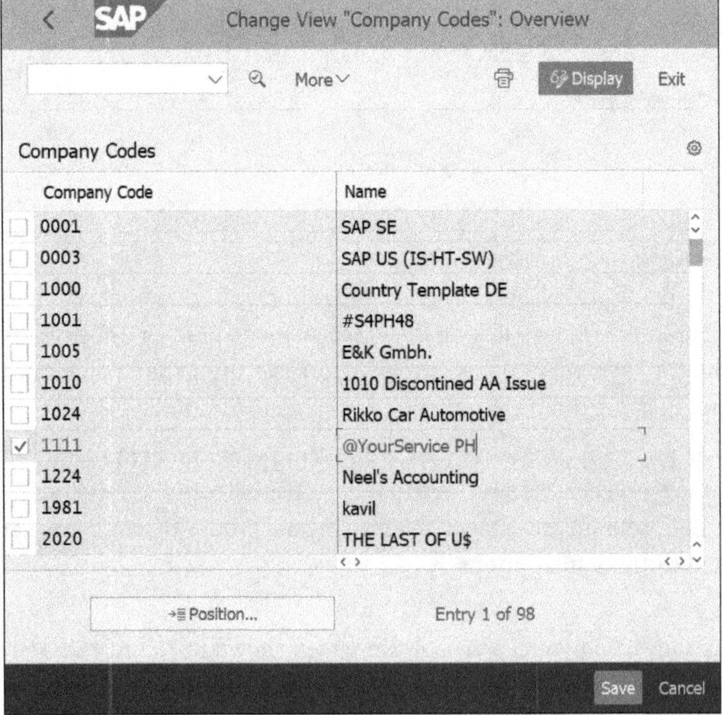

Figure 4.34 Selecting Company Codes

Another screen will appear, as shown in Figure 4.35 and Figure 4.36, where you can define the relevant settings for the selected company code— such as the paying company, which normally matches the company code in the header of the screen, as well as cash discounts and tolerances for payments relevant to early payment discounts. After you've entered all the required fields, click **Save** to save the data.

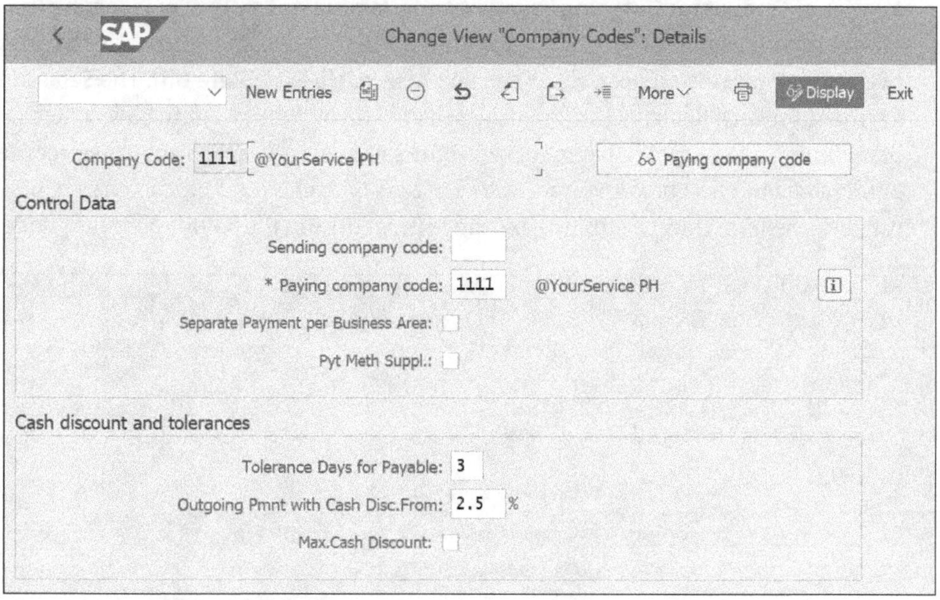

Figure 4.35 Company Codes Settings (Part 1)

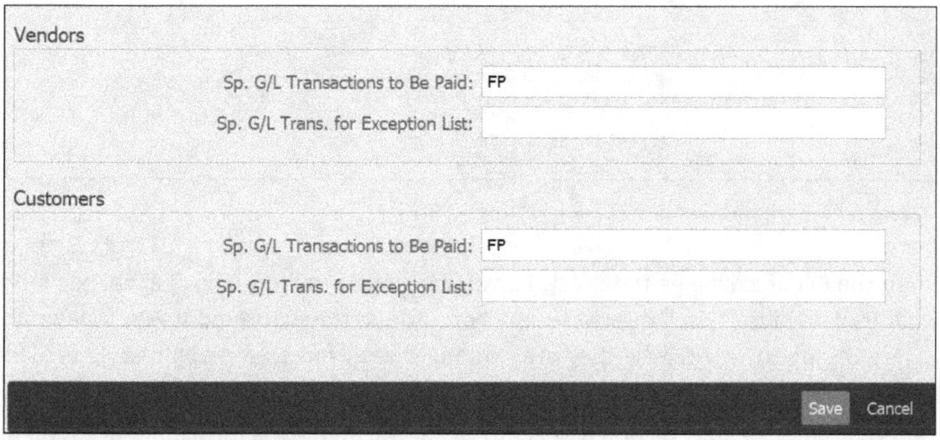

Figure 4.36 Company Codes Settings (Part 2)

- **Paying company codes**
  Here, you perform the following settings:
  - **Control Data**
    Minimum amounts for incoming and outgoing payment.

- **Bill of Exchange**
  Bill of exchange parameters.
- **Forms**
  Forms for payment advice and electronic data interchange (EDI).

Let's start with the **Control Data** section, shown in Figure 4.37. Here you can enter a **Minimum Amount for Outgoing Payment** and a **Minimum Amount for Incoming Payment** for your customer. Further, if you don't want to have any automatic postings of exchange rate differences, select the box next to **No Exch. Rate Diff.** The **Separate Payment for Each Ref.** checkbox may be selected if you don't want to have a netting of incoming and outgoing payments with the same reference. You should select the **Bill/Exch Pymt** checkbox if you want to use bills of exchange, bill of exchange payment requests, or the check/bill of exchange procedure in the paying company code.

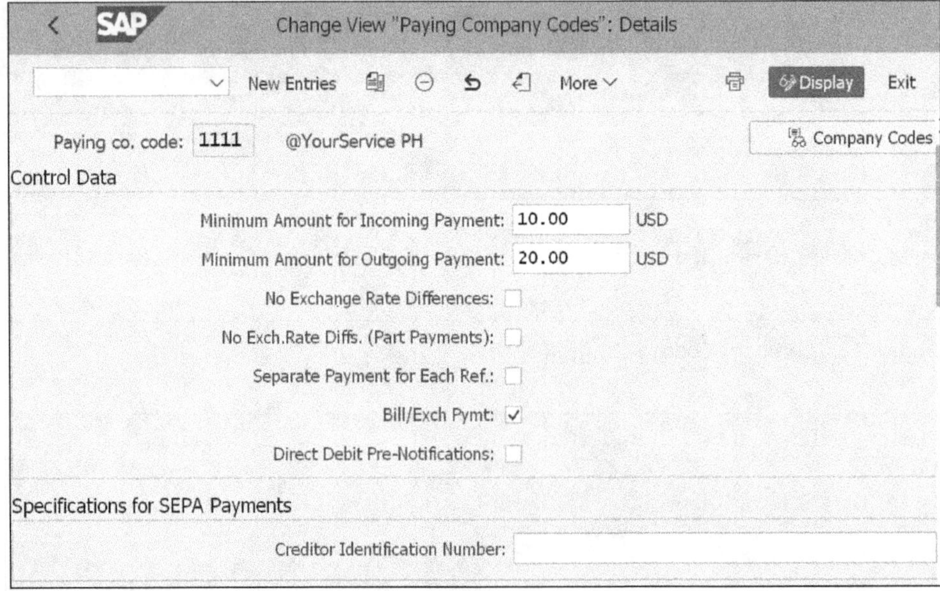

Figure 4.37 Payment Company Codes Settings (Part 1)

In the **Bill of Exchange Data** section, you can configure how bills of exchange look. Although bills of exchange were very common in the 1980s and 1990s, today they are disappearing because they are no longer relevant, so we won't explain them here.

Finally, in the **Forms** section (see Figure 4.38), you define the format of the payment advice you generate with the payment run. In addition, if you generate an EDI file, you can select a specific data format. SAP has its own format called SAPscript. An SAP consultant can customize this format to add **Letter Header**, **Footer**, **Signature Text**, and **Sender** details to it and change them as needed.

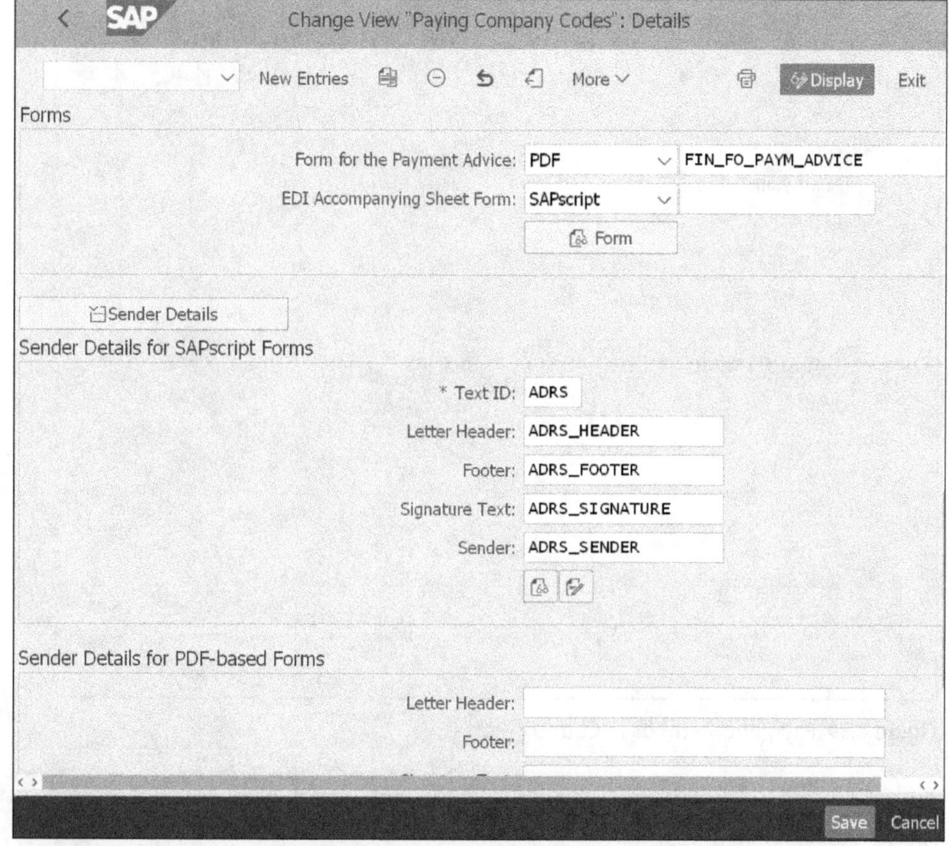

Figure 4.38 Paying Company Codes Settings (Part 2)

After you've filled in the required fields, click the **Save** button, and you will receive the message that the data is saved.

- **Pmnt methods in country**

  Here, you configure the methods of payment, settings for individual payment methods, document types for posting, print programs, and permitted currencies for each country relevant for your organization.

  Figure 4.39 shows how payment method B (bank transfers for outgoing payments) for the US (see the **Country** field) is configured.

  Scroll down to see further settings for payment type B (bank Transfer; see Figure 4.40). This configuration is very country-specific, and these settings are normally preconfigured in so-called country templates when SAP S/4HANA is implemented. To explain all the fields and implications would go far beyond the scope of this book. Note that any changes here would require country-specific expertise from a local SAP consultant.

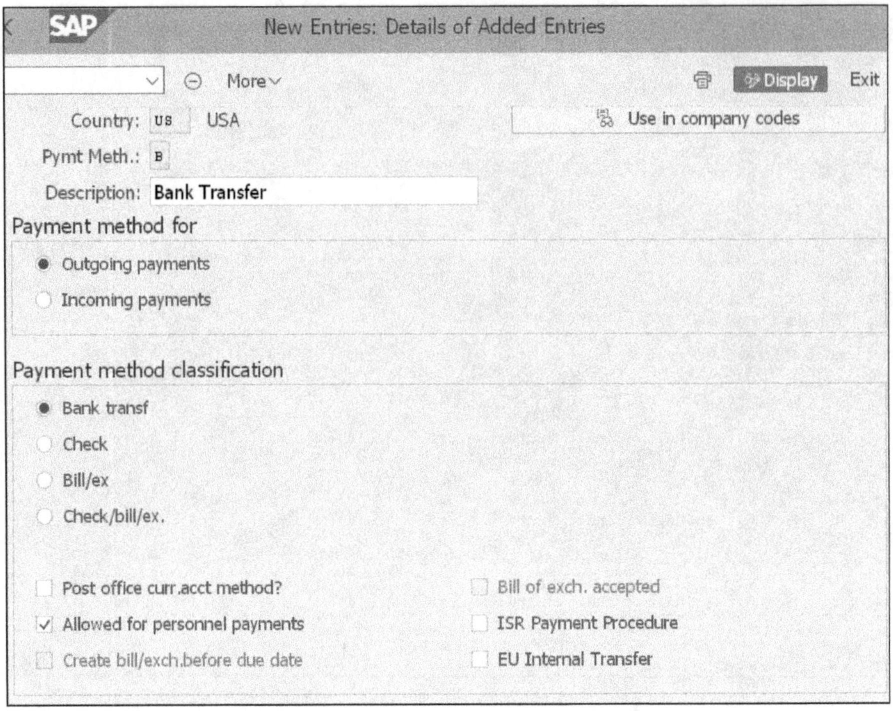

Figure 4.39 Payment Methods in Country (Part 1)

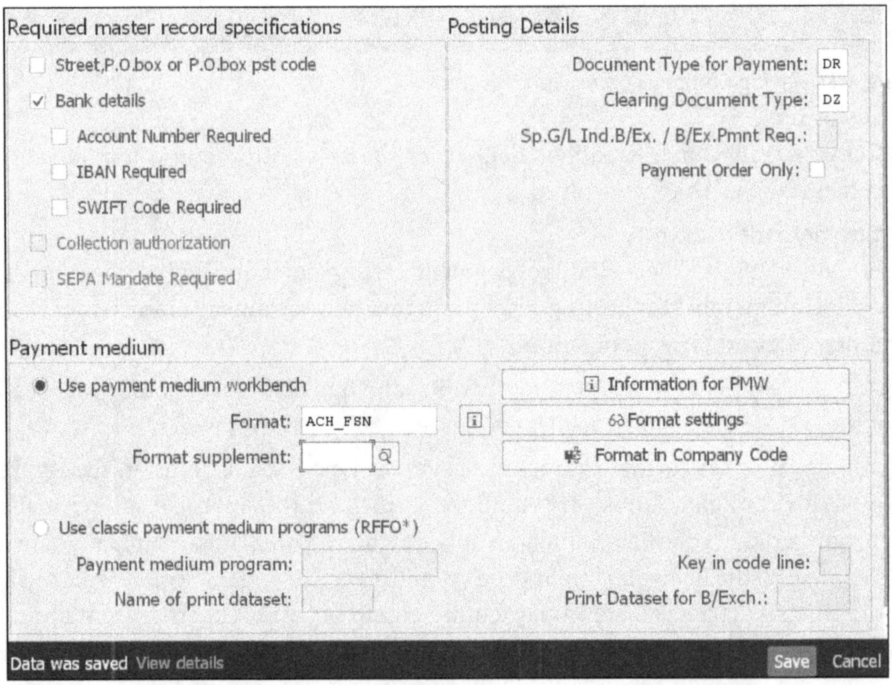

Figure 4.40 Payment Methods in Country (Part 2)

- **Pmnt methods in company code**
  Here you perform settings such as minimum and maximum payment amounts, grouping possibilities, bank optimization, forms for payment media, and so on.

  Figure 4.41 shows the configuration of payment method **T** (bank transfer) for company code 1111. The payment program will select this payment method for any payments above the minimum amount and below the maximum amount. The payment items can be grouped per day or shown individually. You can further configure whether foreign business partners, foreign currencies, or foreign bank accounts are allowed through the payment method assigned to the specific company code.

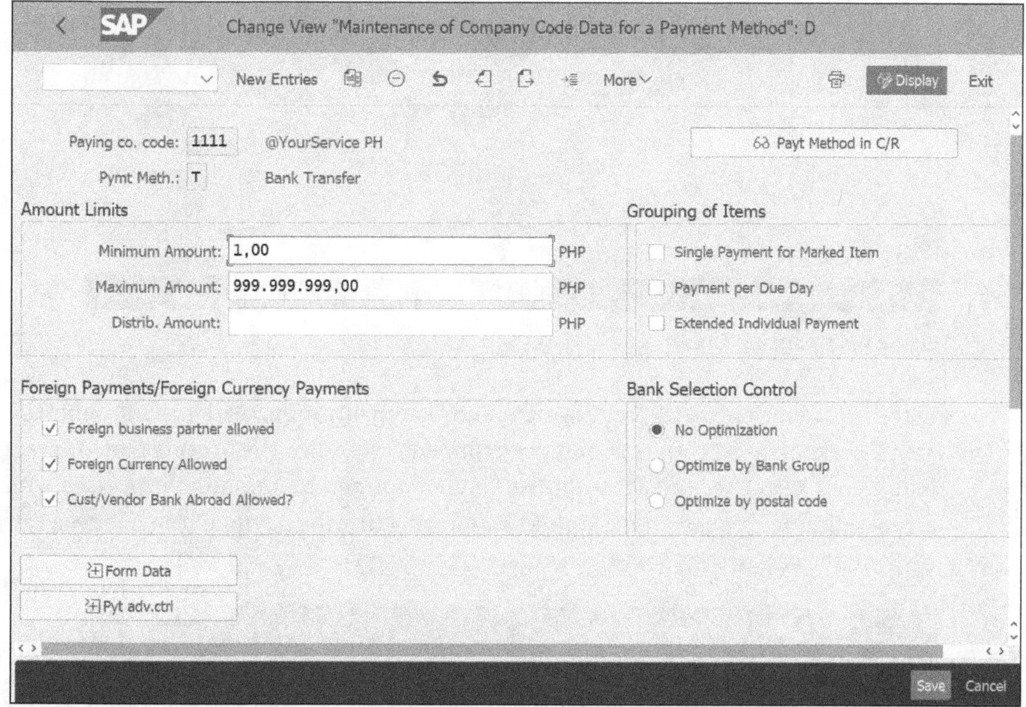

Figure 4.41 Payment Methods in Company Code

- **Bank determination**
  Here, you preconfigure your bank preferences by choosing which bank account is used for the defined payment methods. In our example, shown in Figure 4.42, payment method **C** (check) uses currency (**Crcy**) **USD** with house bank (**House Bk**) **USBK1**. On the second rank for the same payment method, you find house bank **USBK2**.
- **House bank**
  Here, you check the assignment of house banks to company codes.

After you've finished with the configuration steps for the payment process, then you can continue with the execution of the payment program. We'll start back at the **Status** screen shown earlier in Figure 4.30, but now we will enter the relevant parameters.

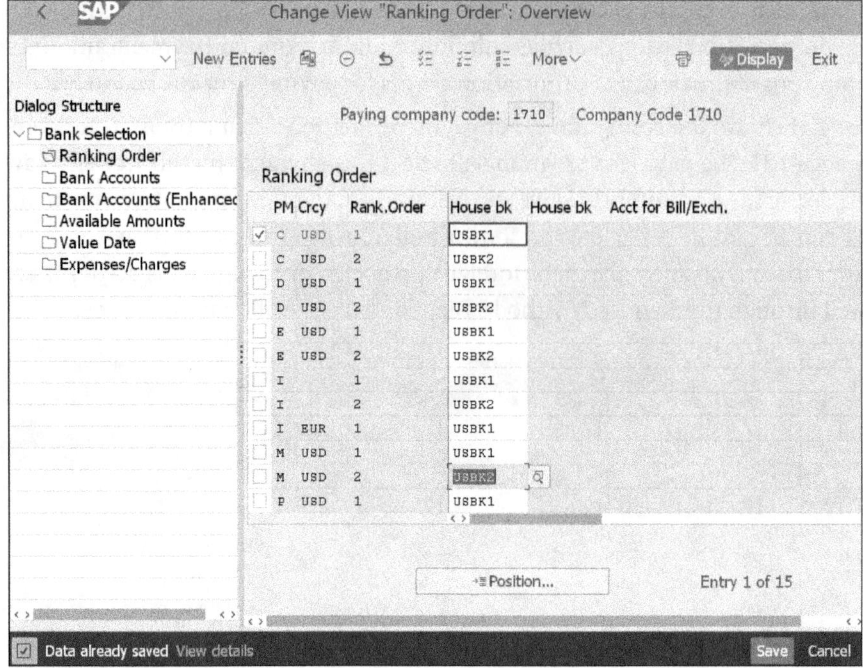

Figure 4.42 Ranking Order

In the top menu, choose **More • Payment Run • Reorganization**. Mark the **Start Immediately** checkbox to start the payment run immediately. Click the **Execute** icon or press the `Enter` key. After the payment has been performed successfully, a message with green indicators appears in the **Status** tab, showing that the payment run has been carried out, the parameters have been entered, and so on (see Figure 4.43).

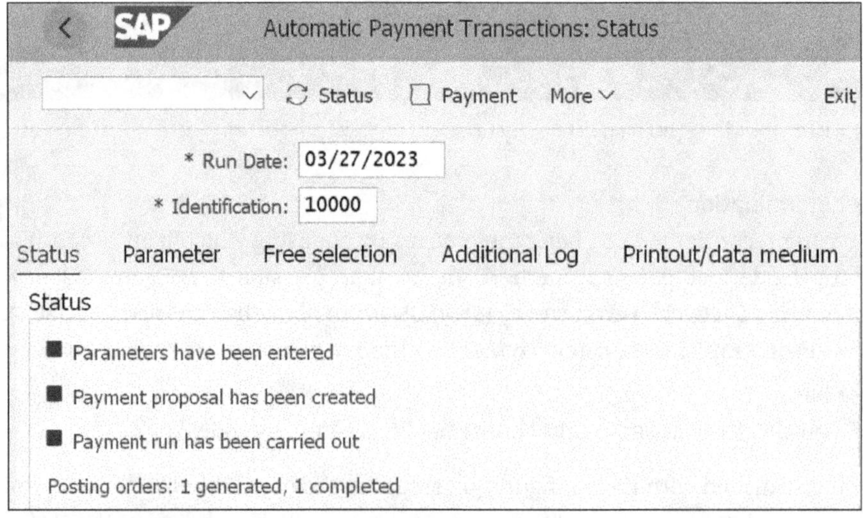

Figure 4.43 Automatic Payment Run Message

Now that you have seen how the automatic payment process works, we will move on to a very specific topic arising from legal requirements: withholding taxes.

4.6 Withholding Taxes

Withholding taxes are used by many countries to collect income tax at the source of income. Such a tax requires a person or entity making a payment (the *withholding agent*) to deduct a certain percentage of each payment and remit it directly to the government on behalf of the recipient. This system is often applied to various types of payments, including salaries, interest, dividends, royalties, and payments to contractors or freelancers.

4.6.1 Business Process Overview

SAP S/4HANA offers two distinct withholding tax mechanisms—simple withholding tax and extended withholding tax—each serving specific statutory or tax-related requirements. We'll discuss each of these in the following sections.

Simple Withholding Tax

Simple withholding tax calculation provides limited but mostly adequate functionality for withholding tax calculation. You can use the simple withholding tax functionality if taxes are to be withheld only from accounts payable, and if a payee only has one type of payment that is subject to withholding tax.

Extended Withholding Tax

Figure 4.44 offers a schematic diagram of the extended withholding tax process for the sake of facilitating easy comparisons. Unlike simple withholding tax, the extended withholding tax process offers a significantly more intricate set of functionalities for tax withholding calculations. For instance, it enables you to handle situations involving multiple withholdings for the same payee, conduct self-withholding when your business is subject to withholding requirements, calculate withholding tax at the invoicing stage rather than at the payment stage, and much more. This expanded functionality provides greater flexibility and adaptability in addressing diverse tax withholding scenarios.

Furthermore, the extended withholding tax process introduces two critical concepts: the withholding tax country and the withholding tax type. The **Withholding Tax Country** field allows you to designate a distinct country for the purpose of calculating and reporting withholding taxes. For instance, in the United States, taxes withheld on specific income of foreign individuals or corporations are reported on IRS Form 1042, and this withholding tax is contingent upon the payee's country of origin.

Conversely, the **Withholding Tax Type** field empowers you to segregate various types of withholdings, such as federal withholding, state withholding, and more. To manage these aspects, you maintain information related to withholding tax countries, withholding tax types, and withholding tax codes in the payee's master records.

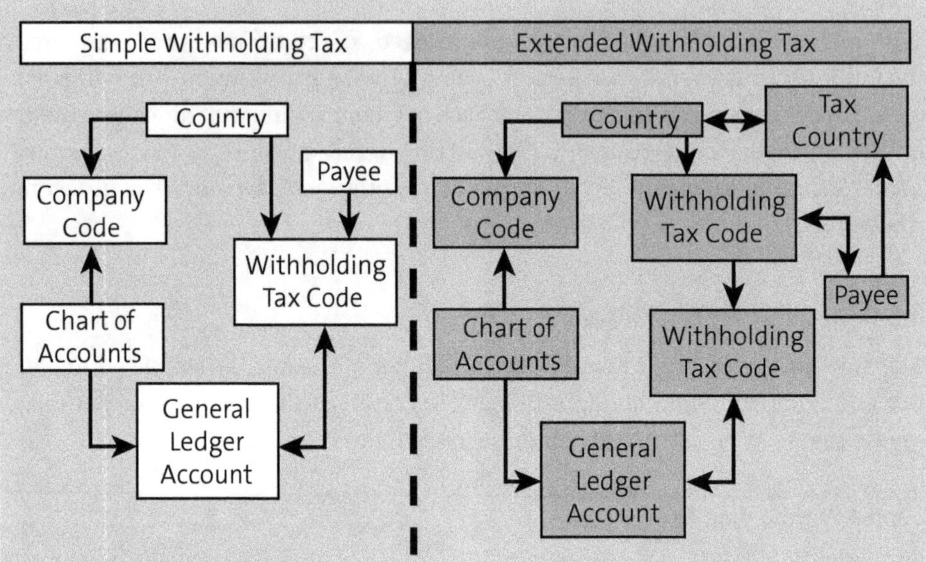

Figure 4.44 Simple Withholding Tax

4.6.2 Withholding Tax Accounts

In this section, we'll discuss withholding tax accounts for both simple withholding tax and extended withholding tax.

Simple Withholding Tax

For simple withholding tax, you assign general ledger accounts directly to withholding tax codes that are associated with the payee account. You carry out this assignment in the configuration activity accessed via menu path **IMG • Financial Accounting Global Settings • Withholding Tax • Withholding Tax • Posting • Define Accounts for Withholding Tax.**

Figure 4.45 shows the account used for withholding tax. In SAP S/4HANA, the system employs transaction key QST for general ledger account determination when dealing with simple withholding taxes. Moreover, for more detailed general ledger account determination, you have the option to utilize account modifiers like debit/credit indicators and withholding tax codes. These modifiers allow for a finer level of granularity in account selection. When you clear open items in customer or vendor accounts, typically during the posting of a payment transaction, the general ledger accounts associated with this transaction key come into play and are posted accordingly. Now, let's delve into the extended withholding tax process in SAP S/4HANA.

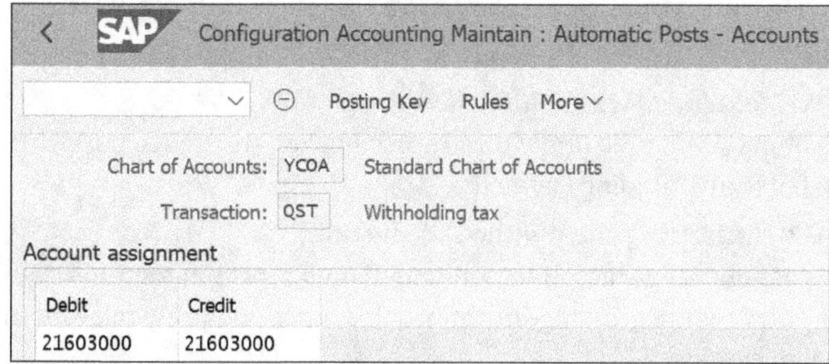

Figure 4.45 Account for Withholding Tax

Extended Withholding Tax

The assignment of general ledger accounts for extended withholding tax is accomplished through configuration activities found under IMG menu path **Financial Accounting • Financial Accounting Global Settings • Withholding Tax • Extended Withholding Tax • Posting • Accounts for Withholding Tax**. When you push the activity button in the configuration, the **Configuration Accounting Maintain: Automatic Posts—Accounts** screen will appear (see Figure 4.46). This setup ensures precise tracking and reporting of withholding tax obligations.

Configuration Accounting Maintain : Automatic Posts - Accounts

Posting Key Rules More

Chart of Accounts: YCOA Standard Chart of Accounts

Transaction: WIT Extended withholding tax

Account assignment

Withholding ta...	Withholding ta...	Debit	Credit
		21603000	21603000
	01	21603000	21603000
01		21603000	21603000
01	00	21603000	21603000
01	01	21603000	21603000
01	02	21603000	21603000
01	03	21603000	21603000
01	04	21603000	21603000
01	10	21603000	21603000
01	11	21603000	21603000
01	20	21603000	21603000
01	30	21603000	21603000
02		21603000	21603000
02	01	21603000	21603000
02	02	21603000	21603000
02	03	21603000	21603000
02	04	21603000	21603000

Figure 4.46 Configuration of Extended Withholding Tax (Transaction Key WIT)

You configure general ledger account determination for the following types of transactions:

- **Transaction key WIT (extended withholding tax)**
  General ledger accounts configured under this transaction key are used to post withholding tax deducted from outgoing payments.
- **Transaction key GRU (offsetting entry without deduction)**
  SAP S/4HANA uses general ledger accounts configured under this transaction key to post an offsetting entry for the grossing-up option. This function is only relevant in Argentina for foreign vendors.
- **EE transaction key OPO (self-withholding)**
  SAP S/4HANA uses general ledger accounts configured under this transaction key to post withholding tax calculated and deducted from incoming payments.
- **EE transaction key OFF (offsetting entry with deduction)**
  In SAP S/4HANA, the system utilizes general ledger accounts configured under this specific transaction key to record the offsetting entry. This approach becomes applicable when the configuration for the withholding tax code mandates SAP S/4HANA to generate two withholding tax line items with opposite debit and credit signs.

For each of these transactions, there are available account modifiers, including the debit/credit indicator, withholding tax type, and withholding tax code. Enabling all of these rule modifiers allows you to perform general ledger account determination at the most granular level of detail.

However, it's important to keep in mind that implementing such intricate general ledger account determination comes with its complexities. Maintaining this level of configuration can be challenging on an ongoing basis and may require significant effort during month-end reconciliation processes. Hence, striking the right balance between granularity and practicality is a key consideration.

4.7 Interest Calculation

The SAP S/4HANA system offers extensive features for computing interest on overdue items and account balances. It's worth noting that you can apply this interest calculation functionality not only to AR and AP but also to general ledger accounts.

In this section, we explain all relevant elements to configure the automatic interest calculation and we explain the account determination objects in detail.

4.7.1 Business Process Overview

Figure 4.47 illustrates the intricate connections between various account determination elements that play a role in the interest calculation process. All the objects shown, their meanings, and how they interact with each other will be explained in the following section.

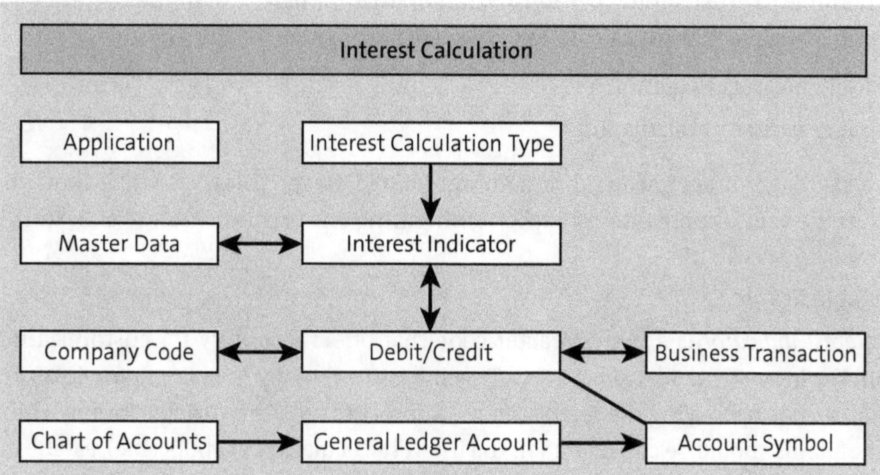

Figure 4.47 Interest Calculation Account Determination Objects

4.7.2 Account Determination Objects

Apart from the more apparent elements like a company code and a chart of accounts, the interest calculation process relies on several additional account determination components, which we'll describe in the following sections.

Application

In this context, an *application* is a task domain associated with a particular business procedure. All configuration activities related to general ledger account determination for interest calculation are performed within a task domain specifically tailored for that application.

Table 4.2 lists the values pertinent to the current topic. It's essential to note that these values come predefined and cannot be altered. Depending on the configuration task at hand, the characteristic's value is automatically chosen.

Application	Transactions Allowed for the General Ledger Account
0002	Interest calculation on AR arrears
0009	Interest calculation on AP arrears
0005	Interest calculation on AR balances
0006	Interest calculation on AP balances

Table 4.2 Application Values for Interest Calculation

Interest Calculation Type

The interest indicator type is responsible for determining whether the interest calculation pertains to arrears (on a per-item basis) or account balances. In the current context, the following interest calculation types are pertinent:

- P, item interest calculation
- S, balance interest calculation

These values are predefined and remain immutable. Nevertheless, as you'll soon discover, these values come into play when configuring one or more interest indicators.

Interest Indicator

An *interest indicator* is a two-character code that offers flexibility for customization within the interest calculation process. This indicator allows you to configure intricate details, such as specifying the reference interest rate to be utilized, determining which types of items should be considered in the interest calculation (if applicable), establishing the frequency at which interest should be calculated, and more.

To define these interest indicators, you can access the configuration through IMG path **Financial Accounting • Accounts Receivable and Accounts Payable • Business Transactions • Interest Calculation • Interest Calculation Global Settings • Define Interest Indicator**. When you start the configuration activity, you first need to define interest indicators, as shown in Figure 4.48.

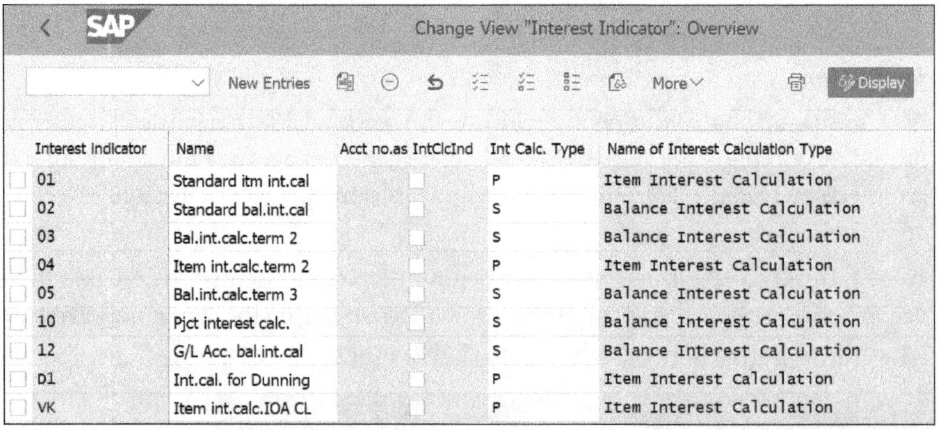

Change View "Interest Indicator": Overview

New Entries More Display

Interest Indicator	Name	Acct no.as IntClcInd	Int Calc. Type	Name of Interest Calculation Type
01	Standard itm int.cal		P	Item Interest Calculation
02	Standard bal.int.cal		S	Balance Interest Calculation
03	Bal.int.calc.term 2		S	Balance Interest Calculation
04	Item int.calc.term 2		P	Item Interest Calculation
05	Bal.int.calc.term 3		S	Balance Interest Calculation
10	Pjct interest calc.		S	Balance Interest Calculation
12	G/L Acc. bal.int.cal		S	Balance Interest Calculation
D1	Int.cal. for Dunning		P	Item Interest Calculation
VK	Item int.calc.IOA CL		P	Item Interest Calculation

Figure 4.48 Define Interest Indicator

Within the same menu area, you will discover configuration transactions that enable you to fine-tune the specifics of interest rate indicators, whether for arrears calculation or balance interest calculation. The system allows you to create multiple interest indicators to accommodate various scenarios. For instance, you can create distinct interest indicators for different reference interest rates, varying interest calculation frequencies, and so forth.

It's important to note that each interest indicator can be associated with only one type of interest calculation, which can be either item-based or account balance-based. These interest indicators are then assigned to master data records for effective utilization in the interest calculation process.

Master Data

To ensure that the automatic interest calculation processes operate effectively, it is imperative to assign the relevant interest indicators to all account master records that should be encompassed within these processes. In this context, master data pertains to three distinct categories: the general ledger account master, the customer account master, and the vendor account master. To facilitate this assignment, here is a detailed breakdown of where you can assign the interest indicator for each type of master record:

- **General ledger account master records**
  Within the general ledger account master records, you can designate the appropriate interest indicator to govern how interest calculations apply to general ledger accounts. This ensures that the system considers the predefined criteria for interest calculation specific to each general ledger account.
- **Customer account master records**
  For customer accounts, the interest indicator can be assigned at the customer account master level. By doing so, you define how interest calculations are to be carried out in relation to individual customer accounts. This allows for flexibility in tailoring interest calculations to suit different customer agreements or payment terms.
- **Vendor account master records**
  Similarly, vendor account master records provide a space for assigning the relevant interest indicator. This allocation ensures that interest calculations pertaining to vendor accounts adhere to the prescribed criteria, which may vary based on vendor-specific agreements or payment terms.

By assigning interest indicators at the appropriate master record levels, you establish a clear framework for interest calculation processes, enabling the system to apply interest rates, item inclusion criteria, and frequency settings consistently across general ledger accounts, customer accounts, and vendor accounts in alignment with your business requirements.

Business Transaction

Within this context, a *business transaction* has a predefined identifier consisting of four characters. This identifier serves as a distinctive label for categorizing various types of business transactions that pertain to interest posting. When configuring the general ledger account determination for interest calculation, it is imperative to account for all potential business transactions that hold relevance for your company's financial operations.

Table 4.3 lists important business transaction values within the context of configuring interest calculation posting. These values are provided as samples to help you grasp the concept. In the next section, we will delve into a discussion of the remaining account determination components, shedding further light on the intricate web of factors involved in the interest calculation process.

Value	Name
1000	Interest received posting
1020	Debit interest—value date in the past
2000	Interest paid posting
2020	Credit interest—value date in the past

Table 4.3 Business Transaction Values for Interest Calculation

4.7.3 Interest Accounts

Configuration options for interest calculation can be accessed within designated menu area **IMG • Financial Accounting • Accounts Receivable and Accounts Payable • Business Transactions • Interest Calculation • Interest Posting.**

Within this menu area, you'll encounter various configuration transactions tailored to the specific requirements of general ledger account determination, depending on whether you're configuring the interest calculation for AR or AP and whether the interest calculation is based on arrears or account balances. It's worth noting that SAP ERP employs the symbolic account technique for general ledger account determination within the interest calculation process.

In our previous discussion, we explored the account determination objects, including the application, business transaction, and interest indicator. To successfully configure any aspect of general ledger account determination, you'll typically follow these key steps:

1. **Define account symbols**
   Create account symbols that correspond to different general ledger accounts. These symbols serve as representations of the accounts involved in the interest calculation process.
2. **Assign account symbols**
   Pair two account symbols for each combination of business transaction, interest indicator, company code, and business area characteristics. One symbol is designated for debit postings, while the other is designated for credit postings. This assignment links the appropriate symbols to specific transaction scenarios.
3. **Assign general ledger accounts**
   Associate general ledger accounts with their corresponding account symbols. This

step ensures that the system knows which general ledger accounts to utilize for debit and credit postings under varying circumstances.

Figure 4.49 provides an illustrative example of AR account determination configuration. It showcases how you can conveniently access commands for defining account symbols and assigning AR accounts from a single transaction interface, streamlining the configuration process for efficient setup and management.

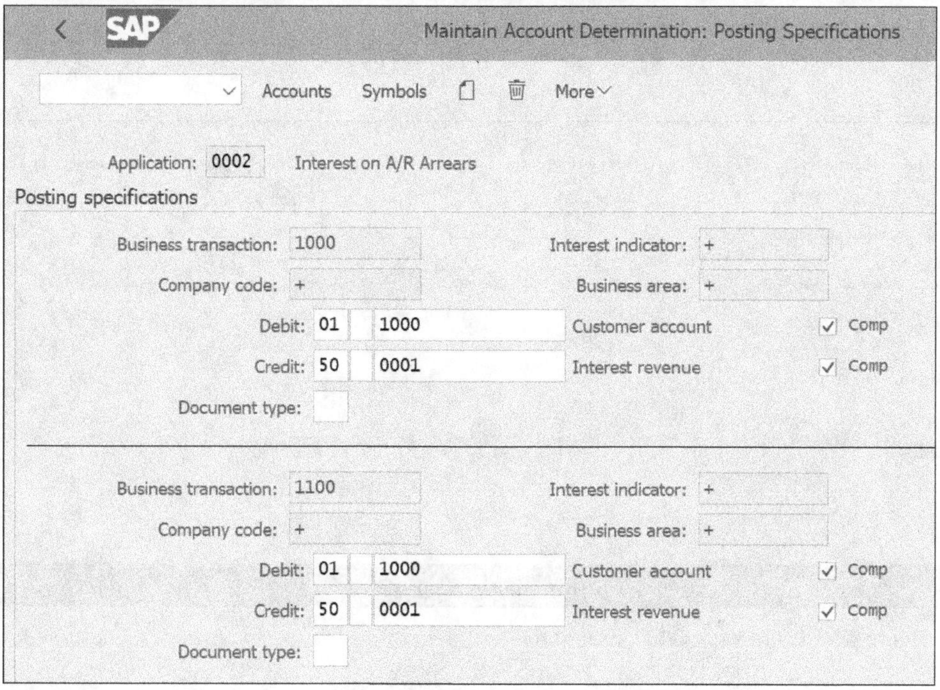

Figure 4.49 Account Determination for AR Accounts Based on Items

4.8 GR/IR Analysis

Goods receipt/invoice receipt analysis is a crucial part of financial and inventory management in SAP that helps organizations ensure the accuracy of their financial records, particularly related to procurement and inventory.

In this section, we explain what the GR/IR account in SAP is good for and how GR/IR automated postings are set up.

4.8.1 Business Process Overview

Here's how GR/IR analysis works:

1. **Goods receipt**
   A GR transaction typically follows placing an order. When an organization receives

goods from a vendor or supplier, they create a GR document in the SAP system. This document acknowledges the receipt of the goods into the company's inventory. It records details such as the quantity, value, and location of the received goods. A GR entry example is shown in Figure 4.50.

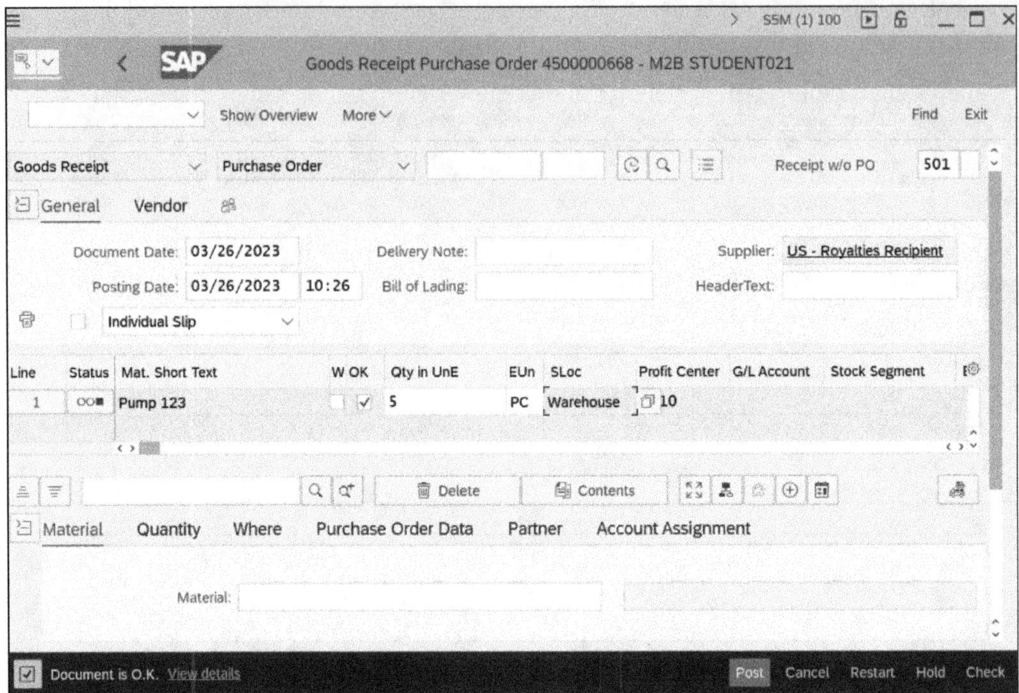

Figure 4.50 Example of GR Transaction

2. **Invoice receipt**
   When a vendor or supplier sends an invoice for the delivered goods, the organization creates an IR document in SAP. This document records the financial aspect of the transaction, including the amount to be paid to the vendor. An IR entry example is shown in Figure 4.51.
3. **Matching GR and IR**
   In an ideal scenario, the quantity and value in the GR document should match those in the IR document. However, discrepancies can occur for various reasons, such as incorrect quantities, prices, or delivery issues. These discrepancies can lead to financial inaccuracies.

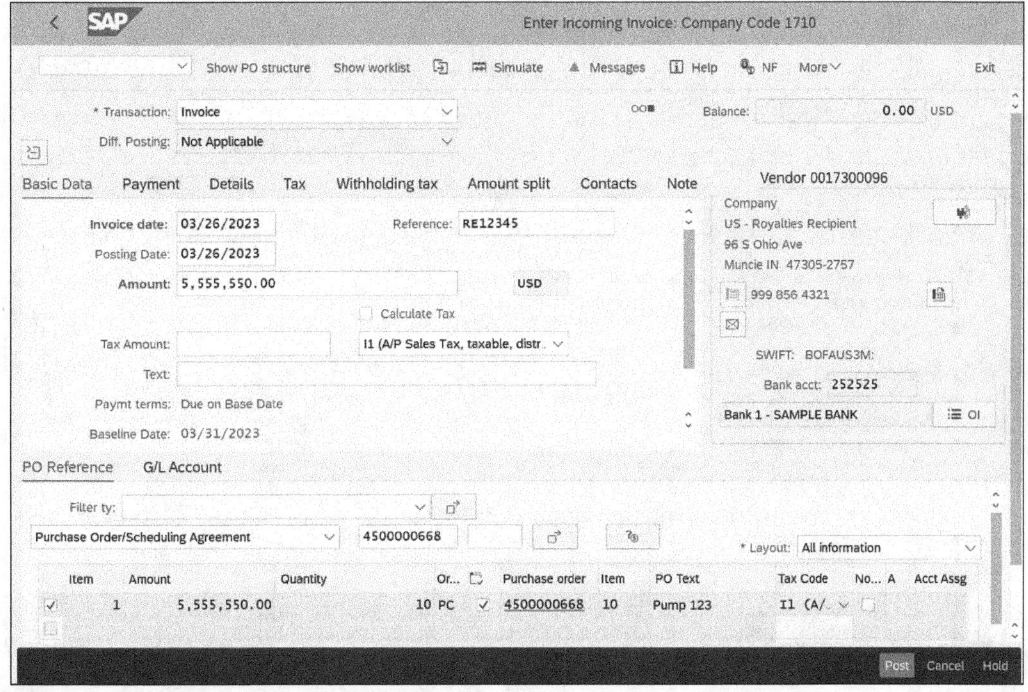

Figure 4.51 Example of IR Transaction

4. **GR/IR account**

   SAP maintains a special account called the GR/IR account to track discrepancies. When you create a GR, the system records the value in the GR/IR account as a credit (liability). When you create an IR, the system records the value as a debit (expense). The goal is to ensure that the GR/IR account balance is zero, signifying that all goods have been received and correctly invoiced. In Figure 4.52, you can see a line item report for a GR/IR account. The lines with WE document types are for GR transactions, and the lines marked with RE document types are for IR transactions. In this case, all WE lines have matching RE lines, which means that the RE transactions can be cleared against the WE transactions. The status of each line will then turn green.

5. **GR/IR analysis**

   GR/IR analysis in SAP involves regularly reconciling the GR/IR account to identify and resolve any discrepancies between GR and IR. This analysis helps ensure the accuracy of financial statements as well as inventory records. If discrepancies are found, they are investigated and resolved, which may involve adjusting financial records or contacting a vendor for clarification. In our example, there is a discrepancy between the GR and IR for purchase transaction 4500000232 (see Figure 4.53). The received quantity is 12 pieces, but the billed quantity is 24 pieces.

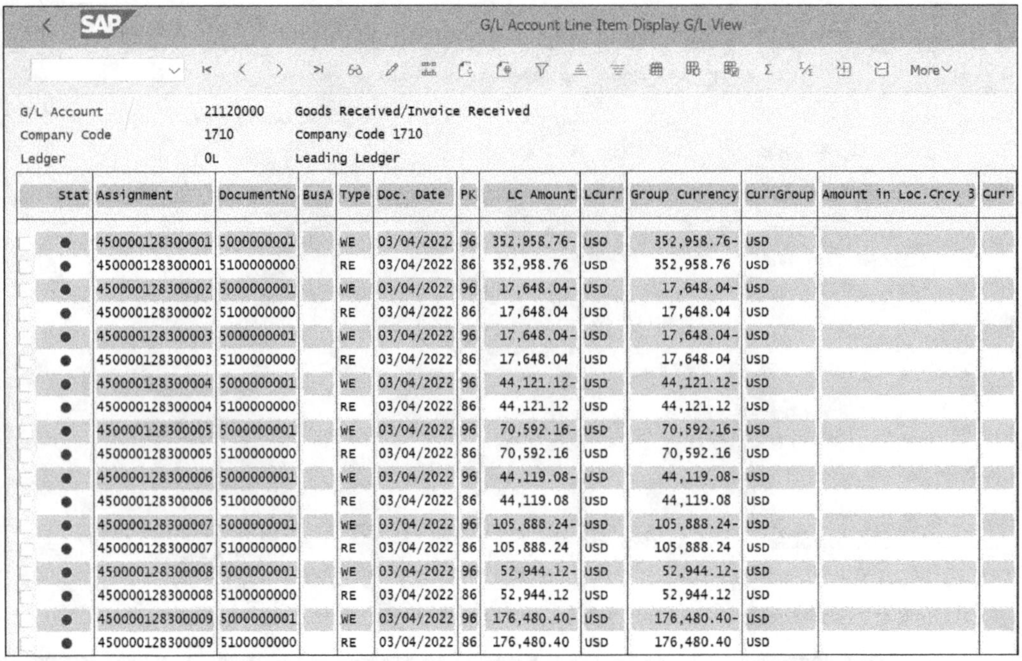

G/L Account Line Item Display G/L View

G/L Account 21120000 Goods Received/Invoice Received
Company Code 1710 Company Code 1710
Ledger 0L Leading Ledger

Stat	Assignment	DocumentNo	BusA	Type	Doc. Date	PK	LC Amount	LCurr	Group Currency	CurrGroup	Amount in Loc.Crcy 3	Curr
●	450000128300001	5000000001		WE	03/04/2022	96	352,958.76-	USD	352,958.76-	USD		
●	450000128300001	5100000000		RE	03/04/2022	86	352,958.76	USD	352,958.76	USD		
●	450000128300002	5000000001		WE	03/04/2022	96	17,648.04-	USD	17,648.04-	USD		
●	450000128300002	5100000000		RE	03/04/2022	86	17,648.04	USD	17,648.04	USD		
●	450000128300003	5000000001		WE	03/04/2022	96	17,648.04-	USD	17,648.04-	USD		
●	450000128300003	5100000000		RE	03/04/2022	86	17,648.04	USD	17,648.04	USD		
●	450000128300004	5000000001		WE	03/04/2022	96	44,121.12-	USD	44,121.12-	USD		
●	450000128300004	5100000000		RE	03/04/2022	86	44,121.12	USD	44,121.12	USD		
●	450000128300005	5000000001		WE	03/04/2022	96	70,592.16-	USD	70,592.16-	USD		
●	450000128300005	5100000000		RE	03/04/2022	86	70,592.16	USD	70,592.16	USD		
●	450000128300006	5000000001		WE	03/04/2022	96	44,119.08-	USD	44,119.08-	USD		
●	450000128300006	5100000000		RE	03/04/2022	86	44,119.08	USD	44,119.08	USD		
●	450000128300007	5000000001		WE	03/04/2022	96	105,888.24-	USD	105,888.24-	USD		
●	450000128300007	5100000000		RE	03/04/2022	86	105,888.24	USD	105,888.24	USD		
●	450000128300008	5000000001		WE	03/04/2022	96	52,944.12-	USD	52,944.12-	USD		
●	450000128300008	5100000000		RE	03/04/2022	86	52,944.12	USD	52,944.12	USD		
●	450000128300009	5000000001		WE	03/04/2022	96	176,480.40-	USD	176,480.40-	USD		
●	450000128300009	5100000000		RE	03/04/2022	86	176,480.40	USD	176,480.40	USD		

Figure 4.52 Example of GR/IR Account, Displayed by General Ledger Line Item Report

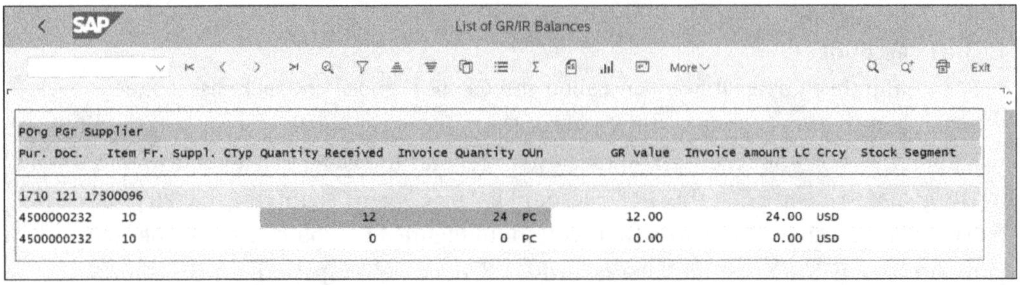

List of GR/IR Balances

POrg PGr Supplier

Pur. Doc.	Item	Fr. Suppl.	CTyp	Quantity Received	Invoice Quantity	OUn	GR value	Invoice amount LC	Crcy	Stock Segment
1710 121 17300096										
4500000232	10			12	24	PC	12.00	24.00	USD	
4500000232	10			0	0	PC	0.00	0.00	USD	

Figure 4.53 Example of Discrepancy between GR and IR

GR/IR analysis is an integral part of maintaining accurate financial records and controlling costs within an organization's procurement and inventory processes. It helps prevent overpayment, underpayment, and inaccuracies in financial reporting, ultimately contributing to better financial management and compliance with accounting standards.

4.8.2 GR/IR Analysis Accounts

The GR/IR account serves as a reconciliation account in which debit entries represent goods received and credit entries correspond to invoices received. These two types of transactions can offset and reconcile only if they pertain to the same business transaction. However, if goods receipts and invoice receipts are associated with different transactions, both types need to be reclassified into two separate accounts: one for items that have been invoiced but not yet delivered and another for items that have been delivered but not yet invoiced. To facilitate this automatic reclassification, it is essential to maintain the relevant accounts in the system's configuration settings, as described in the next section. We'll also discuss the specific settings for BNG and GNB accounts.

GR/IR Configuration Options

Configuration options for GR/IR accounts can be accessed within designated menu area **IMG • Financial Accounting • General Ledger Accounting • Business Transactions • Periodic Processing • Reclassify • Define Adjustment Accounts for GR/IR Clearing.**

There are two main settings for two different procedures (Figure 4.54):

- **BNG**: Invoiced but not yet delivered
- **GNB**: Delivered but not yet invoiced

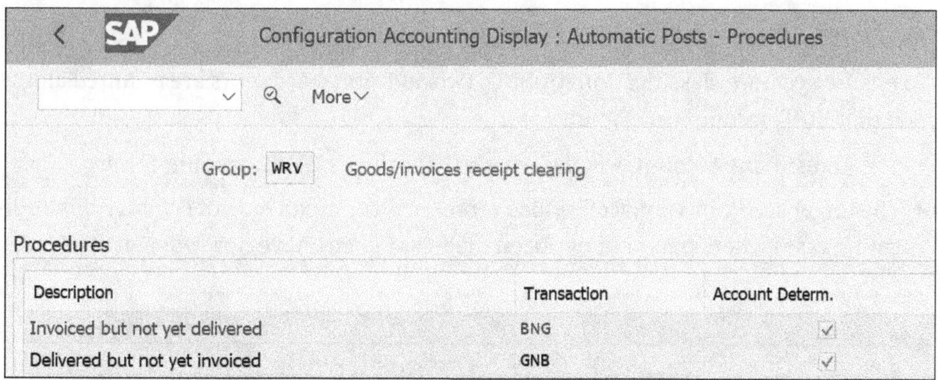

Figure 4.54 Procedures for GR/IR Automatic Postings

For both of these processes, distinct account determinations are utilized. The GR/IR account remains unaffected throughout. Instead, any necessary adjustments to the GR/IR accounts are made at the end of the accounting period following the initiation of the GR/IR run.

Accounts for GNB

To accurately classify transactions, two additional accounts are required alongside the GR/IR account (see Figure 4.55):

- The **Adjustment Account**, which is used to offset and balance the GR/IR account to zero.
- The target account (**Targ.acct**), which represents the debit side of transactions, indicating instances where goods have been received but payment has not yet been made.

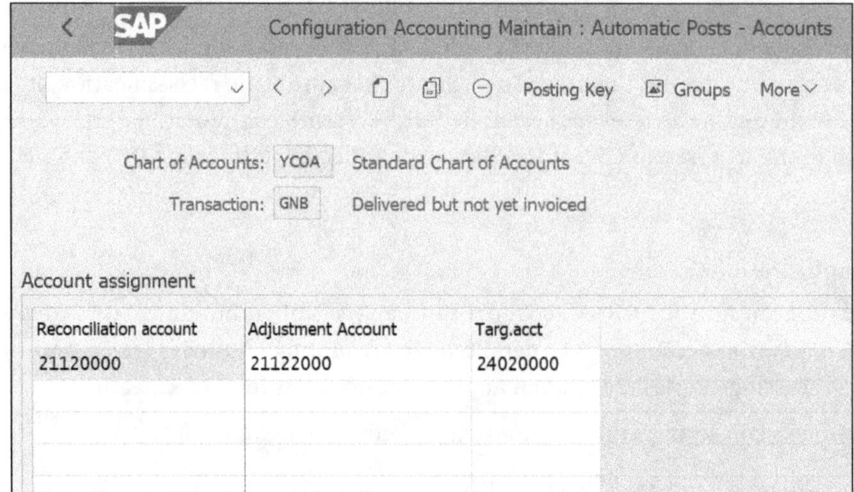

Figure 4.55 Account Determination for GNB

Accounts for BNG

To ensure accurate classification for BNG, two additional accounts are required alongside the GR/IR account (see Figure 4.56):

- The **Adjustment Account**, which serves to balance the GR/IR account to zero.
- The target account (**Targ.acct**), which represents the credit side of transactions, indicating cases where payment has been made but goods have not yet been received.

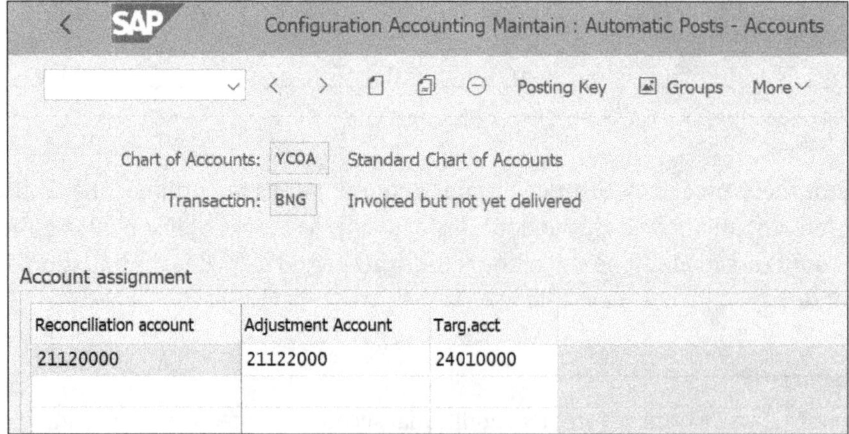

Figure 4.56 Account Determination for BNG

4.9 Summary

The process of general ledger account determination for standard AR and AP transactions is relatively straightforward and user-friendly. All the requisite configuration transactions are thoughtfully organized within the configuration guide, conveniently grouped under distinct business transactions.

It's essential to note that the assignment of alternative reconciliation accounts, as elucidated in this chapter, primarily pertains to documents posted in financial accounting. In contrast, for customer billing documents originating from sales and distribution, the reconciliation account determination follows a different approach known as the *condition technique*, which will be comprehensively covered in Chapter 8.

Given SAP S/4HANA's reputation as a highly integrated system, the general ledger account determination processes for AR and AP exhibit close interconnections with both the purchasing and the sales and distribution areas. As a result, the general ledger account determination procedures detailed in this chapter are generally applicable to transactions initiated not only in AR and AP but also in purchasing and sales and distribution, fostering consistency and ease of management.

Furthermore, another aspect closely intertwined with AR and AP is cash and banking. The next chapter of this book delves into the intricacies of general ledger account determination for cash and bank transactions, shedding light on this essential facet of financial operations within SAP S/4HANA. This comprehensive approach ensures that you have a holistic understanding of how general ledger account determination impacts various aspects of your financial processes.

5

Chapter 5
Cash and Banking

Cash management and banking are crucial parts of the accounting process. In this chapter, we'll walk through the main business processes in these areas that have an impact on accounting and discuss configuring automatic determination for all accounting entries.

In this chapter, we will first explore the core business procedures within the domains of cash management and banking that significantly influence accounting practices. Then we will delve into the crucial aspects of general ledger account master data, offering a detailed walkthrough covering how to set up automated determination for all accounting entries. This comprehensive guide will empower you to effectively manage and align your financial and accounting processes in the context of cash management and banking operations.

5.1 House Bank and Other Subaccounts

Regardless of whether a company utilizes electronic banking features in SAP S/4HANA, the AP functionality is likely to be employed. To conduct payment runs in AP, house banks and house bank accounts must be established. This section will discuss the process of general ledger account determination for house bank accounts and subaccounts.

5.1.1 House Bank

A *house bank* is the financial institution where a company holds its bank accounts. In SAP S/4HANA, house bank identifiers are defined and linked to a company code. Bank account identifiers then are defined for individual bank accounts and assigned to the house bank identifiers. Multiple house banks and multiple bank accounts can be assigned to a company code. As shown in Figure 5.1, the general ledger account field (**G/L Acct**) contains the assignment of a general ledger account to a bank account identifier. This configuration is performed via the path **IMG • Financial Accounting • Bank Accounting • Bank Accounts • Define House Banks**; there, from the menu on the left side, select **Company Code • House Banks • Bank Accounts**.

Figure 5.1 includes a second general ledger account in the **Discount Acct** field for discounted bills of exchange. This field is not the same as the general ledger account for cash discounts. We will discuss bills of exchange in Section 5.8.

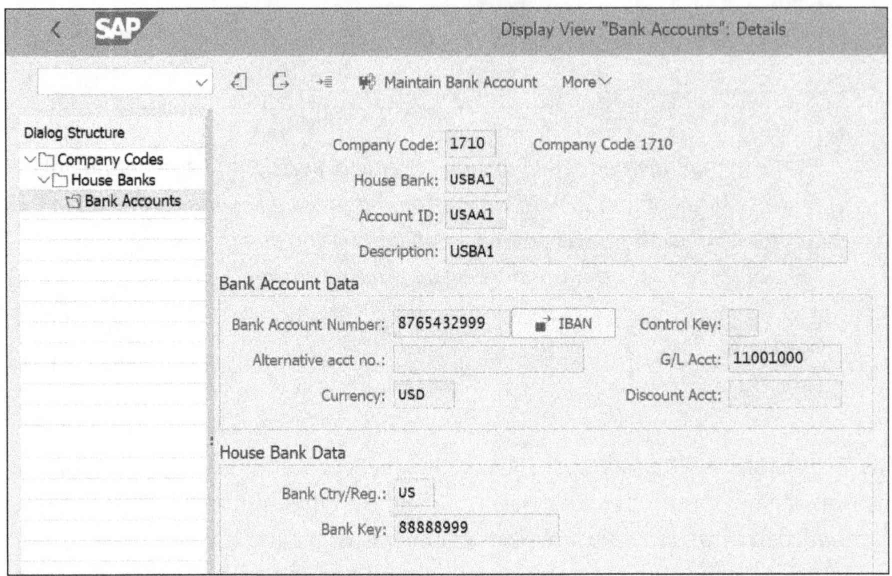

Figure 5.1 Assignment of General Ledger Account to Bank Account

Another important consideration when it comes to bank transactions, especially when dealing with outgoing payments, involves the establishment of bank subaccounts. We'll explore the topic of subaccounts in the next section.

5.1.2 Bank Subaccounts by Payment Method

A standard outgoing payment typically involves debiting a vendor account and clearing the corresponding vendor documents, while crediting the house bank account. However, the actual payment, depending on the method used (e.g., check, wire transfer), may not credit the house bank for several days. To accurately capture this "float" amount, you can establish bank subaccounts for all applicable payment methods (as shown in Figure 5.2). In the case of outgoing payments, this subaccount is credited instead of the primary house bank account. Subsequently, during the bank reconciliation statement process, the credited amount then is transferred from the bank subaccount to the primary house bank account. You can set up this configuration via the following menu path: **IMG • Financial Accounting • Accounts Receivable and Accounts Payable • Business Transactions • Outgoing Payments • Automatic Outgoing Payments • Payment Method/ Bank Selection for Payment Program • Set Up Bank Determination for Payment Transactions**. Once there, from the menu on the left side, select **Bank Selection • Bank Accounts**.

In the example shown in Figure 5.2, for payment methods C, D, E, I, M, P, and T, the outgoing payments procedure will record credits in the respective bank subaccounts instead of the primary house bank account. If you are integrating electronic or manual bank statement reconciliation procedures within the SAP S/4HANA system, this configuration is essential. However, the decision to establish these bank subaccounts depends on the specific requirements of your business processes.

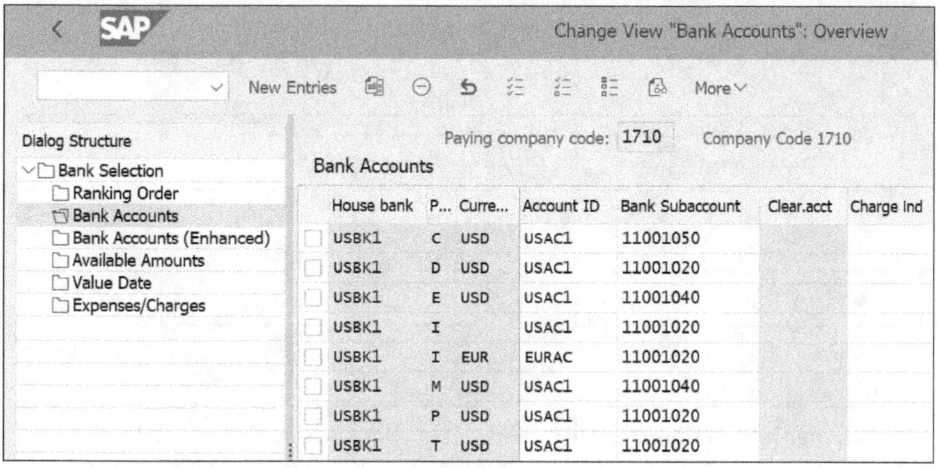

House bank	P...	Curre...	Account ID	Bank Subaccount	Clear.acct	Charge Ind
USBK1	C	USD	USAC1	11001050		
USBK1	D	USD	USAC1	11001020		
USBK1	E	USD	USAC1	11001040		
USBK1	I		USAC1	11001020		
USBK1	I	EUR	EURAC	11001020		
USBK1	M	USD	USAC1	11001040		
USBK1	P	USD	USAC1	11001020		
USBK1	T	USD	USAC1	11001020		

Figure 5.2 Define Bank Subaccounts

In the following section, we will study the account determination elements relevant to the electronic bank statement process.

5.2 Account Determination

The setting of the account determination to process bank statements, deposited check transactions, bills of exchange, and lockbox transactions is quite complex and depends on different objects. In the following sections, we explain all the objects that determine the posting of an incoming or outgoing payment through the different bank transaction methods mentioned previously.

5.2.1 Objects for Automatic Payment Transactions

Figure 5.3 displays the account determination components that pertain to the setup of the configuration for automatic processing of bank statements, deposited check transactions, lockbox transactions, payment requests, and bills of exchange. This setup is a must to ensure efficient, accurate, and complete processing of bank transitions and their integration in financial accounting.

Now let's analyze the intricate aspects of general ledger account determination for electronic bank statement configuration. We start with this complex process because it incorporates all the account determination elements pertinent to banking transactions, as illustrated in Figure 5.3.

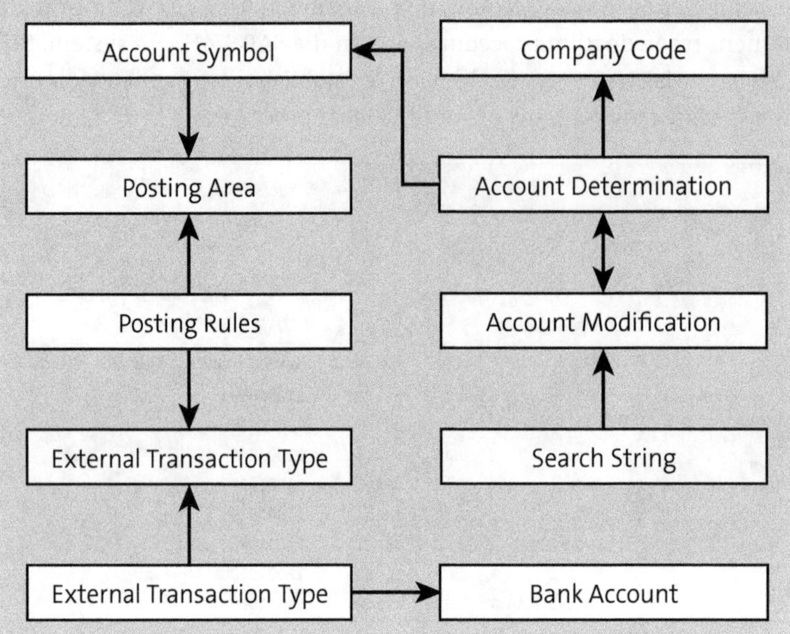

Figure 5.3 Account Determination Objects for Electronic Bank Statement Transaction Processing

Other banking transactions, like manual bank statements and processing of incoming checks, only rely on a subset of these account determination elements. Thus, once you've grasped the interplay among all the account determination components integral to electronic bank statement configuration, comprehending account determination for less complex procedures will become more straightforward.

It's worth noting that the majority of settings pertaining to electronic bank statement configuration can be conveniently managed through a single transaction (as depicted in Figure 5.4). You can access this transaction via menu path **IMG • Financial Accounting • Bank Accounting • Business Transactions • Payment Transactions • Electronic Bank Statement • Make Global Settings for Electronic Bank Statement.**

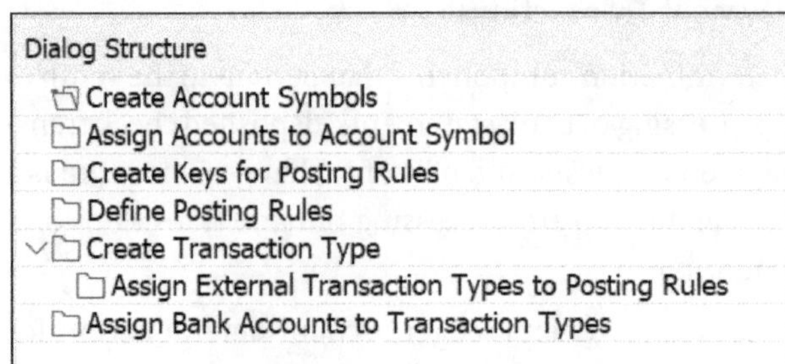

Figure 5.4 Bank Transaction Configuration

Typically, the electronic format(s) in which your bank presents transaction data to its customers are determined by the bank itself. Moreover, various banks across different countries employ distinct data formats for delivering transaction information to their corporate clients. Notable examples of these formats include SWIFT, BAI, MultiCash, and MT940. As anticipated, each format comes with detailed specifications governing the presentation of various transaction data components. In SAP S/4HANA, you establish an internal transaction type that aligns with the data format offered by your designated house bank.

5.2.2 Internal Transactions Type

Within SAP S/4HANA, the data format provided by your house bank is pinpointed by utilizing an account determination element known as the *internal transaction type* (see the **TransTyp** field). As depicted in Figure 5.5, SAP S/4HANA offers a range of transaction types tailored to various formats, with the flexibility to generate additional internal transaction types as needed. It's important to note that you can link only one internal transaction type to a single bank account. However, it's essential to recognize that the internal transaction type is essentially a mnemonic code. For instance, if you're dealing with two bank accounts that both receive bank transaction data in the MultiCash format, but the specifics of the transactions for each account vary slightly, you can establish two distinct internal transaction types, like MC01 and MC02, both associated with the MultiCash format, and then assign them to their respective bank accounts.

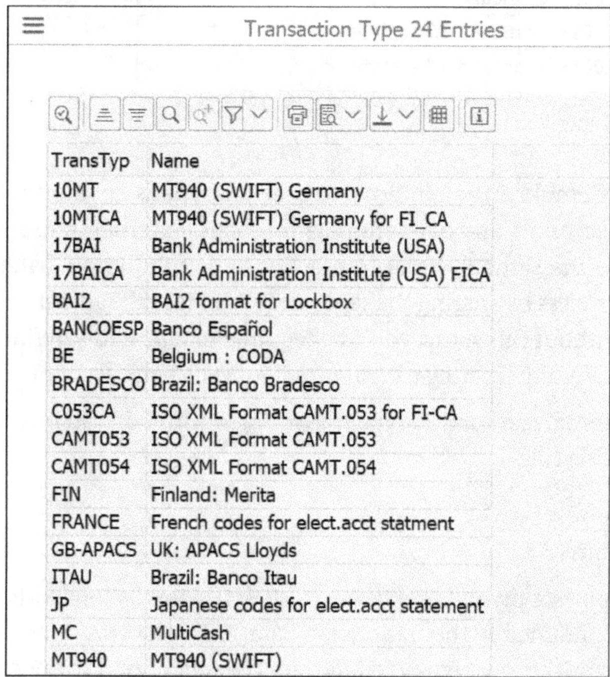

Transaction Type 24 Entries

TransTyp	Name
10MT	MT940 (SWIFT) Germany
10MTCA	MT940 (SWIFT) Germany for FI_CA
17BAI	Bank Administration Institute (USA)
17BAICA	Bank Administration Institute (USA) FICA
BAI2	BAI2 format for Lockbox
BANCOESP	Banco Español
BE	Belgium : CODA
BRADESCO	Brazil: Banco Bradesco
C053CA	ISO XML Format CAMT.053 for FI-CA
CAMT053	ISO XML Format CAMT.053
CAMT054	ISO XML Format CAMT.054
FIN	Finland: Merita
FRANCE	French codes for elect.acct statment
GB-APACS	UK: APACS Lloyds
ITAU	Brazil: Banco Itau
JP	Japanese codes for elect.acct statement
MC	MultiCash
MT940	MT940 (SWIFT)

Figure 5.5 Internal Transaction Types

5.2.3 External Transactions Type

Each internal transaction type in SAP S/4HANA corresponds to a set of external transaction types. These external transaction types serve to categorize the various bank transactions associated with a specific corresponding internal transaction type, aligning with the data format used for receiving bank transaction data.

To effectively manage your bank account, you'll need to establish an external transaction type for each distinct transaction type that is applicable to that account. For instance, you might create external transaction types for incoming payments by check, incoming payments via wire transfer, bank charges, outgoing payments by check, interest income, foreign exchange fees, returned checks, and interbank transfers. Typically, banks maintain an extensive list of transaction codes, each uniquely identifying different transactions the bank processes.

However, within SAP S/4HANA, you have the flexibility to select and create only those external transaction types that are pertinent to your specific bank account. Figure 5.6 shows the setup of an external transaction type for the BAI2 format. Each row in this example represents a distinct transaction, such as 115 (lockbox deposit), 165 (preauthorized ACH credit), 195 (incoming money transfer), and 214 (foreign exchange credit).

Assign External Transaction Types to Posting Rules

External Trans...	+/...	Posting R...	Interpretation Algorithm
115	+	F002	001: Standard Algorithm
165	+	F001	001: Standard Algorithm
195	+	F017	001: Standard Algorithm
214	+	F009	001: Standard Algorithm

Figure 5.6 External Transaction Type

You can now see the benefits of employing an internal transaction type as an intermediary bridge connecting bank accounts and bank transaction codes (external transaction types). When you allocate the same internal transaction type to multiple bank accounts, you streamline the process by sharing the same array of external transaction codes across these accounts. Without this setup, you'd be required to repeatedly define external transaction codes, essentially creating a separate set for each bank account.

In the setup process, every external transaction type is associated with both an interpretation algorithm and a posting rule.

5.2.4 Interpretation Algorithm

The interpretation algorithm plays a pivotal role in how SAP S/4HANA deciphers the reference document number contained in the transaction data. SAP S/4HANA offers a wide array of interpretation algorithms designed to interpret the reference document number in various ways, such as ascertaining it as the check number (011, 012, and 013), accounting document number (020), reference document number within an accounting document (021), document sourced from an invoice list (025), payment order number (029), and so forth (see Figure 5.7). You also have the flexibility to craft up to nine additional custom interpretation algorithms for interpreting bank transaction data.

It's important to note that unlike posting rules (discussed next), the interpretation algorithm primarily exerts an indirect influence on general ledger account determination.

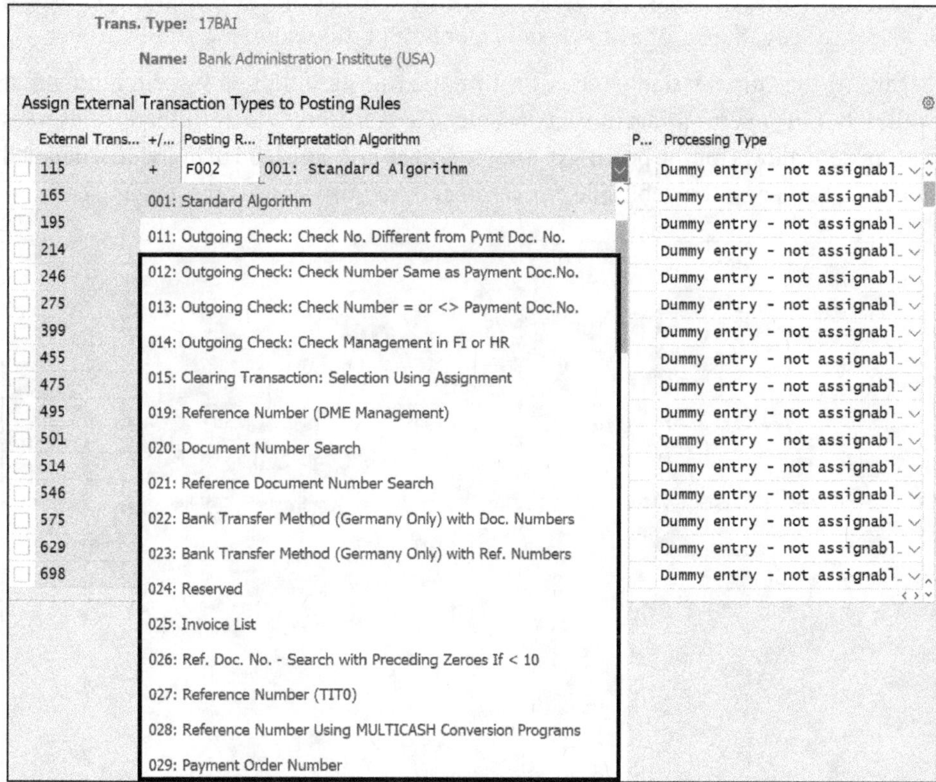

Figure 5.7 Interpretation Algorithms

5.2.5 Posting Rules

The posting rule linked to an external transaction type is in charge of orchestrating the automatic posting of debit and credit entries to the appropriate general ledger accounts for that particular transaction type. A clearer understanding of how this works will emerge once we delve into the configuration of a posting rule.

Let's illustrate this with the example featured in Figure 5.6, as mentioned earlier, in which external transaction type 165 is associated with posting rule F001.

When it comes to a posting rule, you can assign up to two posting areas. Posting area 1 is employed for recording entries in general ledger accounts, encompassing operations like interest income, bank charges, interbank transfers, and the reconciliation of cash-in-transit accounts. In contrast, the settings designated for posting area 2 correspond to subledger postings, which involve tasks like clearing a vendor invoice for an outgoing payment or reconciling a customer invoice for an incoming payment.

Figure 5.8 shows a glimpse of the configuration of a posting rule denoted as *F001*. This rule sets out the posting keys and the account symbol to be employed for both debit and credit postings. To gain a deeper understanding of how these account symbols function in general ledger account determination, you might want to refer back to the earlier discussion of the symbolic account technique in Chapter 1, Section 1.3.2, which outlines the advantages of using account symbols in this context.

Regarding the **Compression** indicator within the rule definition, it's worth noting that while it doesn't exert a direct impact on general ledger account determination, it serves a purpose. It allows for the posting of a summarized entry instead of individual detailed entries, whether on the debit side, the credit side, or both sides of the ledger.

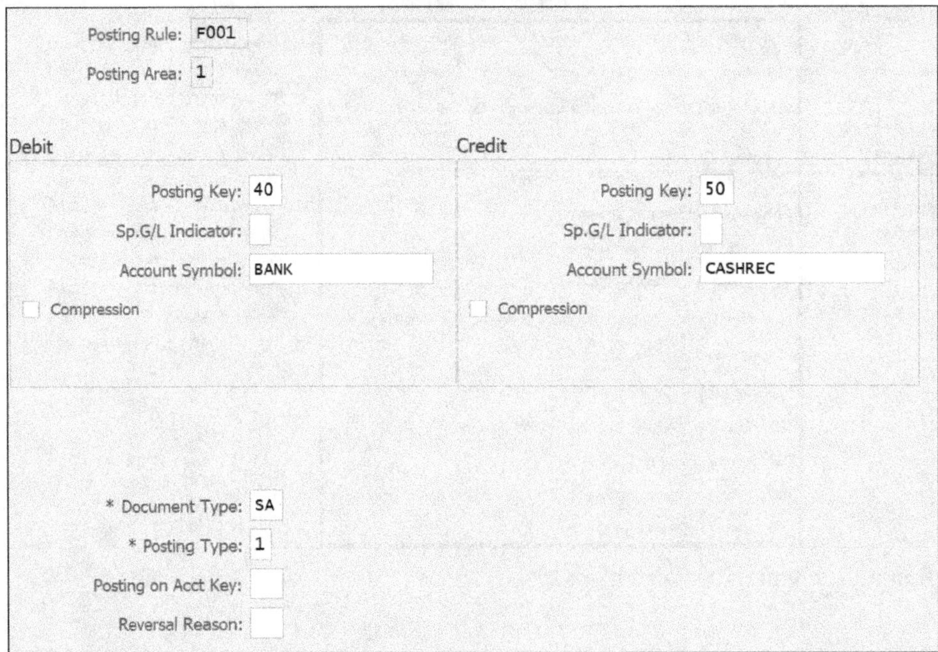

Figure 5.8 Posting Rule Configuration

In the next section, we will discuss the application of these account symbols in the process of determining general ledger accounts.

5.3 Automatic Bank Statements

In this and the next section, we will discuss general ledger account determination for different banking processes between your company and your house bank.

To process electronic bank statements automatically, you have to set up an interface with your bank to electronically receive bank transactions and post those transactions automatically. In subsequent sections, we will discuss other transactions, such as manual bank statements, lockbox statements, and others.

5.3.1 Business Process Overview

In the preceding section, we went over the assignment of the following elements (refer to Figure 5.3):

- Linking external transaction types to an internal transaction type
- Associating posting rules with external transaction types
- Assigning account symbols to the posting rules

The final step in the process of general ledger account determination involves the translation of these account symbols into actual general ledger accounts. This setup is performed through the following menu path: **IMG • Financial Accounting • Bank Accounting • Business Transactions • Payment Transactions • Electronic Bank Statement • Make Global Settings for Electronic Bank Statement**. There, from the menu on the left side, select **Assign Accounts to Account Symbols**.

5.3.2 Automatic Bank Statement Accounts

You have two methods at your disposal for allocating general ledger accounts to account symbols. You can either directly assign a general ledger account to an account symbol, or you can introduce an additional layer of abstraction by employing generic accounts. Figure 5.9 shows an example in which a single general ledger account (in the **G/L Account** column) is linked to an account symbol regardless of the currency involved, with no account modifier in place.

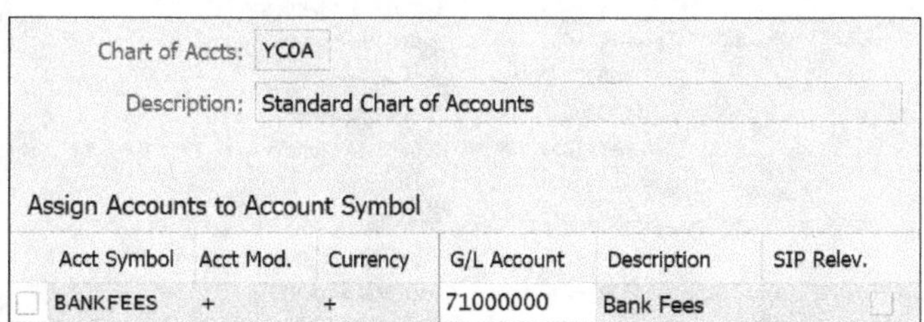

Figure 5.9 Single General Ledger Account Assignment

However, implementing such a straightforward account assignment approach may not always be feasible or advisable within various business contexts. To address the need for increased flexibility in determining general ledger accounts while accommodating business requirements, one can leverage the functionality of generic accounts offered by SAP S/4HANA.

By employing generic accounts, you can streamline your configuration management efforts by substituting common digits in account numbers with plus signs (+). During the posting process, SAP S/4HANA identifies these shared digits from the general ledger account number stored in the house bank master data. It thus becomes possible to allocate postings to distinct accounts based on the transaction context through the use of generic accounts.

Consider a scenario where multiple house bank accounts engage in transactions across various currencies. In such a setting, you can structure your chart of accounts to allocate the last three digits of each bank subaccount number to denote the bank transaction—for example, 040 for direct debit, 020 for outgoing transfer, and so forth. Simultaneously, the primary bank account number retains 000 as its concluding three digits.

Figure 5.10 shows the application of a generic bank account in a configuration. If two house banks are assigned general ledger accounts 1300000 and 1400000, the configuration depicted in Figure 5.10 will channel all direct debit transactions to subaccounts ending in 040 (e.g., 1300040 and 1400040), route all outgoing transfer transactions to subaccounts ending in 020, and handle all other transactions where there is no specific determination within the main bank accounts.

Chart of Accts: YCOA

Description: Standard Chart of Accounts

Assign Accounts to Account Symbol

Acct Symbol	Acct Mod.	Currency	G/L Account	Description	SIP Relev.
DIRECTDEBIT	+	+	+++++++040	Direct Debit	✓
DPCLEAR	+	+	+++++++070	Digital Pymts Clear	
DPEXTCLEAR	+	+	12530100	DP Ext. AR Clearing	
DPEXTCROSS	+	+	12530100	DP Ext.P Cross Clear	
DPEXTIN	+	+	+++++++030	DP Ext.Payment In	
DPEXTOUT	+	+	+++++++030	DP Ext.Payment Out	
DPEXTREFCLEAR	+	+	11001080	DP Ext.Ref.Bank Clea	
DPEXTTRANSFER	+	+	+++++++030	DP Ext.Pymt.Transfer	
DPSETTLEMENT	+	+	+++++++070	Digital Pymts Settle	
DPTRANSFER	+	+	+++++++030	Digital Pymts Trans	
INTPAID	+	+	71100000	Interest Paid	
INTREC	+	+	70100000	Interest Received	
OTHERTRANSF	+	+	+++++++030	Other Bank Transfer	✓
RETURNS	+	+	++++++++45	Bank Returns	
SFVCOUT	+	+	+++++++057	Suppl.Fin. VCard Out	
SUPFOUT	+	+	+++++++056	Suppl. Financ. Out	
TECH	+	+	+++++++090	Technical Account	
TRANSF	+	+	+++++++020	Outgoing Transfer	✓
WIREOUT	+	+	+++++++++1	Wire Out	✓

Figure 5.10 Generic Account Assignment

The success of this automated account determination approach hinges on the appropriate design of the chart of accounts, particularly the numbering logic for both primary bank accounts and bank subaccounts. If you're attempting to implement this process within an already operational SAP S/4HANA system, employing generic account logic may not be feasible. In such a scenario, you'll need to establish distinct posting rules for each of the individual transactions.

To sum up our discussion, here's a list of actions that encapsulates the entire process:

1. A bank account is associated with an internal transaction type.
2. Multiple external transaction types are linked to this internal transaction type.
3. Incoming bank data includes an external transaction type.
4. Each external transaction type is connected to a specific posting rule.
5. These posting rules are linked to debit and/or credit general ledger account symbols.
6. General ledger account symbols are used to determine one or more general ledger accounts.

Once you grasp how general ledger account determination functions in electronic bank statement processing, you'll find that the process becomes comparatively straightforward when dealing with manual bank statement processing.

5.4 Manual Bank Statements

Manual bank statement processing can be integrated with other financial processes within SAP S/4HANA, such as accounts payable and accounts receivable, to ensure that all financial transactions are accurately recorded and reconciled. Instead of developing a complex electronic interface for your house bank, you have the option to utilize the manual bank statement functionality within SAP S/4HANA, as we'll discuss in the following sections. With manual bank statement functionality, you can manually input bank statement data directly into SAP S/4HANA instead of relying on electronic transaction uploads.

5.4.1 Business Process Overview

Processing manual bank statements typically requires six steps:

1. **Create a new bank statement**
   Access the manual bank statement processing transaction (Transaction FF67) and choose the bank account for which you want to process a manual bank statement. This is the bank account from which you have received a paper or electronic bank statement. Specify the date of the bank statement you are processing. This date should match the date of the received statement.
2. **Enter the bank statement items**
   For each transaction on the bank statement, manually complete the **Transaction Date**, **Transaction Amount**, **Currency**, **Reference Numbers** (e.g., check numbers, invoice numbers, payment references), and **Transaction Description** fields.

3. **Assign the bank statement items the specific business transaction codes for the appropriate accounts**
   Assign a business transaction code to each transaction. This code categorizes the transaction type (e.g., customer payment, vendor payment, bank fee) and is used to determine how the transaction should be processed. For each transaction, specify the general ledger accounts to which it should be posted. Depending on the posting rule, you may need to assign a debit account, a credit account, or both.
4. **Post the bank statement**
   Review the entries to ensure accuracy. Validate that all transactions are correctly entered, assigned the appropriate business transaction code and posting rule, and reconciled with internal records. Once you are satisfied with the accuracy of the entries, you can post the transactions to the general ledger. This updates your financial accounts with the bank statement transactions.
5. **Review the posted bank statement and reconcile balances**
   Periodically, reconcile and balance your bank accounts to ensure that the transactions recorded in SAP S/4HANA match the actual bank statement.
6. **Clear the bank statement**
   When you clear a bank statement in SAP S/4HANA, you are marking the bank statement items as reconciled. This means that the bank statement items have been matched to the corresponding open items in the general ledger. Once a bank statement item is cleared, it can no longer be edited or deleted.

To do all of this, you need to set up the configuration. Thankfully, the configuration is much simpler than that of the electronic bank statement.

5.4.2 Manual Bank Statements Accounts

The configuration process for a manual bank statement is similar to that of an electronic bank statement but with some slight differences. Because there are no electronic data transfers from external house bank accounts, there's no need to maintain external transaction types or establish mappings between external and internal transaction types.

Instead, you maintain your own set of business transaction codes and link them to posting rules. This linkage between a business transaction code and a posting rule is accomplished via menu path **IMG • Financial Accounting • Bank Accounting • Business Transactions • Payment Transactions • Manual Bank Statement • Create and Assign Business Transactions**. There, enter "1" into the **Trans. Type** field, as shown in Figure 5.11.

Attributes that were previously associated with an external transaction type in electronic bank statements are now assigned to a business transaction in manual bank statements. For instance, when configuring manual bank statement transactions, attributes such as the account modifier and interpretation algorithms are tied to a business transaction. The remainder of the configuration activities, like setting up a posting rule and associating general ledger accounts with account symbols, follow the same steps as those in the electronic bank statement configuration.

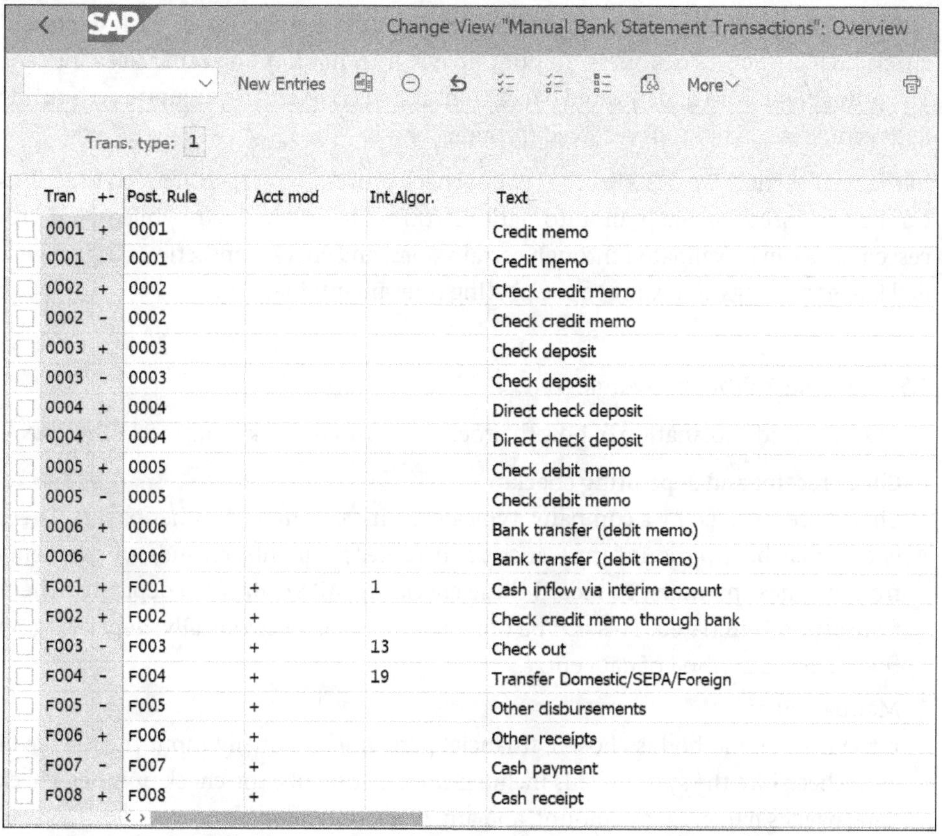

Change View "Manual Bank Statement Transactions": Overview

New Entries More

Trans. type: 1

Tran	+-	Post. Rule	Acct mod	Int.Algor.	Text
0001	+	0001			Credit memo
0001	-	0001			Credit memo
0002	+	0002			Check credit memo
0002	-	0002			Check credit memo
0003	+	0003			Check deposit
0003	-	0003			Check deposit
0004	+	0004			Direct check deposit
0004	-	0004			Direct check deposit
0005	+	0005			Check debit memo
0005	-	0005			Check debit memo
0006	+	0006			Bank transfer (debit memo)
0006	-	0006			Bank transfer (debit memo)
F001	+	F001	+	1	Cash inflow via interim account
F002	+	F002	+		Check credit memo through bank
F003	-	F003	+	13	Check out
F004	-	F004	+	19	Transfer Domestic/SEPA/Foreign
F005	-	F005	+		Other disbursements
F006	+	F006	+		Other receipts
F007	-	F007	+		Cash payment
F008	+	F008	+		Cash receipt

Figure 5.11 Manual Bank Statement Transactions

When manually entering bank statement items in Transaction FF67, each transaction within the bank statement is linked to a specific business transaction code. When the bank statement is eventually posted, SAP S/4HANA determines the appropriate general ledger accounts for each transaction and then posts them accordingly.

In our earlier example, each transaction code corresponds to distinct business transactions, such as credit memos (0001), check credit memos (0002), bearer checks (0003), and so on.

The **Trans. Type** field in Figure 5.11 has a value of 1, indicating that these business transactions are pertinent and configured for manual bank statement processing. If this value is 2, it implies that these business transactions are relevant and configured specifically for deposited check transactions.

5.5 Deposited Check Transaction

A *deposited check transaction* refers to the process of recording and manages checks that have been received by a company or organization and deposited into its bank account. This process involves capturing detailed information about each deposited check, including check numbers, amounts, payer information, and other relevant data. The primary purpose of deposited check transactions is to ensure accurate accounting and reconciliation of funds received through checks.

In the next section, we describe the typical check processing steps. Check transactions can be processed manually or automatically through a file upload. While being processed, data can be validated through certain rules, and check transactions trigger general ledger postings according to the posting rules maintained.

5.5.1 Business Process Overview

Here's a detailed explanation of how to process deposited checks within SAP S/4HANA:

1. **Check receipt and depositing checks**
   Checks are received by a company, typically from customers or other payers. These checks may be in various forms, such as customer payments, refunds, or payments from business partners. To record these checks in SAP S/4HANA, a deposited check transaction is initiated. This can be done manually or electronically, depending on the preferred method of data entry.
2. **Manual entry**
   If the manual method is chosen, financial personnel manually input the details of each check into the system. This includes information like the check number, check date, payer's name, check amount, and any relevant references.
3. **Electronic upload**
   Alternatively, an electronic upload may be used. In this case, a file containing check data is provided by the house bank or a certified service provider. This file is then uploaded into SAP S4HANA, automating the data entry process.
4. **Data validation**
   Regardless of the data entry method, the system validates the information to ensure accuracy and completeness. Any discrepancies or errors are flagged for review and correction.
5. **Check processing and accounting entries**
   Once the check data is recorded, SAP S4HANA processes the checks. This includes updating the company's financial records and reconciling the checks with the corresponding customer accounts. The system generates accounting entries to reflect the checks' impact on the company's financial accounts. For example, a deposit of customer payments would result in a credit to accounts receivable and a debit to the bank account.

6. **Reconciliation and reporting**
   Deposited check transactions are reconciled with the bank statement and other financial records to ensure that the recorded checks match the actual deposits made. SAP S4HANA provides reporting capabilities to track and monitor deposited check transactions. This helps with financial analysis, auditing, and compliance.

5.5.2 Deposited Check Configuration

Much like the process of handling bank statements, you can either manually input the details of deposited checks or electronically upload this data into the SAP S/4HANA system. Typically, the specifics about checks deposited into a bank account are provided by either the house bank itself or a service provider that has been certified by the house bank.

The posting rules for deposited check transactions are set via menu path **IMG • Financial Accounting • Bank Accounting • Business Transactions • Check Deposit • Create and Assign Business Transactions**, as shown in Figure 5.12.

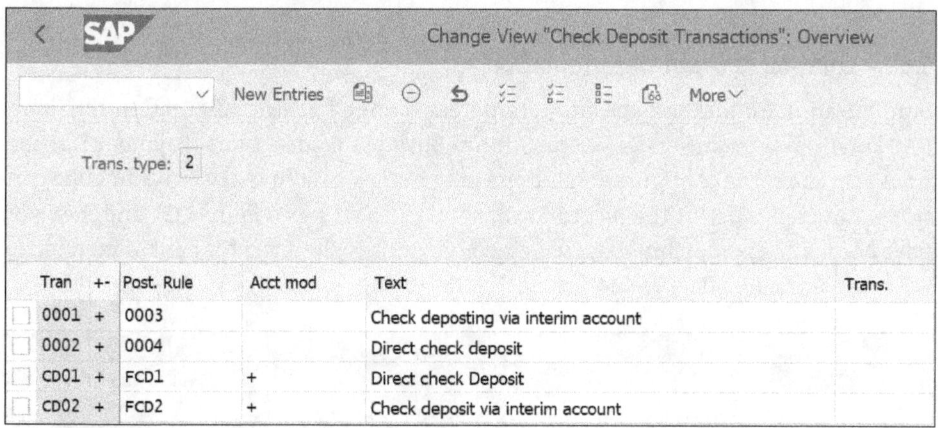

Tran	+-	Post. Rule	Acct mod	Text	Trans.
0001	+	0003		Check deposting via interim account	
0002	+	0004		Direct check deposit	
CD01	+	FCD1	+	Direct check Deposit	
CD02	+	FCD2	+	Check deposit via interim account	

Figure 5.12 Deposited Check Configuration

Starting with the posting rule, the process of determining the general ledger accounts is consistent with the process for electronic bank statements described in Section 5.3.

5.6 Lockbox

Banks in the United States provide a service known as *lockbox processing* in lieu of sending payments to your company and then having someone in your company deposit them to your bank. Using a lockbox is a fairly common banking process in many large and small US companies.

A *lockbox* is a secure United States Postal Service (USPS) box that is used by businesses to receive payments from customers. The lockbox is typically located in a central location, such as a major metropolitan area. When a customer makes a payment, they send it to the lockbox address. USPS then delivers the payments to the business's bank.

In the next two sections, we first show the business process flow of a lockbox process step by step and then show where the settings for the lockbox process can be found in SAP.

5.6.1 Business Process Overview

Here is a step-by-step overview of the lockbox process:

1. The customer makes a payment to the business.
2. The customer sends the payment to the lockbox address.
3. USPS delivers the payment to the business's bank.
4. The bank deposits the payment into the business's account.
5. The bank sends the business a remittance report that lists the payments that were received.
6. The business uses the remittance report to update its accounts receivable records.

5.6.2 Lockbox Account Determination

Configuration for lockbox-specific parameters is done in the configuration activity at **IMG Financial Accounting • Bank Accounting • Business Transactions • Payment Transactions • Lockbox**. The configuration is done in two steps. You first have to define the control parameters, such as the record format and the postings (Figure 5.13), and then you have to define the posting data, which compromises the lockbox bank data and the posting parameters (Figure 5.14).

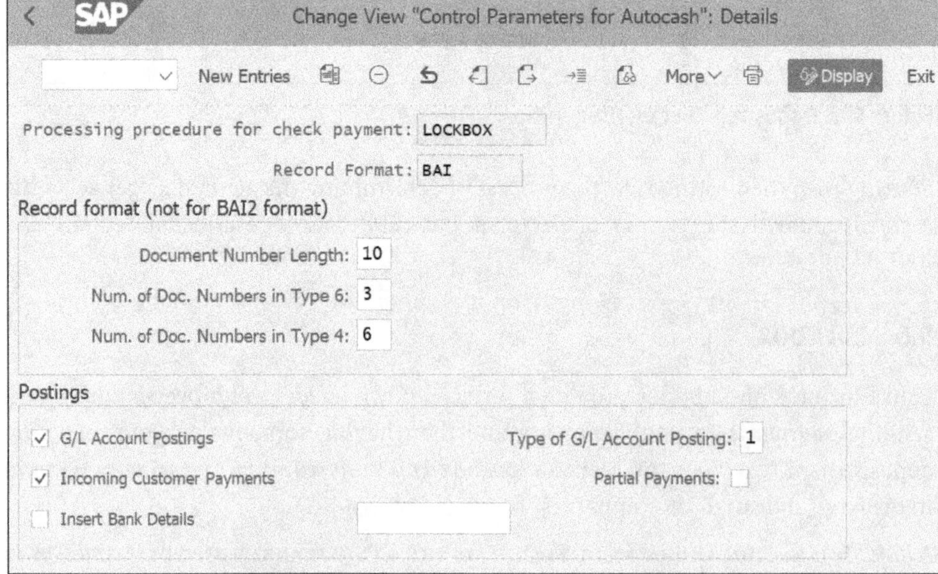

Figure 5.13 Control Parameters for Lockbox

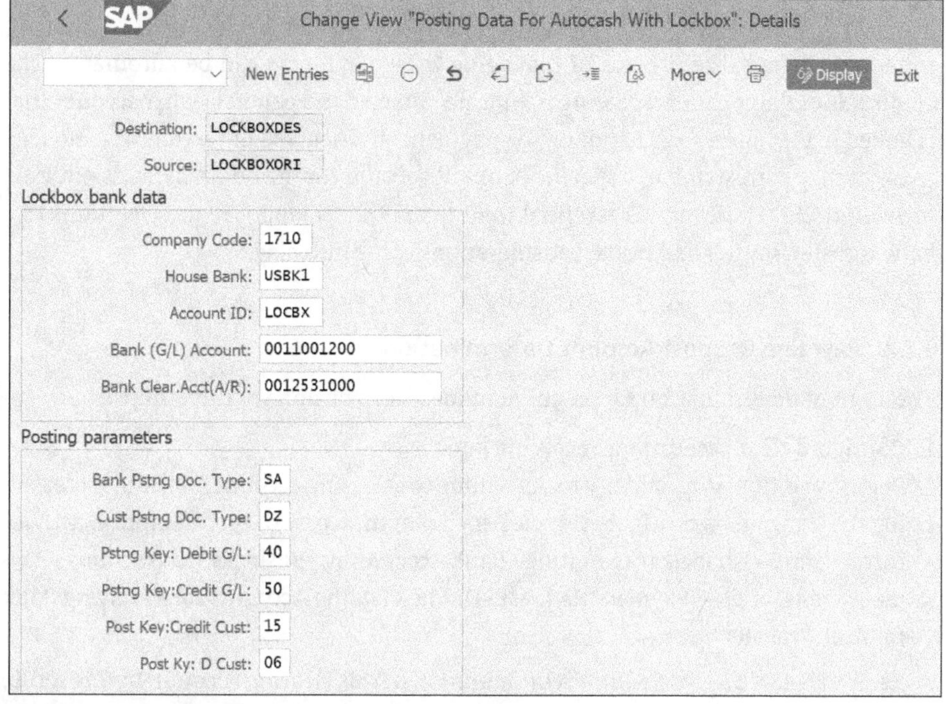

Figure 5.14 Posting Data Configuration for Lockbox

Regarding the process of determining general ledger accounts, the configuration is executed identically to the description provided in Section 5.3, utilizing the same configuration transactions as outlined previously.

5.7 Payment Request Process

One of the most widely utilized transactions in SAP ERP for managing outgoing payments is Transaction F110. This transaction handles outgoing payments, primarily driven by open items present on customer or vendor accounts. Payment requests are an alternative approach to facilitate outgoing payments. Within certain areas of SAP S4HANA, such as treasury, the system automatically generates payment requests whenever an accounting document with outstanding payments is posted. Subsequently, an automatic payment transaction, Transaction F111, is executed to generate outbound payments based on these payment requests rather than generating them directly from accounting documents.

In the following sections, we first outline the business process and then we explain the account determination for payment requests.

5.7.1 Business Process Overview

In most instances, the process of managing these payments can be automated. The credit side of the outbound posting is typically directed to a general ledger account that is linked to the house bank handling the payment, as discussed in Section 5.1.1. Alternatively, it may correspond to a specific bank subaccount for the house bank, as outlined in Section 5.1.2. However, it's essential to note that if a payment request pertains to a bank transfer, several additional considerations come into play.

5.7.2 Payment Request Account Determination

The account determination for payment requests can be done in three steps:

1. **Configure clear account for receiving bank**
   Begin by setting up an activity to determine the clearing account for the bank receiving the transfer. To configure this clearing account, access the configuration activity through **IMG • Financial Accounting • Bank Accounting • Business Transactions • Payment Transactions • Payment Request • Define Clearing Accounts for Receiving Bank for Acct. Transfer.**

   Refer to Figure 5.15 for a sample configuration of this clearing account. In this setup, note that the country (**Cntry/Reg.**) and the payment method (**Payt Meth.**) fields pertain to the paying company code. Other fields, including company code (**CoCode**), **House Bk**, **Currency**, and **Account ID**, relate to the receiving company code.

   For example, any payment requests associated with transfers into house bank USBK1 of company code 1710 or 9100 with any payment method will be posted to clearing account 11001080; transfers into house bank USBK1 will be posted to clearing account 11002080.

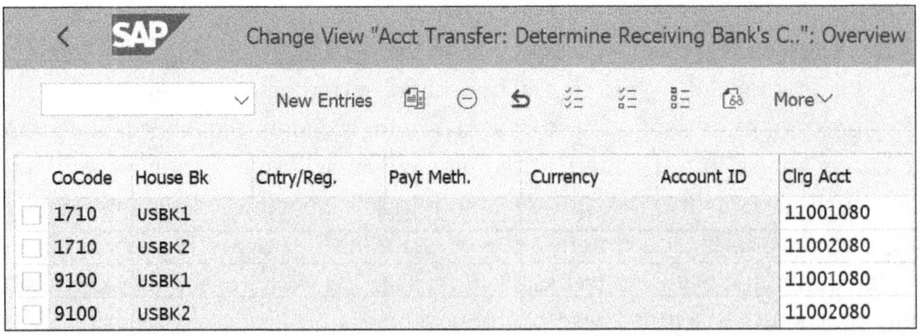

Change View "Acct Transfer: Determine Receiving Bank's C..": Overview

New Entries More

CoCode	House Bk	Cntry/Reg.	Payt Meth.	Currency	Account ID	Clrg Acct
1710	USBK1					11001080
1710	USBK2					11002080
9100	USBK1					11001080
9100	USBK2					11002080

Figure 5.15 Payment Request Clearing Account

2. **Cross-country bank transfers**
   When bank transfers occur between accounts in different countries, an additional layer of complexity arises. To address this, configure a separate clearing account

under **IMG • Financial Accounting Bank Accounting • Business Transactions • Payment Transactions • Payment Request • Define Clearing Accounts for Cross-Country Bank Account Transfers.**

Configure this general ledger account for each company code that may receive cross-country bank transfers. SAP S/4HANA treats cross-country bank transfers as transfers between bank accounts in different countries or those belonging to company codes with varying local currencies.

For instance, company code 1710 maintains cross-company clearing account 21900000 (see Figure 5.16).

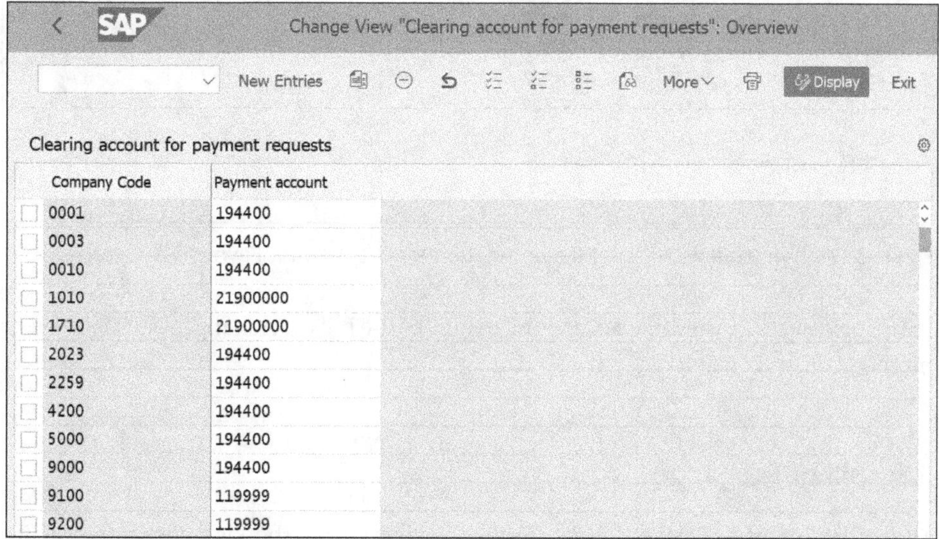

Company Code	Payment account
0001	194400
0003	194400
0010	194400
1010	21900000
1710	21900000
2023	194400
2259	194400
4200	194400
5000	194400
9000	194400
9100	119999
9200	119999

Figure 5.16 Change View “Clearing Account for Payment Requests”: Overview

3. **Configure bank subaccount for the paying bank**
   The final step involves configuring the bank subaccount settings for the paying bank. You can perform this configuration through **IMG • Financial Accounting Bank Accounting • Business Transactions • Payment Transactions • Payment Handling • Bank Clearing Account Determination • Define Account Determination.**

It's worth noting that this subaccount determination setup is similar to the subaccount determination, which we will explain in Section 5.8.2. However, there are two crucial distinctions: this subaccount determination is exclusively used for payment requests related to bank transfers, and only if the originating application hasn't already included a bank subaccount in the payment request.

For instance, house bank USBK1 of company code 1710 maintains subaccount 11001050 for payment requests only (see Figure 5.17).

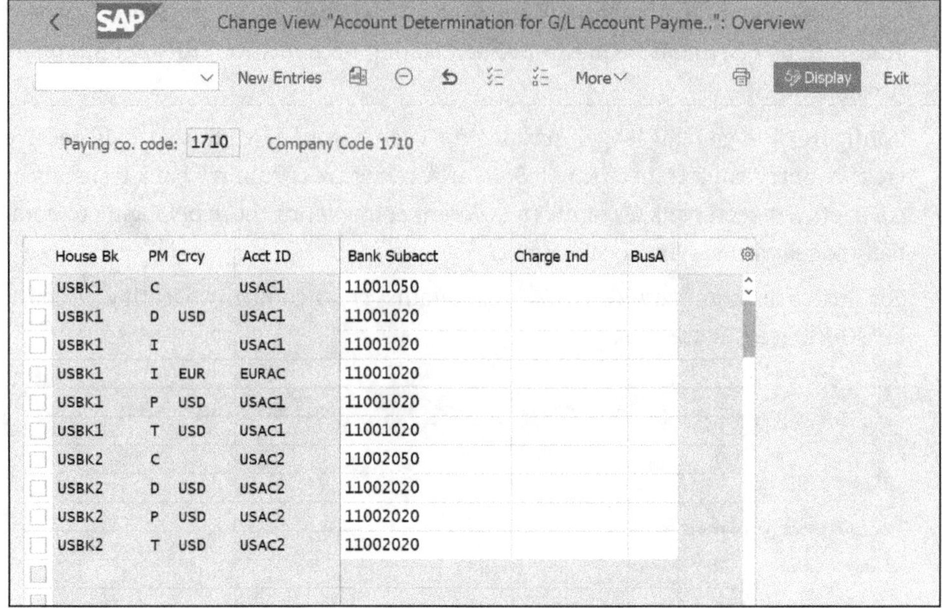

Figure 5.17 Change View "Account Determination for G/L Account Payments": Overview

In the next section, we will examine the details of the bill of exchange special payment method.

5.8 Bill of Exchange

The utilization of a bill of exchange (BoE) as a payment method is a common practice in European countries. When a BoE is employed for payment, the payer is granted a specified timeframe, as indicated in the BoE terms, to fulfill their payment obligation to the payee.

In the next two sections, we first discuss the main steps of the business process relating to the BoE, and then we explain the configuration for both AR and AP transactions.

5.8.1 Business Process Overview

The BoE payment method encompasses several distinct business transactions:

- **Discounting**
  The payee has the option to sell the BoE to a bank before its maturity date. This practice is known as discounting.
- **Presentment**
  The payee can choose to present the BoE to the payer for payment on the agreed-upon due date, known as presentment.

- **Collection**
  Alternatively, the payee can present the BoE to a bank, which then facilitates the collection of funds from the payer.
- **Protest and expiry**
  If the term of the BoE has concluded, and the protest period specified in the BoE has passed, the payee may declare the BoE expired.

In the SAP S/4HANA system, the configuration of BoE transactions involves treating them as special general ledger transactions. Importantly, you have the flexibility to configure BoE as both an incoming (receivable) and an outgoing (payable) payment method. This configuration allows businesses to effectively handle BoE transactions as part of their financial processes within the SAP S/4HANA system.

5.8.2 Bill of Exchange Accounts Receivable Transaction Accounts

SAP S/4HANA treats BoE transactions as special general ledger transactions, which necessitates the management of BoE receivables through distinct reconciliation accounts. To enable this functionality, reconciliation accounts must be configured for each type of BoE transaction, such as BoE payment requests or rediscountable BoE. This configuration is carried out using the following menu path: **IMG • Financial Accounting • Bank Accounting • Business Transactions • Bill of Exchange Transactions • Bill of Exchange Receivable • Post Bill of Exchange Receivable • Define Alternative Reconcil. Accts for Bills/Exch. Receivable.**

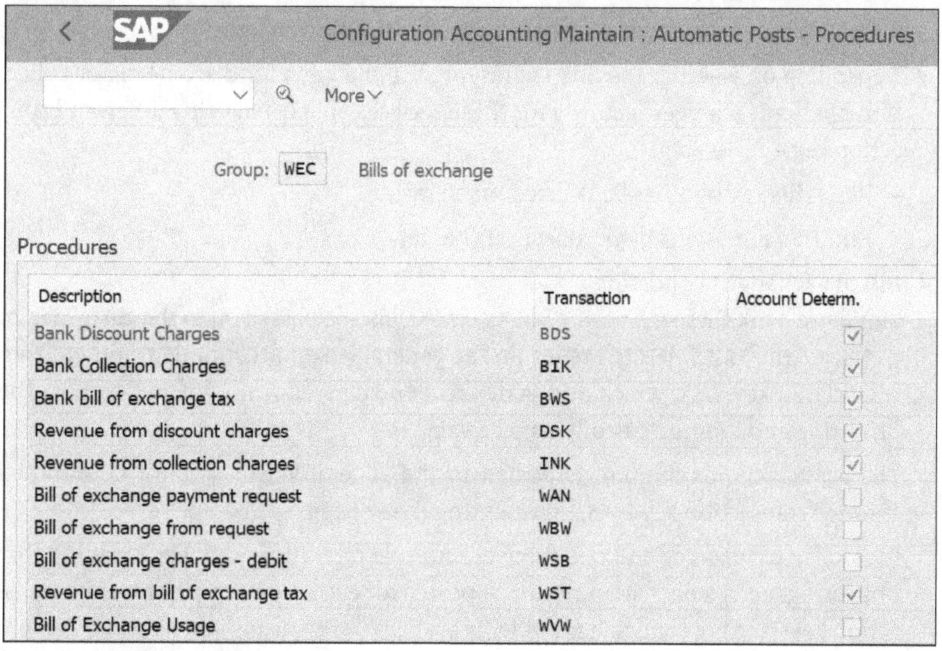

Configuration Accounting Maintain : Automatic Posts - Procedures

Group: WEC Bills of exchange

Procedures

Description	Transaction	Account Determ.
Bank Discount Charges	BDS	☑
Bank Collection Charges	BIK	☑
Bank bill of exchange tax	BWS	☑
Revenue from discount charges	DSK	☑
Revenue from collection charges	INK	☑
Bill of exchange payment request	WAN	☐
Bill of exchange from request	WBW	☐
Bill of exchange charges - debit	WSB	☐
Revenue from bill of exchange tax	WST	☑
Bill of Exchange Usage	WVW	☐

Figure 5.18 Bill of Exchange General Ledger Account Determination

Furthermore, general ledger accounts need to be assigned to various fees and charges associated with BoE transactions. This assignment can be completed through menu path **IMG • Financial Accounting • Bank Accounting • Business Transactions • Bill of Exchange Transactions • Bill of Exchange Receivable • Post Bill of Exchange Receivable • Define Accounts for Bill of Exchange Transactions.** SAP S/4HANA employs the transaction key technique to determine the appropriate general ledger accounts for BoE transactions. The transaction keys relevant to BoE transactions in the WEC (bills of exchange) group are shown in Figure 5.18. This meticulous configuration ensures the accurate management and accounting of BoE transactions within the SAP S/4HANA system.

In Figure 5.18, you'll also notice the presence of transaction keys WAN, WBW, WSB, and WVW. It's important to note that these particular transaction keys serve the sole purpose of determining the posting key and do not play a role in general ledger account determination. Now, let's delve into the explanation of the remaining six transaction keys:

- **Bill of exchange charge posting**
  When engaging in BoE transactions, such as discounting and collection, banks typically impose charges and fees. As the recipient of these charges, you (the payee) have the option to pass approximate charges on to the customer (the payer).

  During the posting of these customer receivable entries, corresponding offsetting entries are generated, and the general ledger accounts for these entries are determined based on specific transaction keys:
  - DSK: This key represents revenue from discount charges.
  - INK: This denotes revenue generated from collection charges.

  Depending on whether the BoE is presented to the bank for discounting or collection, the actual charges incurred for these services are posted using different transaction keys:
  - BDS: This key signifies bank discount charges.
  - BIK: This represents bank collection charges.
- **Bill of exchange tax posting**
  Similar to bank charges, specific BoE taxes can also be passed on to the customer. In this scenario, SAP S/4HANA relies on the general ledger account determined from transaction key WST, which designates revenue derived from BoE tax, to post the offsetting entry for the customer's receivable.

  The actual BoE tax amount is posted to the general ledger account determined through transaction key BWS, representing the BoE tax.

  For the actual charges and tax, there are no rule modifiers available, meaning that only one general ledger account can be posted for each. However, for revenue transaction keys, the tax code rule modifier is available. This means you have the flexibility to post to different general ledger accounts based on the specific tax code applied. In addition to defining the general ledger accounts, two other configuration activities are relevant for the determination of general ledger accounts for receivable BoE transactions:

- **Bill of exchange liability posting**

  When you, as the payee, present a BoE to the bank for discounting, it's imperative to account for the associated liability on your financial records. This step is taken to ensure that if the payer fails to fulfill their obligations under the BoE, the bank has legitimate recourse to collect from you.

  This liability is distinctly represented through dedicated bank subaccounts. These subaccounts can be configured and managed using the following menu path: **IMG • Financial Accounting • Bank Accounting • Business Transactions • Bill of Exchange Transactions • Bill of Exchange Receivable • Present Bill of Exchange Receivable at Bank • Define Bank Subaccounts.**

  For comprehensive management, it's essential to define bank subaccounts for each individual bank account that is utilized for BoE transactions. As shown in Figure 5.19, while it's a requirement to define at least one subaccount for each bank account, you also have the flexibility to determine various bank subaccounts based on combinations of factors such as the purpose of the BoE, special general ledger indicators, and customer accounts receivable reconciliation accounts. This tailored approach ensures that your financial records accurately reflect the liabilities associated with BoE transactions, providing transparency and accountability in the process.

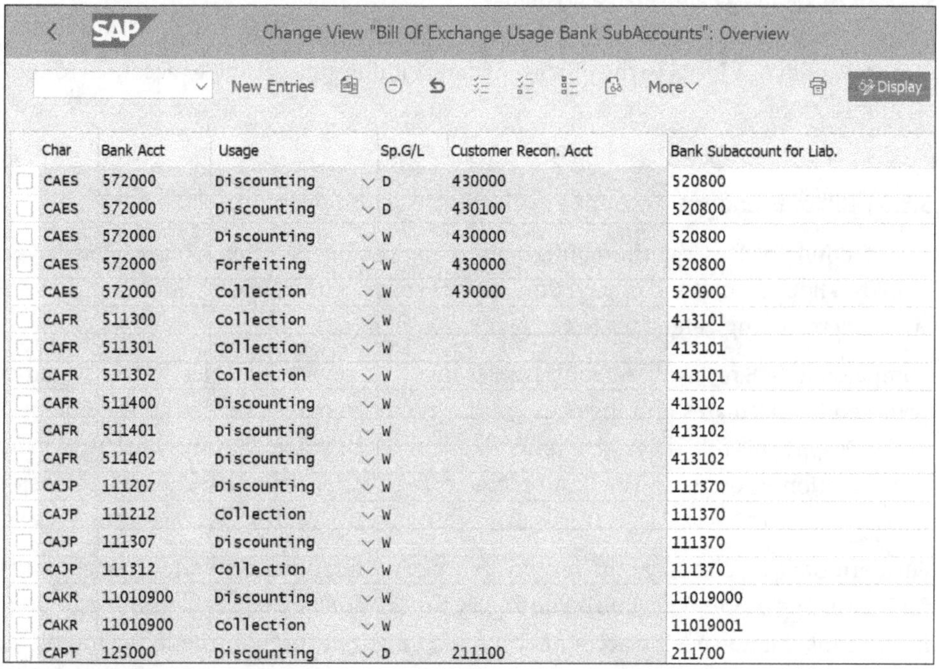

Change View "Bill Of Exchange Usage Bank SubAccounts": Overview

New Entries More Display

Char	Bank Acct	Usage	Sp.G/L	Customer Recon. Acct	Bank Subaccount for Liab.
CAES	572000	Discounting	D	430000	520800
CAES	572000	Discounting	D	430100	520800
CAES	572000	Discounting	W	430000	520800
CAES	572000	Forfeiting	W	430000	520800
CAES	572000	Collection	W	430000	520900
CAFR	511300	Collection	W		413101
CAFR	511301	Collection	W		413101
CAFR	511302	Collection	W		413101
CAFR	511400	Discounting	W		413102
CAFR	511401	Discounting	W		413102
CAFR	511402	Discounting	W		413102
CAJP	111207	Discounting	W		111370
CAJP	111212	Collection	W		111370
CAJP	111307	Discounting	W		111370
CAJP	111312	Collection	W		111370
CAKR	11010900	Discounting	W		11019000
CAKR	11010900	Collection	W		11019001
CAPT	125000	Discounting	D	211100	211700

Figure 5.19 Define Bank Subaccounts

- **Bill of exchange usage posting**

  Upon presenting the BoE to the bank, often referred to as *presentment*, various charges imposed by the bank are applied. These charges are recorded in distinct general ledger accounts, and this setup can be configured through the following menu path: **IMG • Financial Accounting • Bank Accounting • Business Transactions • Bill of Exchange Transactions • Bill of Exchange Receivable • Present Bill of Exchange Receivable at Bank • Maintain Account Determination.**

  When defining these bank charge accounts, you have the flexibility to specify them based on a combination of factors, including the house bank, the specific bank account, and the intended purpose of the BoE (see Figure 5.20).

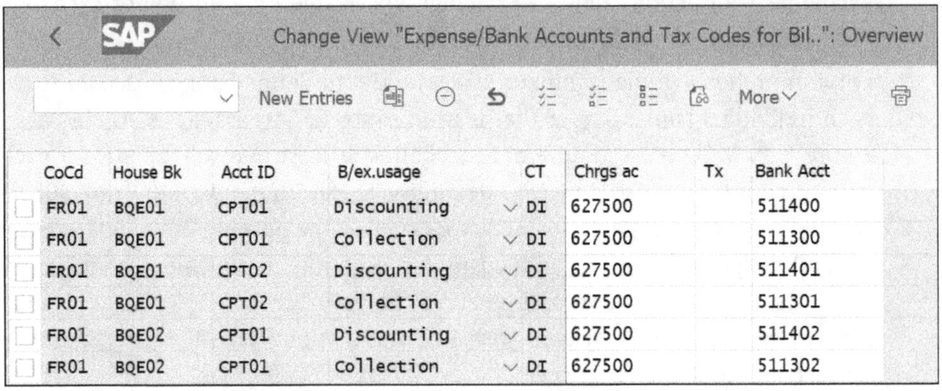

Change View "Expense/Bank Accounts and Tax Codes for Bil..": Overview

New Entries More

CoCd	House Bk	Acct ID	B/ex.usage	CT	Chrgs ac	Tx	Bank Acct
FR01	BQE01	CPT01	Discounting	DI	627500		511400
FR01	BQE01	CPT01	Collection	DI	627500		511300
FR01	BQE01	CPT02	Discounting	DI	627500		511401
FR01	BQE01	CPT02	Collection	DI	627500		511301
FR01	BQE02	CPT01	Discounting	DI	627500		511402
FR01	BQE02	CPT01	Collection	DI	627500		511302

Figure 5.20 Defining Bank Charge Accounts

Now, let's shift our focus to the process of determining general ledger accounts for BoE AP transactions.

5.8.3 Bill of Exchange Account Payable Transaction Accounts

This section focuses on the configuration requirements when you, as the payer, employ a BoE as a method of payment to your vendors (the payees). For a comprehensive understanding, refer back to Chapter 3.

Compared to BoE receivables, BoE payables involve a reversal of roles. In this scenario, your vendor calculates and invoices you for the BoE charges related to the BoE payment. Despite this role reversal, it remains essential for you to specify alternative bank reconciliation accounts and bank subaccounts to effectively manage these BoE transactions.

Furthermore, you also have the responsibility of maintaining general ledger accounts for processing a BoE that your vendor or your vendor's bank has presented to your house bank. These specific accounts can be configured through the following menu path: **IMG • Financial Accounting • Bank Accounting • Business Transactions • Bill of**

Exchange Transactions • Bill of Exchange Payable • Returned Bill of Exchange Payable • Define Account for Returned Bill of Exchange. This configuration is crucial to ensure that BoE payables are accurately managed and accounted for in your financial system.

This account configuration, as depicted in Figure 5.21, is established to accommodate various combinations of the house bank, bank account, and BoE payment methods.

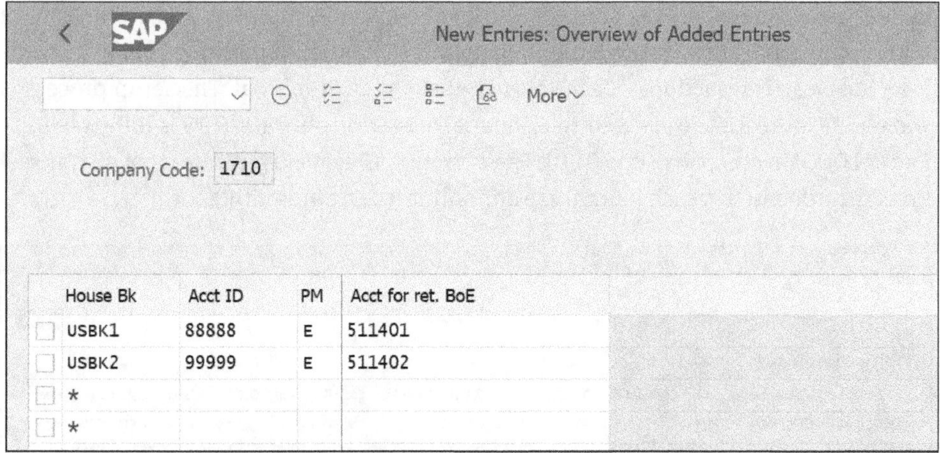

Figure 5.21 General Ledger Accounts for Returned Bill of Exchange

In the next section, we will discuss general ledger account determination for cash journals.

5.9 Cash Journal

A *cash journal* provides a detailed record of all cash inflows and outflows. This includes receipts from sales, cash purchases, expenses paid in cash, and cash withdrawals or deposits. Due to automation and compliance, cash transactions are gradually losing their significance in daily business transactions but are still relevant. For this purpose, SAP offers a cash subledger, which mainly is used to record petty cash transactions in a compliant and structured way.

5.9.1 Business Process Overview

You can leverage the cash journal feature within SAP S/4HANA to document cash transactions. In the operational context of every company, there is typically a petty cash account designated for handling small-value transactions conducted in cash. The definition of what constitutes a petty cash transaction may vary from one organization to another. However, it often encompasses expenses such as acquiring limited quantities of office supplies or covering dinner orders for diligent SAP consultants.

For businesses that engage in transactions with retail consumers, a significant portion of their operations may involve cash payments.

5.9.2 Cash Journal Accounts

To make use of the cash journal functionality, the initial step involves establishing the cash journal general ledger account within the specific company code. This can be achieved by navigating through menu path **IMG • Financial Accounting • Bank Accounting • Business Transactions • Cash Journal • Set Up Cash Journal**. The setup process is shown in Figure 5.22. As an example, general ledger account 100000 is linked to cash journal 0001 in company code 0001. The currency (**Crcy**) field indicates that all transactions recorded in this cash journal are denominated in euros (**EUR**).

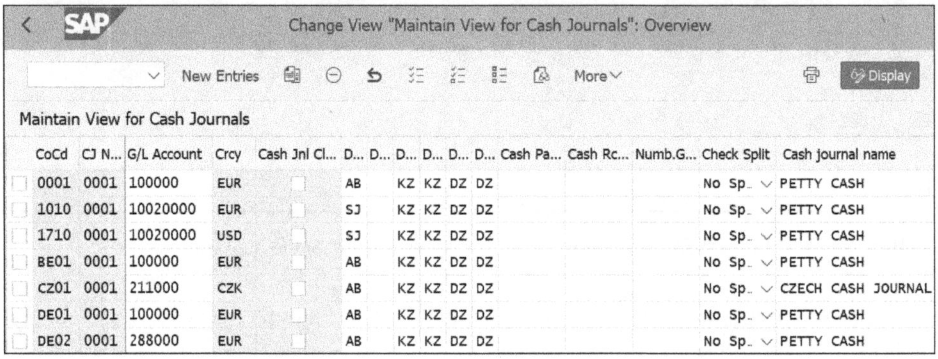

Change View "Maintain View for Cash Journals": Overview

New Entries More Display

Maintain View for Cash Journals

CoCd	CJ N...	G/L Account	Crcy	Cash Jnl Cl...	D...	D...	D...	D...	D...	D...	Cash Pa...	Cash Rc...	Numb.G...	Check Split	Cash journal name
0001	0001	100000	EUR		AB		KZ	KZ	DZ	DZ				No Sp..	PETTY CASH
1010	0001	10020000	EUR		SJ		KZ	KZ	DZ	DZ				No Sp..	PETTY CASH
1710	0001	10020000	USD		SJ		KZ	KZ	DZ	DZ				No Sp..	PETTY CASH
BE01	0001	100000	EUR		AB		KZ	KZ	DZ	DZ				No Sp..	PETTY CASH
CZ01	0001	211000	CZK		AB		KZ	KZ	DZ	DZ				No Sp..	CZECH CASH JOURNAL
DE01	0001	100000	EUR		AB		KZ	KZ	DZ	DZ				No Sp..	PETTY CASH
DE02	0001	288000	EUR		AB		KZ	KZ	DZ	DZ				No Sp..	PETTY CASH

Figure 5.22 Cash Journal Setup

In the subsequent phase, you must establish the transactions designated for posting within the cash journal and determine the corresponding general ledger account that will receive an offsetting entry from the cash journal. To a certain extent, these transactions resemble the business transactions you would typically define for manual bank statements, as discussed in Section 5.4. These transactions can encompass various expenses, such as dinner expenses or postage expenses.

Figure 5.23 shows an example configuration for these cash journal transactions. This can be configured through menu path **IMG • Financial Accounting • Bank Accounting • Business Transactions • Cash Journal • Maintain Business Transaction**.

It's important to bear in mind that the general ledger accounts set up in this activity serve as the offsetting accounts. Credit entries for payments and debit entries for receipts are recorded in the general ledger account that is linked to the cash journal setup, ensuring accurate accounting and financial tracking within the system.

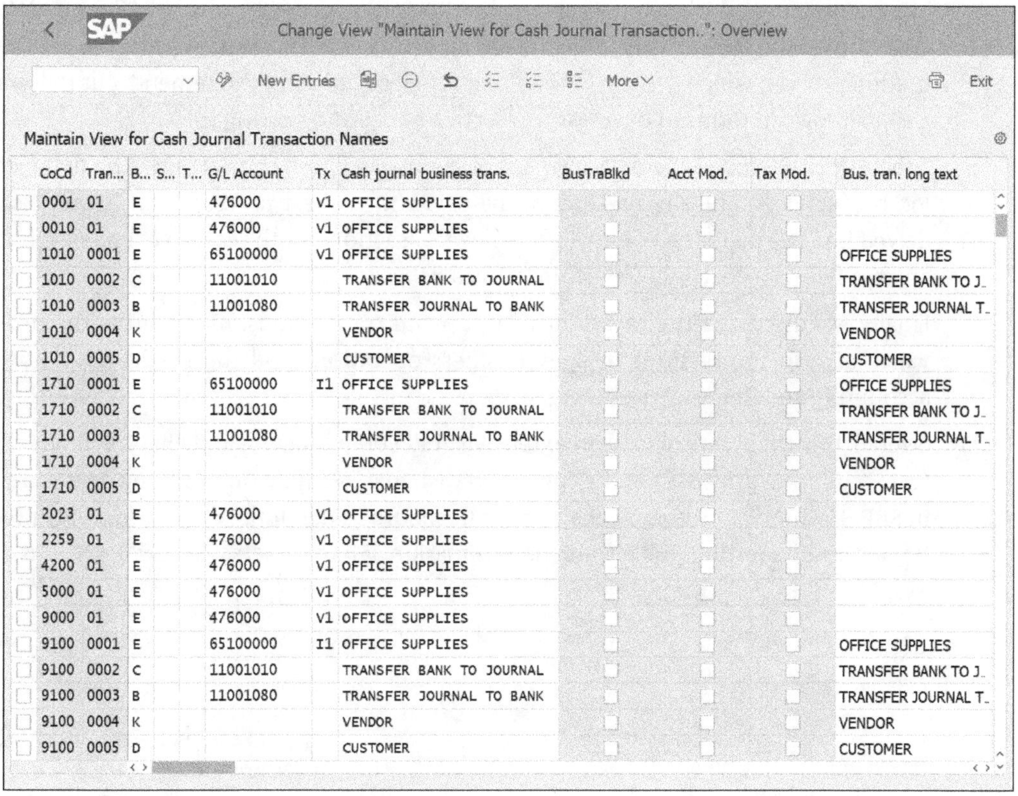

Change View "Maintain View for Cash Journal Transaction..": Overview

New Entries More Exit

Maintain View for Cash Journal Transaction Names

CoCd	Tran...	B...	S...	T...	G/L Account	Tx	Cash journal business trans.	BusTraBlkd	Acct Mod.	Tax Mod.	Bus. tran. long text
0001	01	E			476000	V1	OFFICE SUPPLIES				
0010	01	E			476000	V1	OFFICE SUPPLIES				
1010	0001	E			65100000	V1	OFFICE SUPPLIES				OFFICE SUPPLIES
1010	0002	C			11001010		TRANSFER BANK TO JOURNAL				TRANSFER BANK TO J..
1010	0003	B			11001080		TRANSFER JOURNAL TO BANK				TRANSFER JOURNAL T..
1010	0004	K					VENDOR				VENDOR
1010	0005	D					CUSTOMER				CUSTOMER
1710	0001	E			65100000	I1	OFFICE SUPPLIES				OFFICE SUPPLIES
1710	0002	C			11001010		TRANSFER BANK TO JOURNAL				TRANSFER BANK TO J..
1710	0003	B			11001080		TRANSFER JOURNAL TO BANK				TRANSFER JOURNAL T..
1710	0004	K					VENDOR				VENDOR
1710	0005	D					CUSTOMER				CUSTOMER
2023	01	E			476000	V1	OFFICE SUPPLIES				
2259	01	E			476000	V1	OFFICE SUPPLIES				
4200	01	E			476000	V1	OFFICE SUPPLIES				
5000	01	E			476000	V1	OFFICE SUPPLIES				
9000	01	E			476000	V1	OFFICE SUPPLIES				
9100	0001	E			65100000	I1	OFFICE SUPPLIES				OFFICE SUPPLIES
9100	0002	C			11001010		TRANSFER BANK TO JOURNAL				TRANSFER BANK TO J..
9100	0003	B			11001080		TRANSFER JOURNAL TO BANK				TRANSFER JOURNAL T..
9100	0004	K					VENDOR				VENDOR
9100	0005	D					CUSTOMER				CUSTOMER

Figure 5.23 Cash Journal Transactions

5.10 Summary

This chapter focused on cash management and banking procedures in SAP S4HANA, emphasizing their impact on accounting practices. Key areas covered include the setup of house banks and bank subaccounts, plus the importance of general ledger account determination. The chapter detailed the process for setting up house banks and linking them to company codes, along with configuring bank account identifiers.

Significant attention was given to bank subaccounts, particularly their role in capturing the float amount in outgoing payments and their importance in the bank reconciliation statement process. We thoroughly discussed account determination, highlighting the complexity of setting up configurations for different banking transactions, such as bank statements, lockbox transactions, and bills of exchange.

A comprehensive explanation was provided for various elements like internal and external transaction types, interpretation algorithms, and posting rules. Each element plays a crucial role in the automated processing and posting of bank transactions to the appropriate general ledger accounts.

The chapter also addressed manual bank statement processing within SAP S/4HANA. We outlined the steps involved in processing manual bank statements and the configuration process, which is simpler than that for electronic bank statements but follows a similar logic in terms of business transaction codes and posting rules.

Other topics included the management of deposited check transactions, the usage of lockbox services, and the specifics of the payment request process. The chapter also delved into the handling of bill of exchange transactions, both receivable and payable, and the necessary account determinations for these.

Finally, we discussed the cash journal setup and transactions, highlighting the relevance of cash transactions in daily business operations and the need for accurate accounting and financial tracking.

Overall, this chapter provided an extensive guide on managing and aligning financial and accounting processes in the context of cash management and banking operations in SAP S/4HANA. In the next chapter, we'll explore in detail the characteristics, functionalities, and settings for account determination of fixed assets within SAP S/4HANA.

6

Chapter 6
Fixed Assets

In this chapter, you'll get a comprehensive look at how the process of determining general ledger accounts works within asset accounting. While asset accounting is typically associated with managing fixed assets like machinery, computers, and office equipment, along with tracking their capital costs and calculating depreciation, asset accounting in SAP S/4HANA offers a broader range of functionalities.

This area allows you to handle intangible assets such as patents and goodwill, assess replacement values of assets for insurance purposes, reevaluate assets in countries susceptible to inflation, and calculate intricate details of capital lease payments. What unites all these capabilities is their shared ability to independently track multiple values associated with a single asset and to post periodic increases or decreases in these values.

Within SAP S/4HANA, you have the flexibility to define and manage any type of asset you wish to track, and you can monitor various values for the same asset, all the while implementing highly customizable calculation formulas to record decreases or increases in these values.

A notable enhancement in asset accounting in SAP S4HANA is the introduction of parallel valuation. Assets can be maintained in multiple ledgers for different accounting purposes, such as International Financial Reporting Standards (IFRS) and US Generally Accepted Accounting Principles (GAAP). This allows companies to comply with multiple accounting standards without having to maintain separate sets of asset books.

This chapter is broken down into 10 sections. Each section discusses one major topic in fixed asset accounting. We start with account determination logic, and then we explore all kinds of transactions that occur and can be recorded in fixed asset accounting, such as asset acquisitions, asset depreciations, asset retirements, asset scraps and disposals, asset transfers, asset under construction, asset revaluations, investment support, and imputed interest calculations.

6.1 Fixed Asset Account Determination Logic

Figure 6.1 shows various account determination elements that play a pivotal role in the process of general ledger account determination within asset accounting (AA). These elements serve as the building blocks that empower you to assign general ledger accounts from various perspectives.

For instance, the *asset class* element allows you to categorize assets by their specific types, while *depreciation areas* help you categorize assets based on the methods used for calculating their depreciation. When you combine these account determination elements with the chart of accounts, you establish a well-structured framework for determining general ledger accounts in the realm of asset accounting. Now let's delve into each of these account determination elements in greater detail.

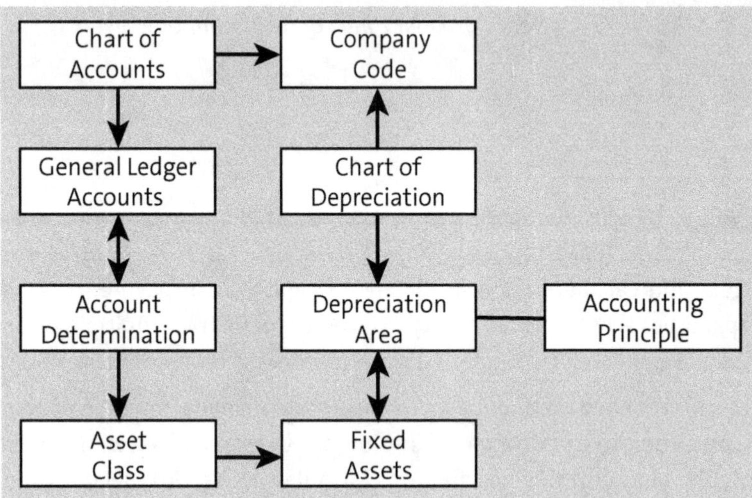

Figure 6.1 Account Determination Logic for Asset Accounting

A proper design and decisions about the chart of depreciation, account determination rules, and asset class can make it significantly easier to expand the future scope of an asset accounting implementation. For example, if the chart of depreciation is designed well, it can be easily adapted to accommodate new asset types or depreciation methods. Similarly, if the account determination rules are carefully defined, they can be used to automate the posting of transactions for new business processes or organizational structures.

Another key benefit of proper asset accounting design is that it can greatly reduce the effort required to implement asset accounting for additional company codes. This is because account determination rules are linked to the chart of accounts, rather than to individual company codes. As a result, if company codes share the same chart of accounts, the implementation effort is reduced to merely maintaining some basic settings.

Here are some specific examples of how proper asset accounting design can benefit an organization:

- **Improved reporting and analysis**
  A well-designed asset accounting system can provide organizations with more detailed and accurate information about their asset holdings and financial performance. This information can be used to improve decision-making in areas such as investment planning, budgeting, and forecasting.
- **Reduced compliance risk**
  A well-designed asset accounting system can help organizations comply with complex accounting and financial reporting regulations. This can help to reduce the risk of fines and penalties.
- **Increased operational efficiency**
  A well-designed asset accounting system can automate the posting of financial transactions, which can reduce the workload on accounting staff and improve the accuracy and efficiency of financial operations.

Overall, proper asset accounting design is essential for any organization that wants to get the most out of its SAP S/4HANA investment.

6.1.1 Asset Class

An asset class functions as a way to categorize assets that share common characteristics, and these categories are determined by the user. For example, you might choose to establish an asset class named *office equipment* to group all forms of office equipment within your company. Alternatively, you could create multiple asset classes like *office furniture*, *telecommunication equipment*, and *office security system* to segregate different types of office equipment.

It's crucial to note that when designing asset classes in SAP S/4HANA, these definitions apply uniformly across all company codes and all charts of depreciation. This standardized approach simplifies the creation of an enterprise-wide asset catalog, ensuring consistency and uniformity in asset management throughout the organization.

An alternative approach to conceptualize the creation of asset classes is to examine the fixed asset section of a balance sheet. You can establish one or more asset classes corresponding to each general ledger account within that section. Figure 6.2 illustrates that the standard SAP S/4HANA system already includes numerous frequently utilized asset classes. If your business requires it, you have the flexibility to generate extra asset classes to align with your specific operational needs.

The process of crafting and managing asset classes is carried out by navigating through menu path **IMG • Financial Accounting • Asset Accounting • Organization Structures • Asset Classes • Define Asset Classes**.

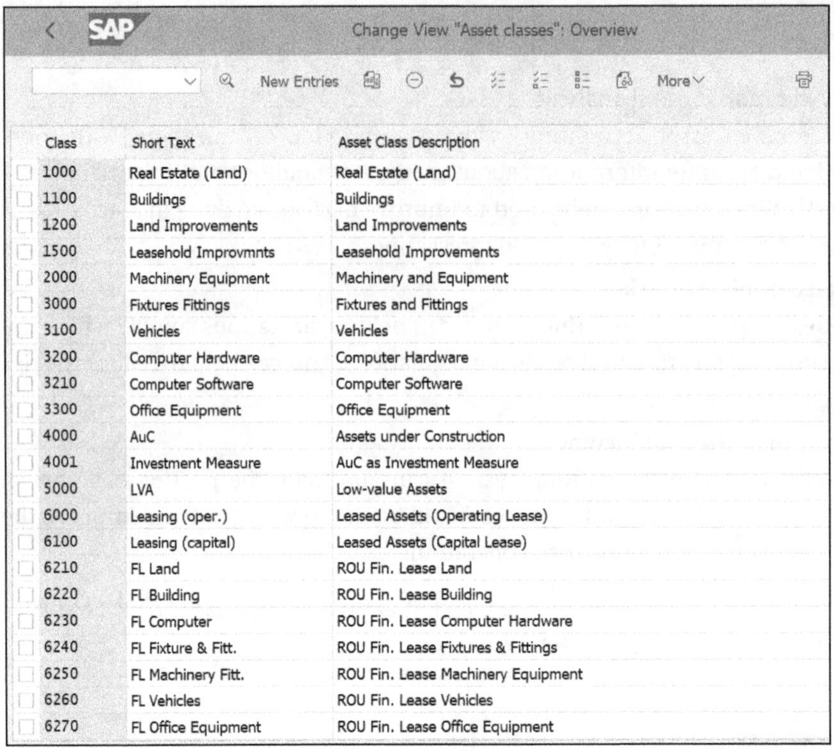

Change View "Asset classes": Overview

New Entries More

Class	Short Text	Asset Class Description
1000	Real Estate (Land)	Real Estate (Land)
1100	Buildings	Buildings
1200	Land Improvements	Land Improvements
1500	Leasehold Improvmnts	Leasehold Improvements
2000	Machinery Equipment	Machinery and Equipment
3000	Fixtures Fittings	Fixtures and Fittings
3100	Vehicles	Vehicles
3200	Computer Hardware	Computer Hardware
3210	Computer Software	Computer Software
3300	Office Equipment	Office Equipment
4000	AuC	Assets under Construction
4001	Investment Measure	AuC as Investment Measure
5000	LVA	Low-value Assets
6000	Leasing (oper.)	Leased Assets (Operating Lease)
6100	Leasing (capital)	Leased Assets (Capital Lease)
6210	FL Land	ROU Fin. Lease Land
6220	FL Building	ROU Fin. Lease Building
6230	FL Computer	ROU Fin. Lease Computer Hardware
6240	FL Fixture & Fitt.	ROU Fin. Lease Fixtures & Fittings
6250	FL Machinery Fitt.	ROU Fin. Lease Machinery Equipment
6260	FL Vehicles	ROU Fin. Lease Vehicles
6270	FL Office Equipment	ROU Fin. Lease Office Equipment

Figure 6.2 List of Available Asset Classes

6.1.2 Depreciation Area

Each depreciation area within asset accounting corresponds to what you can think of as a separate depreciation book. This means you can independently determine the value of an asset within a depreciation area without being influenced by other depreciation areas assigned to the same asset. This independence allows you to manage, for instance, one depreciation area for computing book depreciation according to GAAP requirements and another depreciation area for calculating asset depreciation per local tax regulations.

To configure and review the parameters for a specific depreciation area, follow menu path **IMG • Financial Accounting • Asset Accounting • Organization Structures • Copy Reference Chart of Depreciation/Depreciation Area • Copy/Delete Depreciation Areas**.

Figure 6.3 illustrates that a depreciation area encompasses several attributes related to maintaining values, which control the types of asset transactions you can post to that specific depreciation area. While it's not common to encounter general ledger account determination errors due to incorrect attribute settings, it can lead to errors in asset posting transactions.

A critical aspect of a depreciation area involves how SAP S/4HANA handles the posting of its values to general ledger (GL) accounting. For each depreciation area, you have the flexibility to define this attribute by selecting one of the following options:

- **Area Does Not Post**
  This depreciation area does not post any values to the general ledger.
- **Area Posts in Real Time**
  Values from this depreciation area are posted immediately to the general ledger and are typically used for book depreciation, which contributes to financial statement preparation.
- **Area Posts Depreciation Only**
  This depreciation area posts depreciation values only, not the asset purchase cost.
- **Area Posts APC Immediately, Depreciation Periodically**
  In this case, the depreciation area posts the asset purchase cost immediately and depreciation periodically.

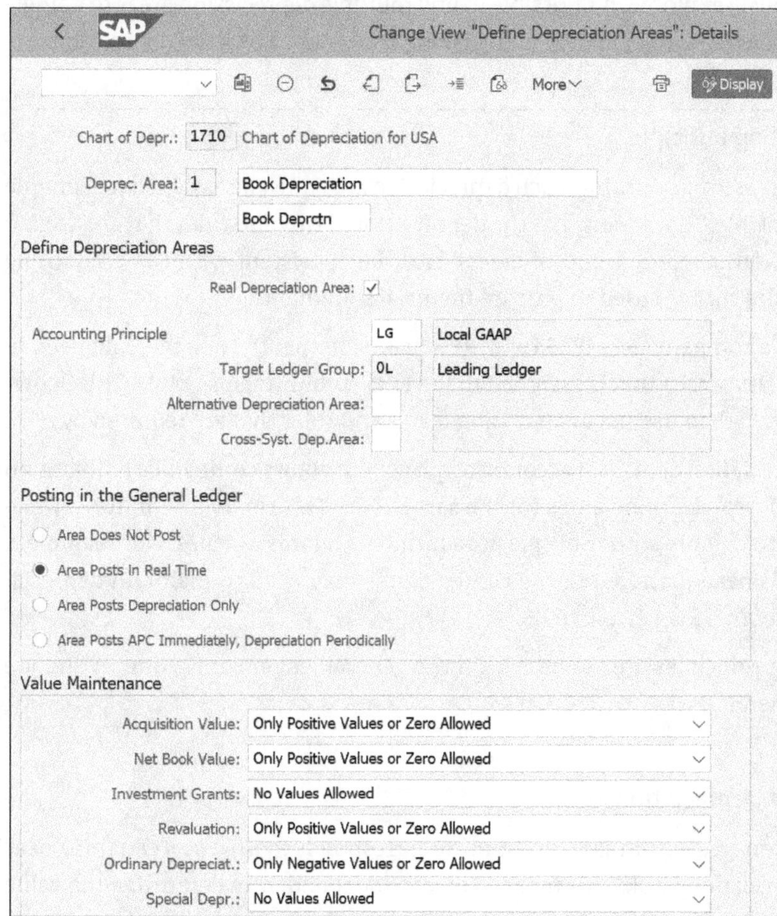

Figure 6.3 Define Depreciation Area Details

It's important to note that you can designate only one depreciation area to post in real time. Typically, this is the depreciation area responsible for book depreciation, which is integral for generating financial statements. During the month-end depreciation run, which is accomplished through Transaction AFAB, only the depreciation values in that specific depreciation area are posted to the general ledger.

Note

In the case of depreciation areas that don't post to the general ledger in real time but rather do so periodically, you should establish and maintain an account determination configuration akin to what has been explained in this chapter. Values from these depreciation areas are transmitted to the general ledger when the periodic posting program is executed.

As we've previously explored, it's possible to link multiple depreciation areas to a single asset. This leads us to our next point of focus: the concept of a chart of depreciation.

6.1.3 Accounting Principle

In SAP S/4HANA asset accounting, each depreciation area must be assigned to a unique accounting principle. This means that the depreciation terms and rules that are used to calculate depreciation for a group of assets must be consistent with the accounting standards or rules that are used to prepare financial statements.

For example, if a company uses IFRS for its financial statements, then it would need to create a depreciation area that is assigned to the IFRS accounting principle. This depreciation area would then use the depreciation terms and rules that are required by IFRS.

The use of depreciation areas and accounting principles allows companies to maintain multiple sets of depreciation values for the same asset. This can be useful for companies that need to comply with multiple accounting standards or rules. For example, a company that is headquartered in the United States may need to maintain depreciation values for both US GAAP and IFRS.

The accounting principles are set up within the general ledger configuration and are not discussed here.

6.1.4 Chart of Depreciation

Each depreciation area within asset accounting corresponds to what you can think of as a separate depreciation book. This means you can independently determine the value of an asset within a depreciation area without being influenced by other depreciation areas assigned to the same asset. This independence allows you to manage, for instance, one depreciation area for computing book depreciation according to GAAP requirements and another depreciation area for calculating asset depreciation per local tax regulations.

Within asset accounting, multiple depreciation areas are organized into a framework known as a *chart of depreciation*. This chart isn't limited to just book depreciation; it can encompass other depreciation areas designed for various purposes, such as group consolidation, tax obligations, insurance requirements, and property tax assessments.

It's worth noting that for many companies operating within the same country, GAAP and tax regulations pertaining to asset depreciation are often consistent. Consequently, the same chart of depreciation is frequently assigned to all company codes operating within that country.

SAP S/4HANA offers a template chart of depreciation for numerous countries worldwide. Each template chart not only includes suggested depreciation areas but also provides associated depreciation calculation parameters. In Figure 6.4, you can see some of the recommended depreciation areas outlined in the template chart of depreciation for the United States. When implementing asset accounting, you have the freedom to retain or remove any of these depreciation areas, except for the book depreciation area.

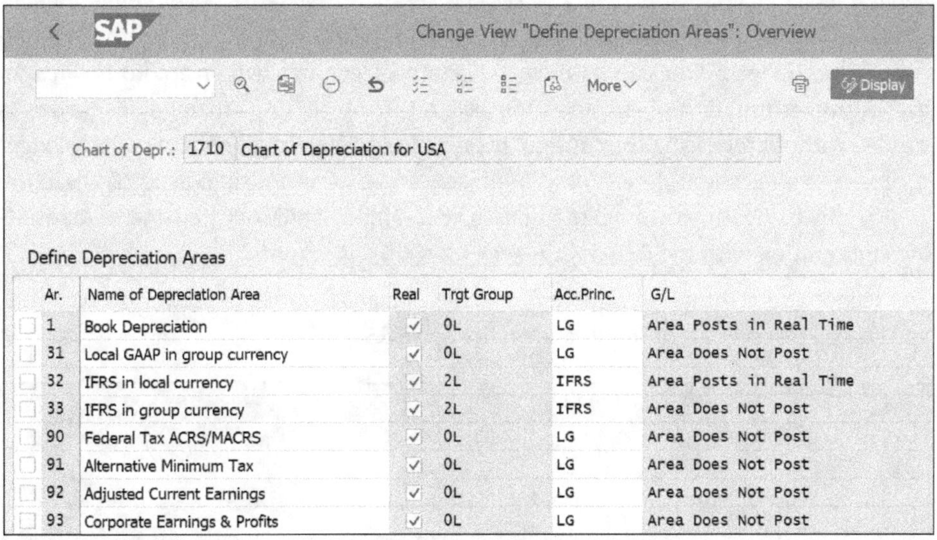

Change View "Define Depreciation Areas": Overview

Chart of Depr.: 1710 Chart of Depreciation for USA

Define Depreciation Areas

Ar.	Name of Depreciation Area	Real	Trgt Group	Acc.Princ.	G/L
1	Book Depreciation	✓	0L	LG	Area Posts in Real Time
31	Local GAAP in group currency	✓	0L	LG	Area Does Not Post
32	IFRS in local currency	✓	2L	IFRS	Area Posts in Real Time
33	IFRS in group currency	✓	2L	IFRS	Area Does Not Post
90	Federal Tax ACRS/MACRS	✓	0L	LG	Area Does Not Post
91	Alternative Minimum Tax	✓	0L	LG	Area Does Not Post
92	Adjusted Current Earnings	✓	0L	LG	Area Does Not Post
93	Corporate Earnings & Profits	✓	0L	LG	Area Does Not Post

Figure 6.4 US Chart of Depreciation

To establish and review the parameters for a chart of depreciation, you can follow this menu path: **IMG • Financial Accounting • Asset Accounting • Organization Structures • Copy Reference Chart of Depreciation • Copy Reference Chart of Depreciation**.

Note

It's advisable to employ a distinct chart of depreciation for each individual country. When it comes to group consolidation, you can consistently utilize a separate depreciation area designated in a group currency, which depreciates assets according to the group's consolidation rules.

Each company code is associated with a specific chart of depreciation. While each company code can have only one chart of depreciation, it's possible for multiple company codes to be linked to the same chart of depreciation. You can execute this assignment through the following menu path within the IMG: **Asset Accounting • Organization Structures • Assign Chart of Depreciation to Company Code**.

6.1.5 Account Determination

In asset accounting, an account determination object corresponds to a collection of general ledger accounts. To create these objects, you can follow menu path **IMG • Financial Accounting • Asset Accounting • Organization Structures • Asset Classes • Specify Account Determination**.

The role of an account determination object is to act as an intermediary linking an asset class to a group of general ledger accounts. As we'll explore in more detail later in this chapter, various asset accounting transactions necessitate postings to different general ledger accounts. If several asset classes share the same set of general ledger accounts, you'd have to configure this separately for each asset class, resulting in repetitive work. Instead, you can associate a group of general ledger accounts with an account determination object and then assign this object to as many asset classes as needed without repeating the configuration process. The assignment of general ledger accounts to an account determination object can be accomplished using Transaction AO90 or by navigating through menu path **IMG • Financial Accounting • Asset Accounting • Integration with the General Ledger • Assign G/L Accounts**.

The account determination is done for balance sheet accounts, depreciation accounts, and special reserves accounts, as shown in Figure 6.5.

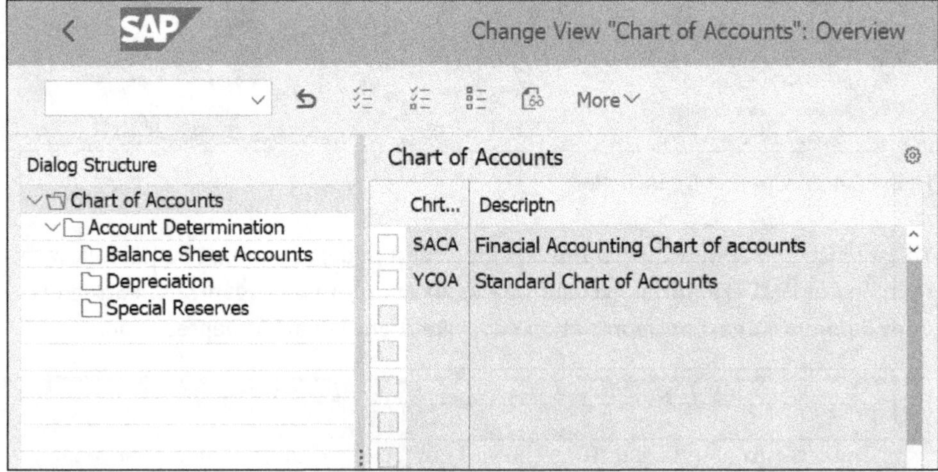

Figure 6.5 Assign General Ledger Accounts: Balance Sheet Accounts, Depreciation, Special Reserves

6.1.6 Other Asset Accounting Objects

Before we dive into the details of general ledger account determination for asset accounting transactions, let's walk through a concise overview of other key asset accounting objects. While these objects may not directly impact general ledger account determination, they play a vital role in the overall functionality of asset accounting. Any discussion of asset accounting would be incomplete without at least touching on these components.

Depreciation Keys

A *depreciation key* within asset accounting acts as a control mechanism for depreciation calculations. Furthermore, it is utilized for calculating imputed interest and processing investment support or subsidies received for an asset. Default depreciation keys are configured for each depreciation area within a chart of depreciation. However, you have the flexibility to alter the depreciation key during the creation or processing of an asset. A depreciation key definition is unique within a chart of depreciation and consists of various calculation methods and parameters governing depreciation calculation. Configuration for depreciation keys is carried out using the following menu path: **IMG • Asset Accounting • Depreciation • Valuation Methods • Depreciation Key • Maintain Depreciation Key** (see Figure 6.6).

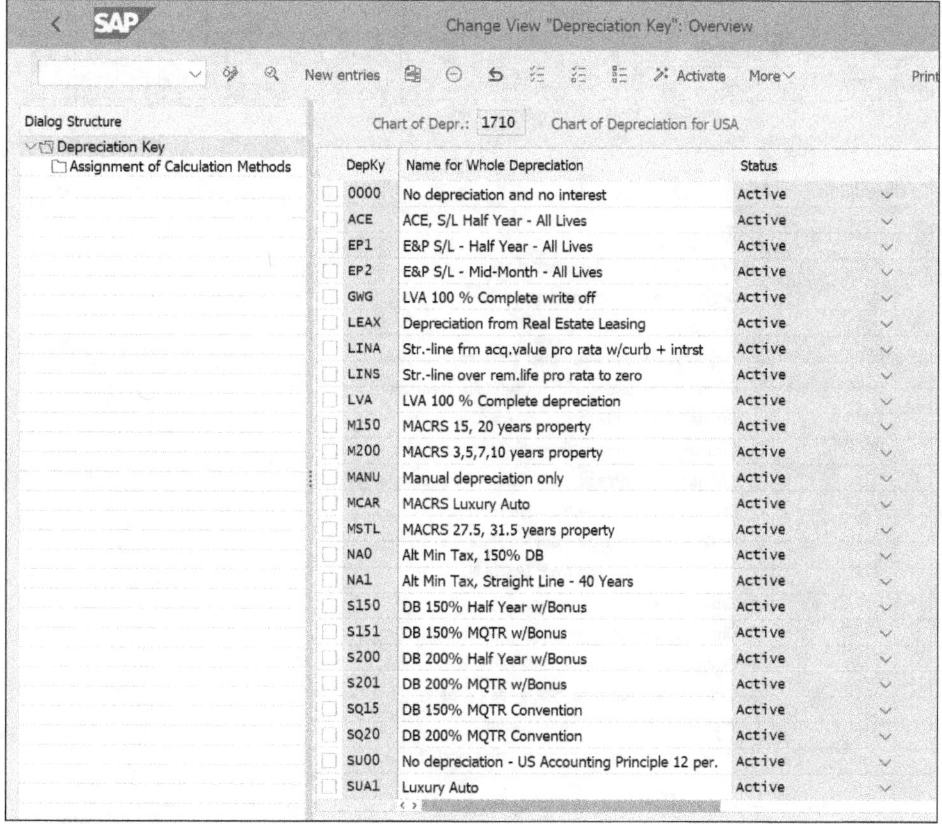

Figure 6.6 Depreciation Keys in Fixed Asset Accounting

Calculation Methods

As the name suggests, *calculation methods* govern calculation parameters for specific types of calculations in asset accounting. These include the following:

- **Base method**
  Specifies depreciation methods such as sum of the year's digits or the percentage of useful life
- **Declining-balance method**
  Defines multiplication factors for depreciation percentages and upper/lower limits
- **Maximum amount method**
  Sets maximum depreciation limits
- **Multi-level method**
  Allows distinct depreciation calculations during different phases of asset depreciation
- **Period control method**
  Determines how depreciation calculations are performed based on periods

Each depreciation key assigns a value to each of these five calculation methods. Configuration for calculation methods is executed through menu path **IMG • Financial Accounting • Asset Accounting • Depreciation • Valuation Methods • Depreciation Key • Calculation Methods**. The predefined base methods are shown in Figure 6.7.

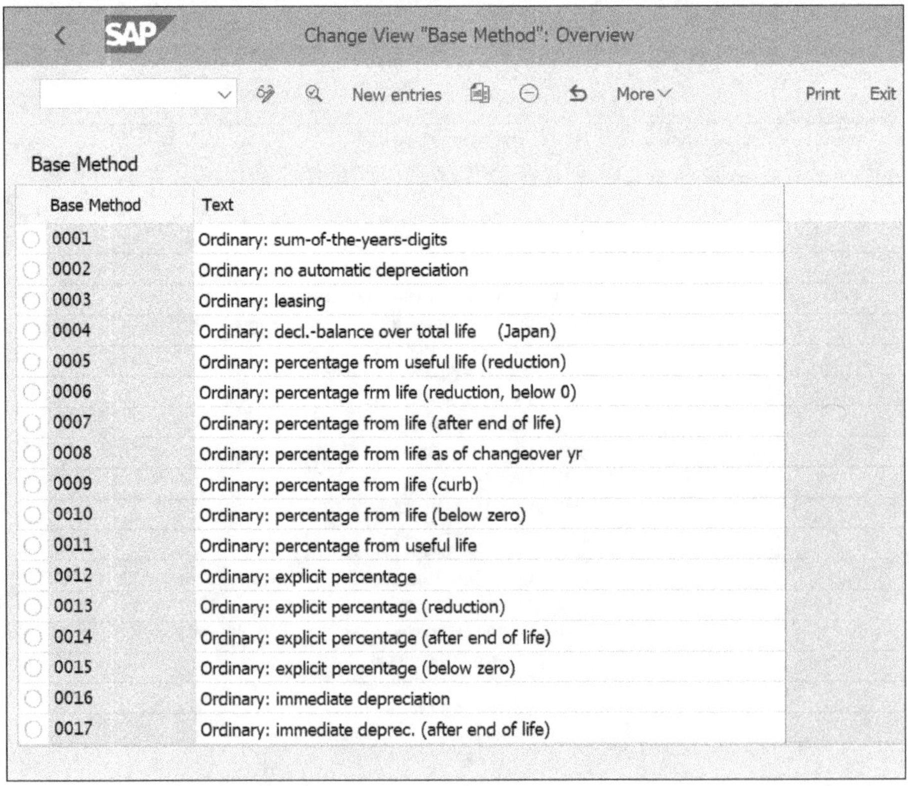

Change View "Base Method": Overview

New entries More Print Exit

Base Method

Base Method	Text
0001	Ordinary: sum-of-the-years-digits
0002	Ordinary: no automatic depreciation
0003	Ordinary: leasing
0004	Ordinary: decl.-balance over total life (Japan)
0005	Ordinary: percentage from useful life (reduction)
0006	Ordinary: percentage frm life (reduction, below 0)
0007	Ordinary: percentage from life (after end of life)
0008	Ordinary: percentage from life as of changeover yr
0009	Ordinary: percentage from life (curb)
0010	Ordinary: percentage from life (below zero)
0011	Ordinary: percentage from useful life
0012	Ordinary: explicit percentage
0013	Ordinary: explicit percentage (reduction)
0014	Ordinary: explicit percentage (after end of life)
0015	Ordinary: explicit percentage (below zero)
0016	Ordinary: immediate depreciation
0017	Ordinary: immediate deprec. (after end of life)

Figure 6.7 Predefined Base Methods

Transaction Types

In asset accounting, a transaction type denotes a specific business transaction, influencing key aspects like whether the transaction debits or credits an asset, the document type used for general ledger posting, and whether it can be manually initiated. For consolidation purposes, transaction types encompass parameters that indicate whether they post to an affiliated company and identify the corresponding consolidation transaction type. SAP S/4HANA provides numerous transaction types covering various asset transactions. However, you can create custom transaction types if necessary. Configuration for asset transaction types can be carried out under menu path **IMG • Financial Accounting • Asset Accounting • Transactions.** There you will find the settings for all fixed asset transactions, like acquisitions, retirements, transfers, and intercompany transfers.

Now, let's shift our focus to explore the process of determining general ledger accounts within asset accounting. Figure 6.1 showed various account determination elements playing a pivotal role in the process of general ledger account determination within asset accounting. These elements serve as the building blocks that empower you to assign general ledger accounts from various perspectives.

For instance, the *asset class* element allows you to categorize assets by their specific types, while *depreciation areas* help you categorize assets based on the methods used for calculating their depreciation. When you combine these account determination elements with the chart of accounts, you establish a well-structured framework for determining general ledger accounts in the realm of asset accounting. In the next section, we'll delve into each of these account determination elements in greater detail.

6.1.7 General Ledger Account Determination

You have two approaches available for configuring general ledger account determination in asset accounting. The first option involves using a single configuration transaction, which enables you to determine general ledger accounts for a wide range of asset accounting business transactions, including depreciation accounts, acquisition and production cost accounts, and imputed interest accounts. The alternative approach allows you to use multiple configuration transactions to determine general ledger accounts for specific asset accounting business transactions, such as special depreciation and ordinary depreciation. To access the single configuration transaction, follow this menu path: **IMG • Financial Accounting • Asset Accounting • Integration with the General Ledger • Assign G/L Accounts.**

In Figure 6.8, general ledger accounts in this configuration are conveniently grouped into balance sheet accounts, depreciation accounts, and special reserves accounts.

To configure general ledger accounts, you'll need to select the chart of accounts, then choose the account determination rule, and finally select one of the three groups displayed on the left side of Figure 6.8.

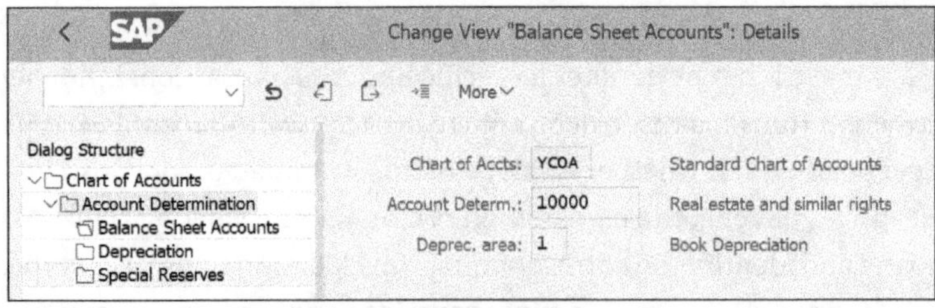

Figure 6.8 General Ledger Account Configuration

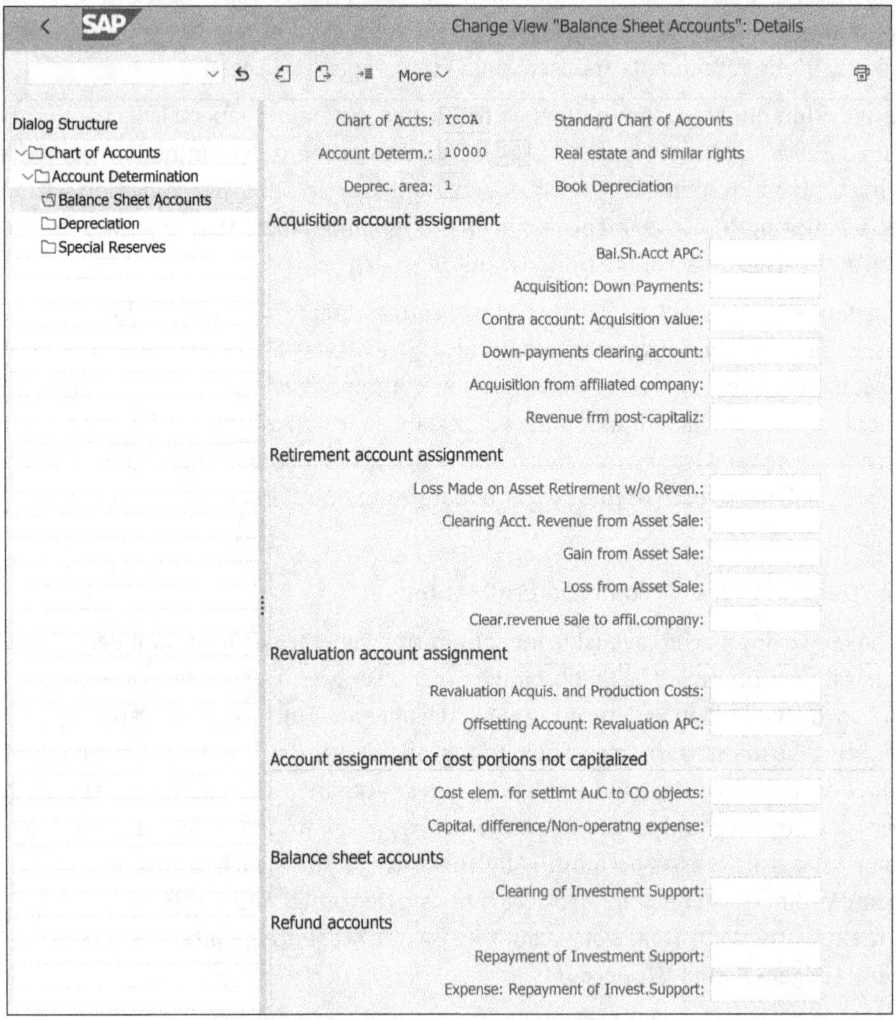

SAP
Change View "Balance Sheet Accounts": Details
More

Dialog Structure
Chart of Accounts
Account Determination
Balance Sheet Accounts
Depreciation
Special Reserves

Chart of Accts: YCOA Standard Chart of Accounts
Account Determ.: 10000 Real estate and similar rights
Deprec. area: 1 Book Depreciation

Acquisition account assignment
Bal.Sh.Acct APC:
Acquisition: Down Payments:
Contra account: Acquisition value:
Down-payments clearing account:
Acquisition from affiliated company:
Revenue frm post-capitaliz:

Retirement account assignment
Loss Made on Asset Retirement w/o Reven.:
Clearing Acct. Revenue from Asset Sale:
Gain from Asset Sale:
Loss from Asset Sale:
Clear.revenue sale to affil.company:

Revaluation account assignment
Revaluation Acquis. and Production Costs:
Offsetting Account: Revaluation APC:

Account assignment of cost portions not capitalized
Cost elem. for settlmt AuC to CO objects:
Capital. difference/Non-operatng expense:

Balance sheet accounts
Clearing of Investment Support:

Refund accounts
Repayment of Investment Support:
Expense: Repayment of Invest.Support:

Figure 6.9 Balance Sheet Accounts for Fixed Assets

Keep in mind that the depreciation area you select within this configuration transaction should be configured to allow posting to the general ledger account, either completely or partially, in real time or via periodic posting. We here take a somewhat distinct approach when discussing general ledger account determination for asset accounting. This is primarily because this particular transaction sets itself apart from other discussions as it empowers you to configure the majority of general ledger accounts for asset accounting.

The rationale behind this distinct approach is the extensive scope of this specific transaction, which enables the configuration of a significant portion of general ledger accounts for asset accounting, setting it apart from conventional discussions. Figure 6.9 shows the balance sheet accounts and Figure 6.10 shows the depreciation accounts for fixed assets. We will discuss almost all of these accounts in detail in the following sections.

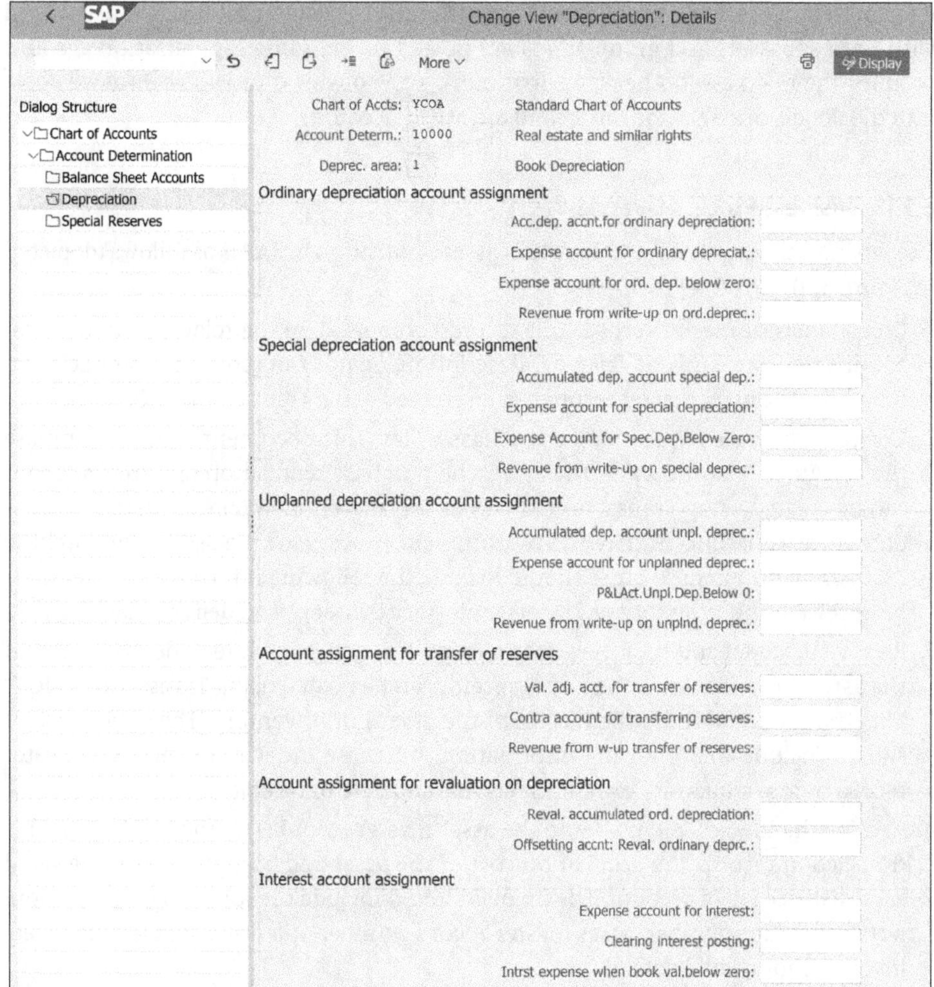

Figure 6.10 Depreciation Accounts for Fixed Assets

It is also important to mention that we will describe the essential general ledger account assignments required for utilizing the core functionality of asset accounting, widely employed in typical scenarios.

6.2 Asset Acquisition and Asset Capitalization

On the asset acquisition and asset capitalization accounts, the acquisition and production costs of assets are recorded. In other words, they capture the capitalized amount of an asset, which is reflected in the general ledger. Whenever an asset is acquired, whether through external procurement or in-house production, the associated capitalization cost is posted to a balance sheet account. This cost is known as the *asset capitalization cost* or the *acquisition* or *production cost*.

All future alterations or augmentations to the value of an asset are likewise recorded in this general ledger account. In your interactions with asset accounting, you might encounter subassets and group assets. Importantly, the capitalization costs associated with both subassets and group assets are posted to the same capitalization account used for the main asset. Ahead, we first outline the business process before we delve into the details of asset acquisition and offsetting accounts.

6.2.1 Business Process Overview

The business process for a simple case of asset acquisition in SAP is as follows (depicted in Figure 6.11):

1. Create an asset master record. This record contains all of the relevant information about the asset, such as its type, description, acquisition cost, and depreciation schedule. The asset master record can be created using Transaction AS01.
2. Create a purchase requisition or purchase order. This document is used to initiate the procurement process for the asset. The purchase requisition or purchase order can be created using Transaction ME51N and Transaction ME21N, respectively. This document is used to initiate the procurement process for the asset. The purchase requisition or purchase order should include the following information: asset master record number, quantity of assets, unit price of assets, and delivery date.
3. Receive the asset and post the goods receipt. This transaction records the receipt of the asset into inventory. The goods receipt can be posted using Transaction MIGO. This transaction records the receipt of the asset into inventory. The goods receipt should include the following information: purchase order number, asset master record number, quantity of assets received, and date of receipt.
4. Receive the invoice and post it to the asset master record (Transaction MIRO). This transaction records the acquisition cost of the asset and capitalizes it to the fixed asset balance sheet account. The invoice should include the following information: purchase order number, asset master record number, quantity of assets invoiced, invoice amount, and date of invoice.

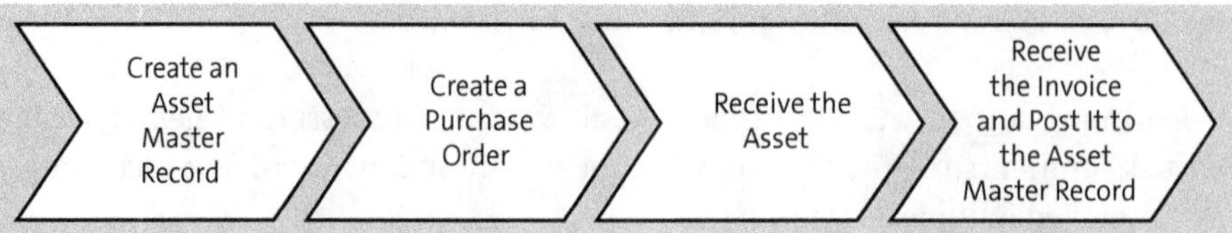

Figure 6.11 Asset Acquisition Process

Once the invoice has been posted, the asset acquisition process is complete.

Note

It is possible to integrate the asset acquisition process with the purchasing process in SAP. This can be done by creating a purchase order with reference to the asset master record. When the goods receipt is posted, the asset acquisition will be automatically posted to the asset master record.

It is also possible to capitalize in-house produced goods or services as assets. This can be done by creating a production order with reference to the asset master record. When the production order is completed, the asset acquisition will be automatically posted to the asset master record.

SAP also offers a number of add-on solutions, such as for SAP Real Estate Management or SAP Enterprise Asset Management (SAP EAM) for asset acquisition management. These solutions can provide additional functionality, such as approval workflows, budget control, and reporting.

6.2.2 Asset Acquisition Account

The asset acquisition account (**Bal.Sh.Acct APC**, ❶) is a temporary account that is used to record the purchase of assets. Once the asset acquisition is posted to a fixed asset account, the asset acquisition account will be cleared. The **Acquisition: Down Payments** ❷ and **Down-payments, clearing account** ❹ accounts are only needed for account allocations which are used in asset classes for assets under construction. The **Contra account: Acquisition value** ❸ account is only needed when the asset acquisition was not posted using the integrated accounts payable (AP) process. Figure 6.12 shows where to configure these accounts.

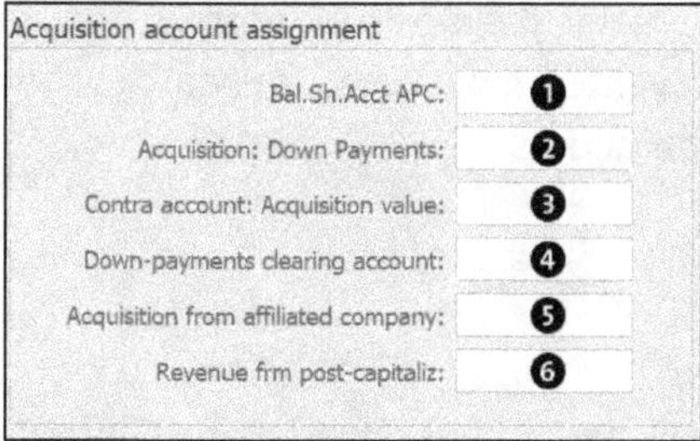

Figure 6.12 Acquisition Accounts for Fixed Assets

6.2.3 Offsetting Entry Accounts

When acquiring assets from a vendor already established within your SAP S/4HANA system, the payable associated with the asset acquisition is directly recorded in that vendor's account. Similarly, during asset sales, if a preexisting customer is specified in the transaction, the receivable is directly posted to the customer's account.

However, if the customer or vendor account hasn't been set up in the SAP S/4HANA system, you have the option to utilize a clearing account where you can post the offsetting entry at the time of asset acquisition, sale, or retirement. These clearing accounts can be configured under the offsetting entry for asset acquisition account ❸ and the offsetting entry for asset sale account. Where to configure the account is described later on in Section 6.4.2.

Furthermore, even when the customer or vendor account involved in the asset transaction is already established, you might still prefer to employ a clearing account. This allows you to post entries from AP or accounts receivable (AR) to the clearing account, and subsequently you can reconcile these entries with the ones posted from the AR or AP functionality.

6.2.4 Offsetting for Post Capitalization

As mentioned earlier, when you directly record an asset acquisition in asset accounting using acquisition transaction types, SAP S/4HANA provides the offsetting entry for the asset acquisition account (03). However, it's important to note that the general ledger account determination differs when you're posting an asset acquisition that falls under the category of post capitalization.

Post capitalization, in this context, refers to the situation where you need to record an asset acquisition in the current fiscal year that should have been posted in one of the previous fiscal years but was not due to certain circumstances. This scenario arises when you have an asset that should have started depreciating from its original capitalization date in a prior fiscal year.

When you enter a post-capitalization transaction in asset accounting, SAP S/4HANA suggests using the general ledger account specified as *revenue from post-capitalization* (06) from your configuration as the account for posting the offsetting entry. It's worth noting that this is merely a default proposal for a general ledger account; you have the flexibility to modify it when posting the asset transaction.

6.3 Asset Depreciation

Except for select asset types like land and goodwill, nearly all assets undergo depreciation over their specified useful lifespan. For assets falling outside these exceptions, a monthly depreciation process is conducted. During this process, depreciation postings are made to lower the asset's book value and record the associated depreciation expense.

In the following sections, we first describe the business process step by step before we dive into the configuration for asset deprecation.

6.3.1 Business Process Overview

The business process for asset depreciation in SAP is as follows (depicted in Figure 6.13):

1. Define the depreciation areas. Depreciation areas can be defined using Transaction AS02 and are used to group assets together for depreciation purposes. For example, you may have a depreciation area for plant and equipment, another depreciation area for vehicles, and a third depreciation area for buildings.
2. Assign assets to depreciation areas. Assets can be assigned to depreciation areas in the asset master data. To do this, go to Transaction AS01 and open the asset master record for the asset you want to assign. In the **Depreciation Area** tab, select the appropriate depreciation area.
3. Define the depreciation keys. Depreciation keys define the depreciation method, calculation period, and other parameters. For example, you may have a depreciation key for straight-line depreciation, another depreciation key for declining-balance depreciation, and a third depreciation key for accelerated depreciation.
4. Assign depreciation keys to depreciation areas. Depreciation keys can be assigned to depreciation areas in the asset master data. To do this, go to Transaction **AS01** and open the asset master record for the asset you want to assign. In the **Depreciation Key** tab, select the appropriate depreciation key.
5. Run the depreciation calculation. The depreciation calculation can be run periodically using Transaction **AFAR**. This transaction will calculate the depreciation for all assets in the system based on the depreciation keys that have been assigned to them.
6. Post the depreciation. The depreciation can be posted periodically using Transaction **AFBP**. This transaction will post the calculated depreciation to the general ledger accounts.

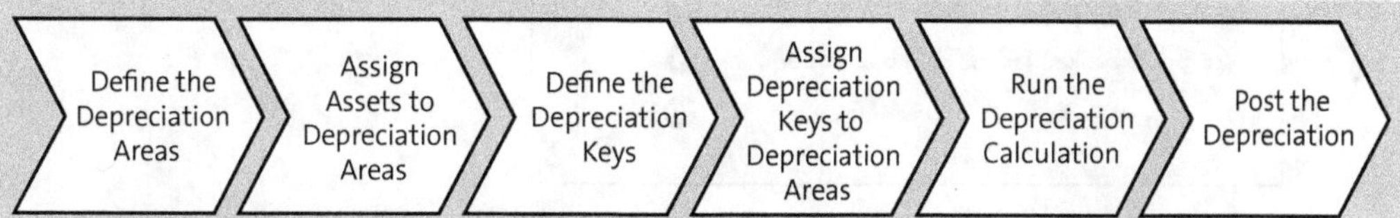

Figure 6.13 Depreciation Process for Fixed Assets

> **Note**
>
> The asset retirement transaction will automatically generate a journal entry to reverse the asset acquisition and depreciation entries.
>
> If the asset is being sold to a customer, the asset retirement transaction will also generate a revenue entry. If the asset is being scrapped, the asset retirement transaction will also generate an expense entry for the scrap value of the asset.

The figures computed for unplanned and special depreciation are not automatically recorded during the routine depreciation process. To account for these amounts, it's necessary to manually post adjustment entries in the financial records to ensure accurate representation of the asset's book value.

6.3.2 Ordinary Depreciation Accounts

The depreciation run carries out this operation by crediting the accumulated depreciation account (**Acc.dep. Accnt.for Ordinary Depreciation**, ❼) to represent the reduction in asset value and debiting the depreciation expense account (**Expense Account for Ord. Dep. below Zero**, ❽). This process is known as *standard depreciation*, often referred to as *ordinary depreciation* within SAP S/4HANA.

Usually, once the net book value of an asset reaches zero, depreciation for local reporting purposes ceases. However, there may be situations where, for business or statutory considerations, you need to continue depreciating the asset even if the net book value falls below zero. To address this, SAP S/4HANA employs the **Expense Account for Ord. Dep below Zero** account ❾) for posting any ordinary depreciation after the net book value of an asset has reached zero.

If, for any reason, the depreciation that has been posted to an asset needs to be reversed, the reversal of ordinary depreciation is recorded in the **Revenue from Write-up on Ord.deprec** general ledger account ❿). Such a write-up in depreciation can occur for various reasons, such as previous depreciation being excessively high, the posting of a credit memo reducing the capitalized cost of an asset, or other pertinent circumstances. The mentioned accounts are shown in Figure 6.14.

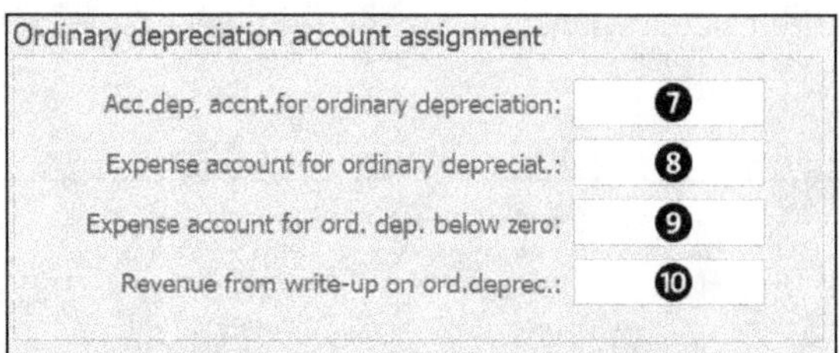

Figure 6.14 Accounts for Ordinary Depreciation

6.3.3 Unplanned Depreciation Accounts

In situations where an asset experiences unforeseen wear and tear, such as damage caused by a natural disaster, you have the option to record unplanned depreciation to decrease the net book value (NBV) of the asset. SAP S/4HANA offers a set of specific general ledger accounts to facilitate the posting of unplanned depreciation. The use of these accounts closely mirrors the approach we've discussed for their counterparts in ordinary depreciation:

- **Accumulated depreciation (Accumulated Dep. Account Unpl.deprec., ⓫)**
  This account is employed to accumulate the depreciation related to the unplanned wear and tear.
- **Depreciation expense (Expense Account for Unplanned Deprec., ⓬)**
  This serves as the account for recognizing the depreciation expense incurred due to the unplanned wear and tear.
- **Depreciation if the net book value is below zero (P&LAct.Unpl.Dep.Below 0, ⓭)**
  When the net book value falls below zero during the depreciation process, this account is used to record any depreciation in that scenario.
- **Revenue from depreciation write-up (Revenue from Write-up on Unplnd. Deprec., ⓮)**
  This account comes into play when you need to write up depreciation for specific reasons, whether it's to rectify previously overstated depreciation or address adjustments, for instance.

The mentioned accounts are shown in Figure 6.15.

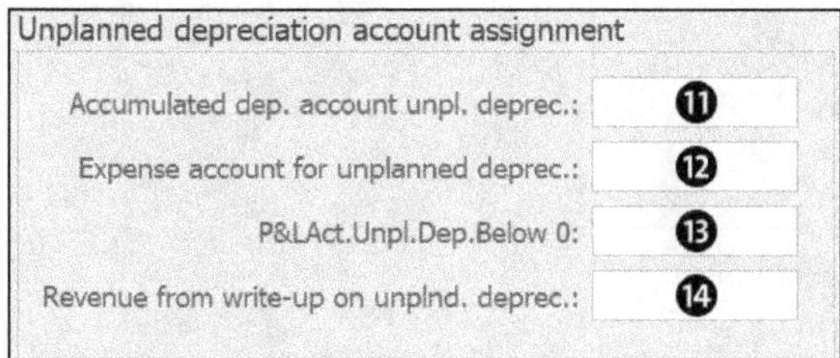

Figure 6.15 Accounts for Unplanned Depreciation

6.3.4 Special Depreciation Accounts

Occasionally, statutory regulations may permit the application of special depreciation, often in the form of tax incentives or other government-backed growth promotion initiatives. A notable example in the United States involves economic stimulus acts over the past decade that granted businesses the ability to claim extra depreciation for assets acquired during specific calendar periods.

While this involves configuring new depreciation keys to calculate special depreciation, it's important to highlight that, in terms of general ledger account determination, the postings for special depreciation can be tracked in distinct general ledger accounts as opposed to the ones used for ordinary depreciation. SAP S/4HANA employs the following general ledger accounts for posting special depreciation, and their usage aligns with the methodology described for corresponding accounts in ordinary depreciation or unplanned depreciation:

- **Accumulated depreciation (Accumulated Dep. Account Special Dep., ⓯)**
  This account is designated for accumulating the special depreciation posted.

- **Depreciation expense (Expense Account for Special Depreciation, ⓰)**
  This serves as the account for recording the depreciation expenses associated with special depreciation.
- **Depreciation if the net book value is below zero (Expense Account for Spec.Dep.Below Zero, ⓱)**
  In situations where the net book value drops below zero due to special depreciation, this account is utilized for posting such depreciation.
- **Revenue from depreciation write-up (Revenue from Write-up on Special Deprec., ⓲)**
  If there's a need to adjust depreciation upward, this account comes into play, addressing various scenarios like correcting previously overestimated depreciation or handling specific adjustments.

The mentioned accounts are shown in Figure 6.16.

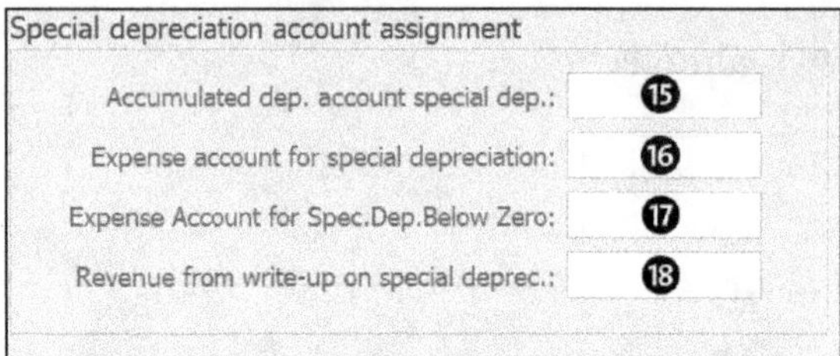

Figure 6.16 Accounts for Special Depreciation

Note

Prior to considering the utilization of distinct general ledger accounts for various forms of depreciation, it's crucial to assess your specific business requirements. From a technical standpoint, it is entirely feasible to allocate different general ledger accounts for ordinary depreciation, unplanned depreciation, and special depreciation.

Nonetheless, it's important to recognize that asset accounting offers multiple views and reporting features that allow you to differentiate between ordinary, unplanned, and special depreciation. This means that you might not necessarily need a separate set of general ledger accounts within your primary chart of accounts to distinguish these different depreciation types. The availability of these views and reports can often obviate the need for an intricate general ledger account structure.

The figures computed for unplanned and special depreciation are not automatically recorded during the routine depreciation process. To account for these amounts, it's necessary to conduct a periodic asset posting run specifically for this purpose.

6.4 Asset Retirement by Sales

When it comes to disposing of an asset, you have the option to do so even if it still retains some book value. The accounting entries you make depend on whether the proceeds from the asset sale exceed or fall short of its book value.

In the following section, we outline the business process in four major steps and then we explain the relevant configuration of asset retirement by sales.

6.4.1 Business Process Overview

The business process for asset retirement by sales in SAP S/4HANA is as follows (see Figure 6.17):

1. Create a sales order for the asset. This is done using Transaction VA01. The sales order should include the following information: asset master record number, quantity of assets being sold, unit price of assets being sold, customer information.
2. Create a delivery note for the asset. This is done using Transaction VL01N. The delivery note should include the following information: sales order number, asset master record number, quantity of assets being delivered, delivery date.
3. Create a billing document for the asset. This is done using Transaction **VF01**. The billing document should include the following information: sales order number, delivery note number, asset master record number, quantity of assets being billed, billing amount, and customer information.
4. Retire the asset from the system. This is done using Transaction **F-92**. The asset retirement transaction should include the following information: asset master record number, retirement date, sales order number, and billing document number.

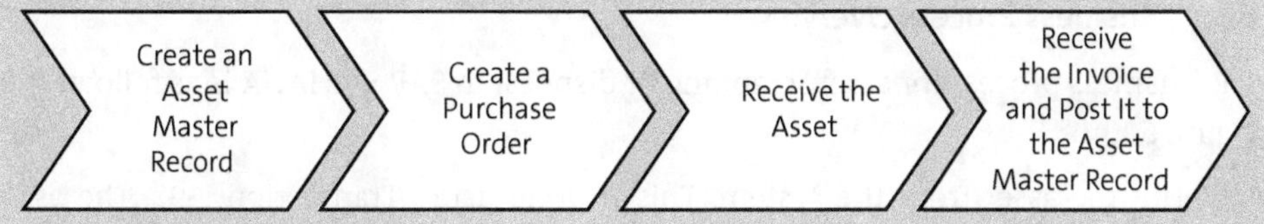

Figure 6.17 Asset Retirement Process

6.4.2 Gain and Losses Accounts from Asset Sales

If the sale of an asset yields more than its net book value, SAP S/4HANA records this surplus as a gain on the asset sale account (**Gain from Asset Sale, ㉑**). Conversely, if the asset is sold for an amount less than its net book value, the difference is posted as a loss from the asset sale account (**Loss from Asset Sale, ㉒**).

The net book value of an asset is calculated as the variance between the initial capitalized value of the asset account (**Bal.Sh.Acct APC**, Figure 6.12 ❶) and the cumulative depreciation recognized on the asset account (**Acc.dep. Accnt.for Ordinary Depreciation**, Figure 6.14 ❼). Figure 6.18 shows where the accounts are configured.

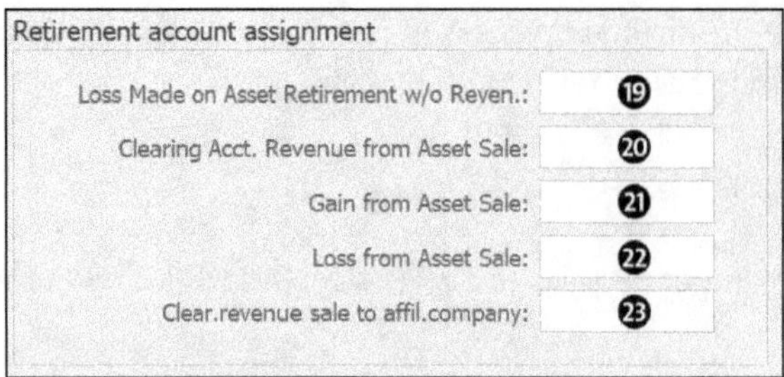

Figure 6.18 Accounts for Asset Retirement

6.5 Asset Scrap or Disposal

On occasion, an asset may become unsuitable for sale due to performance issues, usability concerns, or legal requirements. In such instances, you have the option to retire the asset by recording an asset scrap transaction. An asset scrap transaction is akin to an asset retirement transaction, with the key distinction that there are no proceeds from a sale involved.

In the following section, we outline the business process in three main steps before we examine the configuration for asset scrap and disposal accounts.

6.5.1 Business Process Overview

The business process for asset scrapping or disposal in SAP S/4HANA is as follows (see Figure 6.19):

1. Retire the asset from the system. This is done using Transaction F-92. The asset retirement transaction should include the following information: asset master record number, retirement date, and retirement reason (e.g., scrapped, sold to scrap dealer, donated, etc.).
2. Physically dispose of the asset. This may involve scrapping the asset, selling it to a scrap dealer, donating it, or disposing of it in another way.
3. Record the disposal. This is done using Transaction FB60. The disposal entry should include the following information: asset master record number, disposal date, disposal amount (if any), and disposal account (e.g., scrap income account, donation expense account, etc.).

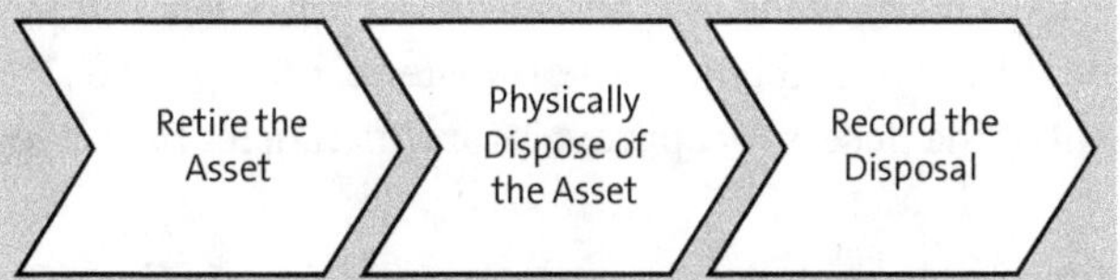

Figure 6.19 Process of Scrapping or Disposal of an Asset

> **Note**
>
> If the asset is being sold to a scrap dealer, the disposal entry should also generate a revenue entry. If the asset is being donated, the disposal entry should also generate an expense entry for the donation. If the asset is being scrapped, the disposal entry should also generate an expense entry for the scrap value of the asset.

6.5.2 Scrap and Disposal Accounts

When an asset is scrapped, the entire net book value is recorded as a loss attributed to the asset scrap account (Figure 6.14 ❼). The calculation of the asset net book value follows the same method as explained for asset sales or retirements. The net book value of an asset is calculated as the variance between the initial capitalized value of the asset account (**Bal.Sh.Acct APC**, Figure 6.12 ❶) and the cumulative depreciation recognized on the asset account (**Acc.dep. Accnt.for Ordinary Depreciation**, Figure 6.14 ❼). Figure 6.18 shows where the accounts are configured.

6.6 Asset Intercompany Transfer

In certain situations, you may find the need to engage in asset acquisitions or asset sales. When such transactions involve affiliated companies within the same corporate group, SAP S/4HANA employs specific accounts for recording these transactions.

6.6.1 Business Process Overview

The business process for asset intercompany transfer in SAP S/4HANA is as follows (see Figure 6.20):

1. Create a transfer order. This is done using Transaction AS09. The transfer order should include the following information: sending company code, receiving company code, asset master record number, and transfer date. The transfer order is used to initiate the asset intercompany transfer process. The transfer order should include all of the relevant information about the transfer, such as the sending and receiving company codes, the asset master record number, and the transfer date.

2. Release the transfer order. This is done using Transaction AS10. Once the transfer order has been created, it must be released before the transfer can be posted. The release process ensures that all of the necessary approvals for the transfer have been obtained.
3. Post the transfer in the sending company code. This is done using Transaction AS11. The transfer posting in the sending company code removes the asset from the sending company's books. The transfer posting should include the asset master record number, transfer date, and receiving company code.
4. Post the transfer in the receiving company code. This is done using Transaction AS12. The transfer posting in the receiving company code adds the asset to the receiving company's books. The transfer posting should include the asset master record number, transfer date, and sending company code.

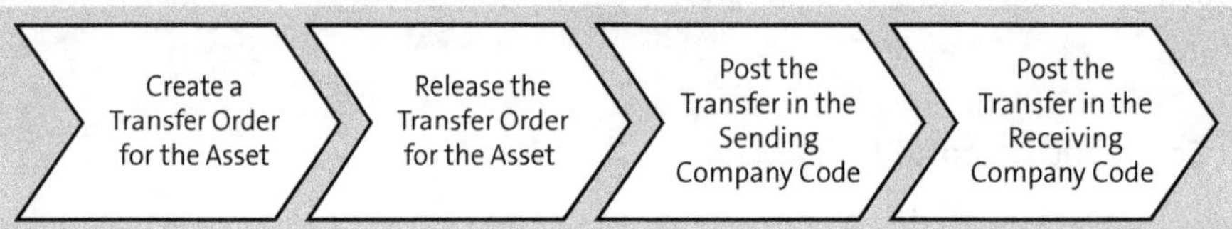

Figure 6.20 Process for Transferring Assets between Two Affiliated Companies

Note

The asset intercompany transfer process can be automated using SAP's automatic intercompany transfer functionality. Automatic intercompany transfer allows you to transfer assets between company codes without having to create and release transfer orders.

The asset intercompany transfer process can also be integrated with SAP Transportation Management, which allows you to track the shipment of an asset during the transfer process.

6.6.2 Acquisition from Affiliates Account and Revenue of Sale to Affiliates Account

An acquisition from an affiliated company is recorded to an *acquisition from the affiliated company* account (05), and a sale to an affiliated company is recorded to a *revenue sale to the affiliated company* account (23). After the acquisition is recorded, the asset follows the same lifecycle as any other asset. It is depreciated through its period of use before it is scraped and disposed of at some point in time.

6.7 Asset under Construction

A straightforward asset acquisition is typically applicable to relatively small assets or those obtainable as single units, such as individual computers, plant equipment, copier machines, or even cars. However, when it comes to substantial, long-term projects like constructing a plant or entering into extensive SAP S/4HANA implementation contracts spanning many years, a different approach is necessary to account for and capitalize these endeavors.

Assets under construction (AuC) serve as a specialized asset class within asset accounting designed to address this need. AuC is a distinct type of asset class that allows you to monitor and report all costs associated with a specific cost collector in SAP S/4HANA. In most cases, this cost collector takes the form of an internal order, a production order, a project defined in project systems, or a combination of these. Nevertheless, if there are no such predefined cost collectors, you can even create a unique asset under the AuC class, serving as a repository for tracking and reporting all incurred costs.

Importantly, AuC assets are not subject to depreciation. Instead, the costs accumulated within an AuC are, as needed, settled into one or more fixed assets. This settlement process can involve a complete settlement for all costs or a partial settlement, and it can be executed periodically, typically at the conclusion of each project phase or based on other business-specific criteria. This allows for the allocation of costs from the AuC to various destinations, such as cost centers or specific general ledger accounts, ensuring that only the relevant expenses are capitalized.

In the following sections, we describe first the business process with SAP S/4HANA and then the accounts for assets under construction and the accounts for capitalization of down payments.

6.7.1 Business Process Overview

The business process for assets under construction in SAP S/4HANA is as follows (see Figure 6.21):

1. Create an asset master record for the asset under construction. This is done using Transaction AS01. The asset master record should include the following information: asset class, asset type, asset description, construction start date, expected completion date, and construction budget. The asset master record for the asset under construction is used to track the asset throughout the construction process. The asset master record should include all of the relevant information about the asset, such as its asset class, asset type, asset description, construction start date, expected completion date, and construction budget.
2. Record the costs incurred on the asset under construction. This is done using Transaction AIAB. The costs can be recorded using a variety of document types, such as purchase orders, invoices, and internal orders. The costs incurred on the asset under construction are also recorded using Transaction AIAB. The costs can be recorded using a variety of document types, such as purchase orders, invoices, and internal orders. The cost postings should be referenced to the asset under construction master record.

3. Periodically calculate the accumulated costs of the asset under construction. This is done using Transaction AFAR. The accumulated costs represent the total cost of the asset up to a certain point in time. The accumulated costs of the asset under construction are calculated using Transaction AFAR. The accumulated costs represent the total cost of the asset up to a certain point in time. The accumulated costs are calculated by summing all of the cost postings that have been referenced to the asset under construction master record.
4. Capitalize the asset under construction to a fixed asset account. This is done using Transaction AS92. The capitalization transaction will create a new fixed asset master record for the asset and transfer the accumulated costs from the asset under construction master record to the fixed asset master record. The asset under construction is capitalized to a fixed asset account using Transaction AS92. The capitalization transaction will create a new fixed asset master record for the asset and transfer the accumulated costs from the asset under construction master record to the fixed asset master record. The asset will then be depreciated over its useful life.

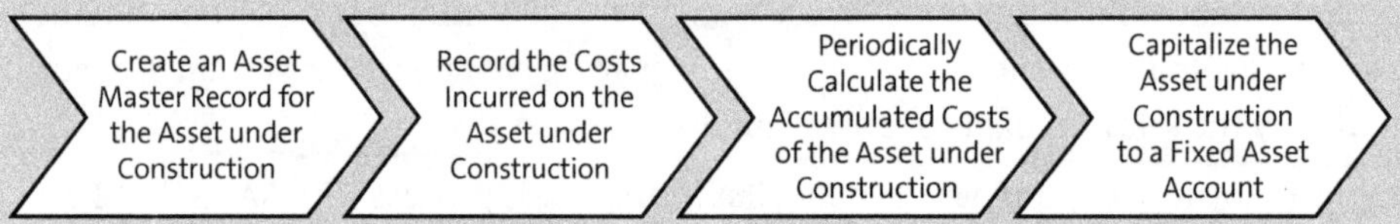

Figure 6.21 Process for Asset under Construction

Note

SAP S/4HANA offers a number of reports, like Capital Investment Reports for Projects, Project Progress and Status Report, Cost Element Reports in Controlling, Asset under Construction Aging Report, and Line Item Reports for Assets, that can be used to track the progress of asset under construction projects.

SAP S/4HANA also offers a number of add-on solutions like SAP Real Estate Management or SAP Portfolio and Project Management for asset under construction management. These solutions can provide additional functionality, such as project budgeting, schedule management, and risk management.

6.7.2 Accounts for Asset under Construction Settlement

Within the asset accounting general ledger account determination configuration, there are two essential cost element accounts to set up for this process (Figure 6.22):

- **Cost element account (Cost Elem. for Settlmt AuC to CO Objects, ㉔)**
  This cost element comes into play when you're settling line items from an asset under construction to a cost center, an order, or a work breakdown structure (WBS) element in a project system. It plays a crucial role during line item settlements in asset accounting.

- **Cost element account (Capital. Difference/Non-operatng Expense, ㉕)**
  This second cost element serves a distinct purpose. It is utilized for posting nonoperating expenses or any disparities in capitalization costs. These differences may occur when an asset is capitalized differently for the local financial records compared to the consolidated financial books.

These two cost elements are integral to managing the financial aspects of asset settlements and ensuring accurate accounting for capitalization variations and nonoperating expenses.

Account assignment of cost portions not capitalized	
Cost elem. for settlmt AuC to CO objects:	㉔
Capital. difference/Non-operatng expense:	㉕

Figure 6.22 Cost Element Accounts for Asset under Construction Settlement

6.7.3 Accounts for Capitalization of Down Payments

An AuC account representing a long-term project that is in the process of being capitalized often involves situations where down payments need to be made even before the asset is fully completed or delivered. When no down payments are involved, the costs are capitalized to the AuC when a full or partial portion of the asset is deemed complete and delivered.

However, in cases where you have made down payments to the vendor before capitalizing any assets, you have the option to also include the capitalization of these down payments. This decision hinges on your specific business requirements.

If you choose to capitalize the down payment, SAP S/4HANA utilizes two accounts for this purpose. The first is the acquisition down payments account (**Acquisition: Down Payments**, Figure 6.12 ❷) used for posting the capitalization of down payments. The second is the down payments clearing account (**Down-Payments, Clearing**, Figure 6.12 ❹), which is employed for recording offsetting entries. When posting a vendor invoice, this entry is subsequently reversed and cleared against the corresponding down payment.

6.8 Asset Revaluation Postings

Asset revaluation is a multifaceted subject, and different companies may undertake it for various reasons using diverse methodologies. For instance, if your company holds assets in a region with a history of inflation, you might consider periodic revaluation to account for inflation adjustments. Alternatively, for insurance purposes, you might utilize the index replacement series to reassess your assets annually.

Each revaluation method has its own advantages and drawbacks, and each caters to specific needs. For example, the index replacement series is ideal when a majority of assets are situated in a high-inflation region with a dependable and widely recognized inflation index. On the other hand, the asset revaluation measure is more suitable for isolated, one-time asset revaluations.

In the following two sections, we first explain the business process in four steps, then we explore the settings for the revaluations of acquisitions and offsetting accounts.

6.8.1 Business Process Overview

The business process for asset revaluation postings in SAP S/4HANA is as follows (see Figure 6.23):

1. Create a revaluation proposal. This can be done using Transaction AR29N. The revaluation proposal should include the following information: asset master record number, revaluation date, revaluation method (e.g., fair value, historical cost), and revaluation amount. The revaluation proposal is used to create a proposed revaluation for the asset. The revaluation proposal can be created using the fair value or historical cost method. The revaluation proposal should include the asset master record number, revaluation date, revaluation method, and revaluation amount.
2. Review and approve the revaluation proposal. This can be done using Transaction AR29N. The revaluation proposal must be reviewed and approved before it can be posted. The review and approval process ensures that the revaluation is necessary and accurate.
3. Post the revaluation. This can also be done using Transaction AR29N. Once the revaluation proposal has been approved, it can be posted. The posting process updates the asset master record with the new revalued amount.
4. Record the revaluation. This is done using Transaction FB60. The revaluation entry should include the following information: asset master record number, revaluation date, revaluation amount, and revaluation account (e.g., asset revaluation account). The revaluation must be recorded in the accounting system so that it is reflected in the financial statements. The revaluation entry should include the asset master record number, revaluation date, revaluation amount, and revaluation account.

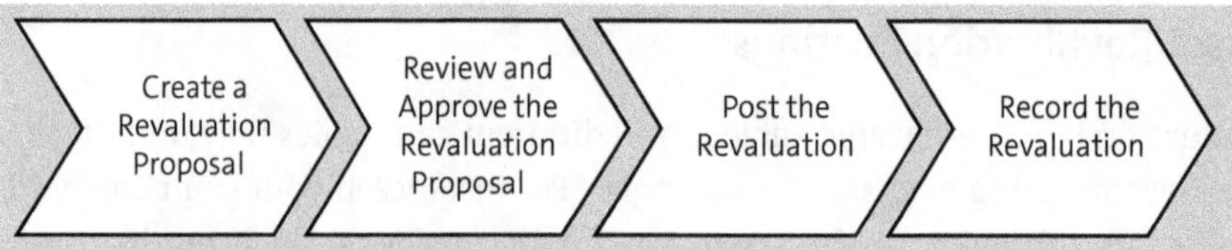

Figure 6.23 Process for Asset Revaluation

Note

Asset revaluations can be performed individually or in bulk.

It is recommended to perform asset revaluations regularly to ensure that the asset values are reflected accurately in the financial statements. SAP offers a number of tools like the Asset Explorer (Transaction AW01N) or Asset Value Adjustment (Transaction AR29N) and reports like the Asset History Sheet to help you with asset revaluations.

6.8.2 Revaluations of Acquisitions and Offsetting Accounts

Exploring the extensive details of revaluation reasons and methods is beyond the scope of this discussion. Instead, let's focus on general ledger account determination for the asset revaluation process. In all such instances, SAP S/4HANA employs the following accounts for posting revaluation adjustments and corresponding entries to the asset purchase cost and accumulated depreciation (Figure 6.24 and Figure 6.25):

- **Revaluation of asset purchase cost adjustments (Revaluation Acquis. and Production Costs, Figure 6.24 ㉖)**
  This account is used to record adjustments in the asset purchase cost due to revaluation.
- **Revaluation of asset purchase cost offsetting (Offsetting Account: Revaluation APC, ㉗)**
  This serves to balance the entries related to APC adjustments during revaluation.
- **Revaluation of accumulated depreciation Figure 6.25 ㉘)**
  This account is used to document the revaluation of accumulated depreciation.
- **Revaluation of accumulated depreciation offsetting ㉙**
  This is utilized to balance the entries associated with revalued accumulated depreciation.

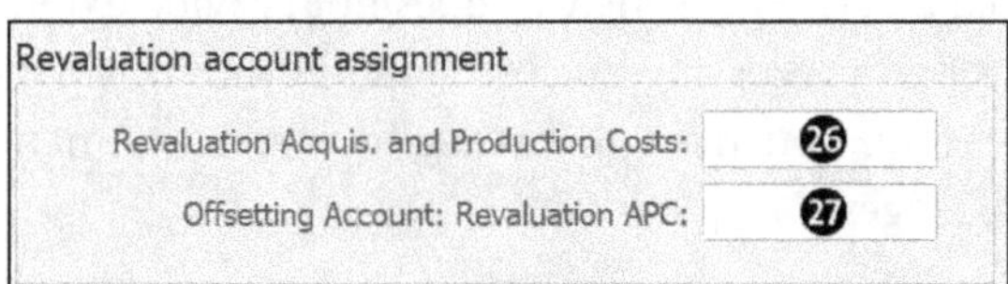

Figure 6.24 Accounts for Revaluation of Asset Purchase Cost Adjustments and Cost Offsetting

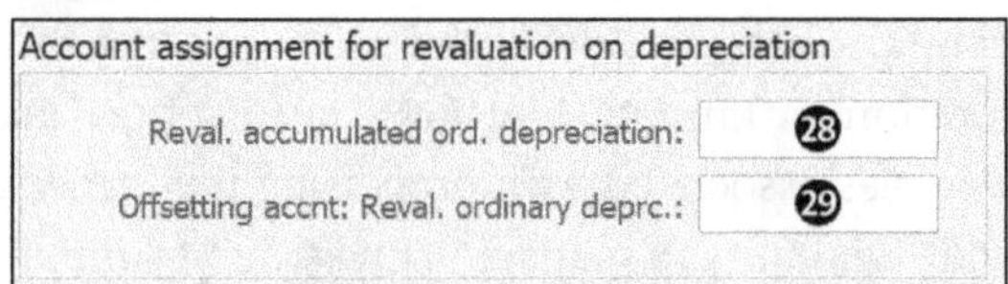

Figure 6.25 Accounts for Revaluation of Accumulated Depreciation and Offsetting

These accounts share similarities with the accounts you configure for the posting of ordinary, unplanned, or special depreciation. However, it's noteworthy that the general ledger account determination for imputed interest calculations does not encompass any configuration related to accumulated depreciation or the write-up of imputed interest back to general ledger accounts.

6.9 Investment Support

When we talk about *investment support*, we are referring to subsidies or loans that may be extended to assist in financing your capital expenditures for asset procurement or production. The extent and availability of this investment subsidy hinge on various factors, including the type of asset, your industry, and the intended purpose of the asset.

In the following sections we explore investment support from a business perspective in the business process overview and we explain the configuration of accounts for investment support transactions.

6.9.1 Business Process Overview

In SAP S/4HANA asset accounting, there are two distinct approaches to manage and track this investment support:

- **Investment support on the asset side**
  In this approach, you account for the investment support as a reduction in the asset acquisition and procurement cost. Essentially, the cost of acquiring the asset is decreased by the amount corresponding to the investment support or subsidy. For depreciation calculations, you utilize the reduced asset cost, considering the investment subsidy. It's important to note that while depreciation calculations for the asset align with those of other fixed assets, SAP S/4HANA does not provide extensive automation for investment support processing, such as handling monthly payments or early termination. Therefore, you need to manually post investment support transactions to the general ledger.
- **Investment support on the liability side**
  Alternatively, you can choose to maintain the investment support on the liability side as a special reserve. Under this method, you depreciate the original asset acquisition cost on the asset side, similar to any other fixed asset. Simultaneously, you "depreciate" the investment support on the liability side. It's worth noting that this approach entails additional configuration within SAP S/4HANA, but it offers the advantage of automation for most entries associated with processing investment support. These two approaches provide flexibility in managing investment support within asset accounting, and the choice between them may depend on the specific needs and automation preferences of your organization.

Note

Several of the general ledger accounts discussed in this section become accessible within the configuration transaction only once you've set up at least one investment measure.

To configure investment support, you can navigate through menu path **IMG • Financial Accounting • Asset Accounting • Special Valuation • Investment Support**.

6.9.2 Accounts for Investment Support Transactions

Figure 6.26 illustrates the general ledger accounts that play a role in general ledger account determination for investment support in asset accounting.

Figure 6.26 Investment Support Accounts

Now let's examine the general ledger account determination that pertains to scenarios where the investment support measure is maintained on the liability side:

- **New investment support entry**
  For a fresh investment support measure in SAP S/4HANA, you need to set up the corresponding investment support clearing account, as illustrated in Figure 6.26 ㉚. It's crucial to remember that these entries are generated exclusively when investment support is maintained on the liability side as a special reserve.
- **Asset termination and retention period**
  If you decide to terminate the procurement or production of an asset before the retention period specified for the investment support, two scenarios emerge based on whether the asset is terminated prior to the retention period and whether the investment support necessitates repayment:
  - **Repayment required—asset termination before retention period**
    If there is a repayment requirement for the investment support, and the asset is terminated before the retention period, the system generates repayment entries. The original support amount is posted to the *repayment of investment support* account ㉛. Simultaneously, the planned write-off amount is recorded in the *expense: repayment of investment support* account ㉜.

- **No repayment required—asset termination before retention period**
  In cases where there's no obligation to repay the investment support and the asset is terminated prior to the retention period, the remaining portion of the investment support is straightforwardly written off.

It's worth noting that the general ledger account determination process in asset accounting is streamlined, as a single configuration transaction grants access to all pertinent scenarios. This simplifies the management of investment support within the system.

In the next section, we will learn about the possibility of setting up accounts for imputed interest that is tied up in fixed assets.

6.10 Imputed Interest Calculation

SAP S/4HANA offers the capability to calculate imputed interest on the capital invested in fixed assets. This information can prove invaluable in cost accounting, particularly for conducting what-if analyses. By leveraging this data, you can assess whether investing in fixed assets represents an optimal strategy for deploying your capital.

In asset accounting, the process for performing imputed interest calculations closely mirrors that of depreciation calculations. It involves configuring a depreciation key specifying the interest rate and then assigning this key to a distinct depreciation area designated for interest calculation. You also need to decide whether to analyze this interest calculation within asset accounting in SAP S/4HANA or to post the interest calculation to general ledger accounts.

In the following sections, we explain the business process in four steps before we delve into the configuration of accounts for imputed interest calculations.

6.10.1 Business Process Overview

The business process for imputed interest calculation in SAP S/4HANA is as follows (see Figure 6.27):

1. Create a loan agreement. This is done using Transaction FB01. The loan agreement should include the following information: loan amount, interest rate, loan term, and payment schedule. The loan agreement is used to document the terms of the loan. The loan agreement should include all of the relevant information about the loan, such as the loan amount, interest rate, loan term, and payment schedule.
2. Define the imputed interest calculation method. This is done using Transaction OBAA. The imputed interest calculation method can be defined using the following options: simple interest, compound interest, or zero-coupon. The imputed interest calculation method is used to calculate the imputed interest on the loan.

3. Calculate the imputed interest. This is done using Transaction AFAR. The imputed interest calculation will be based on the loan agreement and the imputed interest calculation method.
4. Post the imputed interest. This is done using Transaction FB60. The imputed interest posting should include the following information: loan agreement number, imputed interest calculation date, imputed interest amount, and imputed interest account.

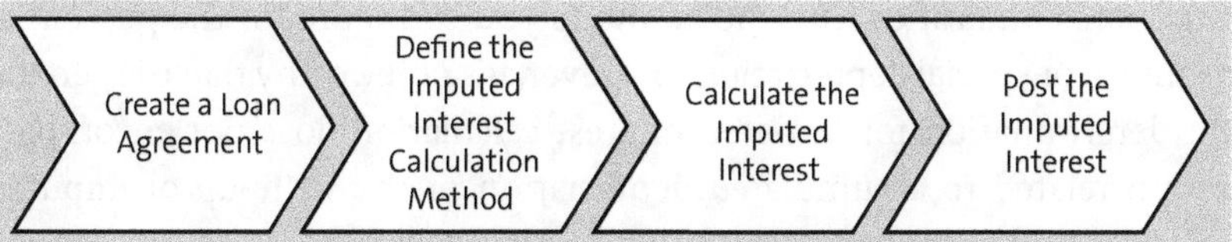

Figure 6.27 Process for Imputed Interest Calculation

> **Note**
>
> Imputed interest is calculated on a monthly basis and is posted to the accounting system at the end of each month.
>
> SAP offers a number of reports, such as the Interest Calculation Report or Asset History Sheet, that can be used to track imputed interest.

6.10.2 Accounts for Imputed Interest Calculation

If you opt to post the results of interest calculations to general ledger accounts, SAP S/4HANA employs the following general ledger accounts for this purpose, as shown in Figure 6.28:

- **Expense interest posting account (Expense Account for Interest, ㉝)**
  This general ledger account serves a parallel function to the depreciation expense account. It is the destination for imputed interest calculated on an asset.
- **Offsetting interest posting account (Clearing Interest Posting, ㉞)**
  This general ledger account is used to post an offsetting entry for the imputed interest calculation, effectively balancing the interest posted to the expense interest posting account ㉝.
- **Expense interest posting when the net book value is below zero (Intrst Expense when Book Val.below Zero, ㉟)**
  Similar to the expense interest posting account ㉝, this general ledger account fulfills the same role, but it comes into play when the net book value of an asset reaches zero.

Figure 6.28 shows where these accounts are configured.

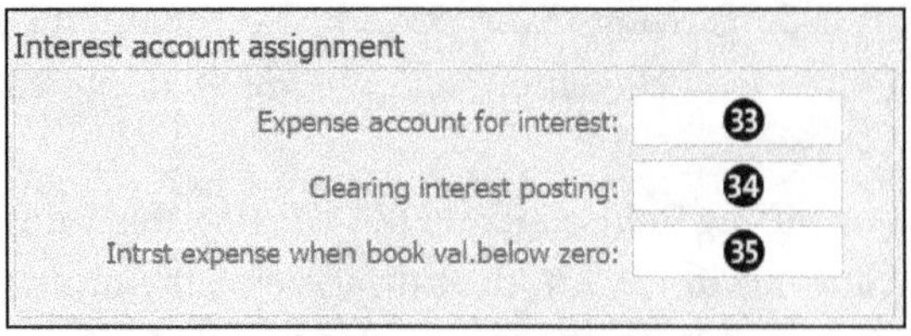

Interest account assignment

Expense account for interest: 33

Clearing interest posting: 34

Intrst expense when book val.below zero: 35

Figure 6.28 Accounts for Interest Calculation

These accounts share similarities with the accounts you configure for the posting of ordinary, unplanned, or special depreciation. However, it's noteworthy that the general ledger account determination for imputed interest calculation does not encompass any configuration related to accumulated depreciation or the write-up of imputed interest back to general ledger accounts.

6.11 Summary

The chapter on fixed assets intricately explored the configurations and account determination within asset accounting in SAP S/4HANA. It commenced with an overview of asset accounting's versatility, highlighting its application in managing not just traditional fixed assets like machinery and office equipment, but also intangible assets, replacement values, and complex capital lease payments.

Delving into the nuances of fixed asset accounting determination logic, we underscored the importance of various elements such as asset classes and depreciation areas. These elements are crucial in structuring general ledger account determination and ensuring robust asset accounting. The integration of asset categorization and depreciation methods with the chart of accounts was emphasized for its critical role in asset management.

The chapter further elaborated on the concept of parallel valuation in SAP S/4HANA, which facilitates the maintenance of assets in multiple ledgers adhering to different accounting standards like IFRS and US GAAP. This was followed by a detailed discussion of linking asset classes to general ledger accounts, the role of depreciation keys, and the importance of a well-designed asset accounting setup for operational efficiency and compliance.

Intricate processes such as asset acquisitions, capitalization, and depreciation were explored in depth, providing insights into the business processes and general ledger account determination involved. Similarly, the accounting entries and configurations necessary for asset retirement, scrapping, or disposal were discussed, along with the methodologies and reasons behind asset revaluation and the associated general ledger account configurations.

The chapter concluded by examining investment support in asset accounting, discussing various approaches for managing subsidies or loans used for asset procurement. Throughout, we maintained a balance between theoretical understanding and practical application, offering comprehensive guidance for professionals involved in fixed asset management and accounting in SAP S/4HANA.

The Authors

Abdullah Galal is a certified accountant and a leading SAP financials and controlling solution architect with extensive experience in SAP implementation projects worldwide. He has a deep passion for teaching complex business processes and their implementation in SAP S/4HANA through easy-to-understand videos and articles, which he regularly shares on professional social networks. Recognized for his skill in simplifying complex topics, Abdullah effectively communicates SAP solutions to consultants and users. His work is highly valued in the SAP community for making complex SAP solutions understandable and accessible to a wide audience.

Jonas Tritschler is a managing partner at FALK, a regional accounting firm based in Heidelberg, Germany. With nearly 25 years of expertise, he is a distinguished IT and business consultant, possessing certifications in accounting and a PhD in accounting, auditing, and taxation.

Jonas oversees the assessment of IT systems and applications, focusing on ensuring their compliance with local and international requirements, particularly in the critical realms of accounting and finance. His extensive experience as an SAP user, honed through years of serving as an interim manager and CFO, greatly enhances his proficiency in this role.

Additionally, he excels in educating accountants and consultants on SAP, offers professional guidance to companies adopting SAP, and holds a position as an accounting professor at VICTORIA University in Berlin.

Index

D

E

F

G

H

I

L

M

N

O

P

Q

R

S

T

U

V